EMPIRE, COLONIALISM, AND THE HUMAN SCIENCES

In this bold reconsideration of the human sciences, an interdisciplinary team employ an expanded theoretical and geographical critical lens centering the notion of the encounter. Drawing insights from Indigenous and Latin American Studies, nine case studies delve into the dynamics of encounters between researchers, intermediaries, and research subjects in imperial and colonial contexts across the Americas and Pacific. Essays explore ethical considerations and knowledge production practices that prevailed in field and expedition science, custodial institutions, and governance debates. They reevaluate how individuals and communities subjected to research projects embraced, critiqued, or subverted them. Often, research subjects expressed their own aspirations, asserted sovereignty or autonomy, and exercised forms of power through interactions or acts of refusal. This book signals the transformative potential of Indigenous Studies and Latin American Studies for shaping future scholarship on the history of the human sciences. This title is also available as Open Access on Cambridge Core.

ADAM WARREN is Associate Professor of History at the University of Washington, Seattle. He is the author of *Medicine and Politics in Colonial Peru: Population Growth and the Bourbon Reforms* (University of Pittsburgh Press, 2010), and the coauthor of *Baptism through Incision: The Postmortem Cesarean Operation in the Spanish Empire* (Penn State University Press, 2020.)

JULIA E. RODRIGUEZ is Professor of History at the University of New Hampshire (USA). She is the author of *Civilizing Argentina: Science, Medicine, and the Modern State* (University of North Carolina Press, 2006) and editor of the open-source website *HOSLAC: History of Science in Latin America and the Caribbean* (www.hoslac.org).

STEPHEN T. CASPER is Professor of History at Clarkson University in Potsdam, New York. His research focuses on the history of the human sciences, neuroscience, and neurology, and his latest monograph, *Punch Drunk and Dementia: A Cultural History of Concussion, 1870–Present*, is under contract with Johns Hopkins Press and explores the cultural history of brain injury and violence in the modern world.

EMPIRE, COLONIALISM, AND THE HUMAN SCIENCES

Troubling Encounters in the Americas and Pacific

Edited by

ADAM WARREN
University of Washington

JULIA E. RODRIGUEZ
University of New Hampshire

STEPHEN T. CASPER
Clarkson University

CAMBRIDGE
UNIVERSITY PRESS

CAMBRIDGE
UNIVERSITY PRESS

Shaftesbury Road, Cambridge CB2 8EA, United Kingdom

One Liberty Plaza, 20th Floor, New York, NY 10006, USA

477 Williamstown Road, Port Melbourne, VIC 3207, Australia

314–321, 3rd Floor, Plot 3, Splendor Forum, Jasola District Centre, New Delhi – 110025, India

103 Penang Road, #05–06/07, Visioncrest Commercial, Singapore 238467

Cambridge University Press is part of Cambridge University Press & Assessment,
a department of the University of Cambridge.

We share the University's mission to contribute to society through the pursuit
of education, learning and research at the highest international levels of excellence.

www.cambridge.org
Information on this title: www.cambridge.org/9781009398138

DOI: 10.1017/9781009398152

First published 2024

A catalogue record for this publication is available from the British Library

Library of Congress Cataloging-in-Publication Data
Names: Warren, Adam, Ph.D., editor. | Rodríguez, Julia, 1967- editor. | Casper, Stephen T., editor.
Title: Empire, colonialism, and the human sciences : troubling encounters in the Americas and
Pacific / Edited by Adam Warren, University of Washington, Julia E. Rodriguez, University of New
Hampshire, Stephen T. Casper, Clarkson University, New York.
Description: New York, NY : Cambridge University Press, 2024. | Includes bibliographical
references and index.
Identifiers: LCCN 2024007277 (print) | LCCN 2024007278 (ebook) | ISBN 9781009398138
(hardback) | ISBN 9781009398169 (paperback) | ISBN 9781009398152 (epub)
Subjects: LCSH: Imperialism–History. | Colonies–America–History. | Colonies–Pacific Area–
History. | Social sciences–History. | Indigenous peoples–Colonization. | America–History | Pacific
Area–History.
Classification: LCC JC359 .E445 2024 (print) | LCC JC359 (ebook) | DDC 325.8–dc23/eng/
20240322
LC record available at https://lccn.loc.gov/2024007277
LC ebook record available at https://lccn.loc.gov/2024007278

ISBN 978-1-009-39813-8 Hardback

CONTENTS

FIGURES

CONTRIBUTORS

MAILE ARVIN is Associate Professor of History and Gender Studies at the University of Utah.

EVE E. BUCKLEY is Associate Professor of History at the University of Delaware.

STEPHEN T. CASPER is Professor in the History of Science at Clarkson University in the Department of Humanities and Social Sciences.

ROSANNA DENT is the Peter Lipton Lecturer in History of Modern Science and Technology at the University of Cambridge.

MARÍA ELENA GARCÍA is Professor in the Comparative History of Ideas Department at the University of Washington, Seattle.

SEBASTIÁN GIL-RIAÑO is Assistant Professor in the History and Sociology of Science Department at the University of Pennsylvania.

ALBERTO ORTIZ DÍAZ is Assistant Professor of History at the University of Texas at Arlington.

JULIA E. RODRIGUEZ is Professor of History at the University of New Hampshire.

KARIN ALEJANDRA ROSEMBLATT is Professor of History at the University of Maryland, College Park.

GABRIELA SOTO LAVEAGA is Professor of the History of Science and Antonio Madero Professor for the Study of Mexico at Harvard University.

LAURA STARK is Associate Professor of Medicine, Health, and Society and History at Vanderbilt University.

ADAM WARREN is Associate Professor of History at the University of Washington, Seattle.

PREFACE

This is a book about how we think about encounters in the history of the human sciences, the forms of knowledge production they engendered, and what they can tell us about the relationship between science, empire, and colonialism. Focusing on case studies from Latin America and the United States Empire, many of which center Indigenous peoples, it asks how we might decolonize the history of the human sciences and develop a more ethical, social justice-oriented approach to writing about past encounters that today seem problematic, or troubling. Its thematic organization moves between a variety of scales for reconstructing interactions between scientists, the human subjects and nonhuman (or in some cases, once-human or human-related) objects they studied, and a variety of other historical actors. Some chapters adopt a local perspective, others a national one, and yet others draw attention to transnational and even global domains. In doing so, they explore the myriad interactions of expedition science, the relationality implied in fieldwork, the logics of settler colonial custodial institutions, the global circulation of ideas about human nature and behavior, and the relationship between science, state power, and governance.

A notable feature of this book, and one that has aided us in thinking about the themes described above, is its engagement of scholarship from Indigenous Studies. This has not occurred by accident. At an information session during the 2017 meetings of the Latin American Studies Association (LASA) in Lima, Peru, two of us (Warren and Rodriguez) listened as a notable intellectual historian of Latin America refused to consider the possibility that Indigenous epistemologies should be recognized specifically as playing a meaningful role within a broader initiative on Latin American contributions to the global history of knowledge. We interpreted his comments as casually dismissing not just the role of past and present Indigenous peoples in global knowledge production, but also Indigenous epistemology as a category of analysis and the contributions of an entire field of study, Indigenous Studies, that seeks to center and advocate for Indigenous ways of knowing, rights, and self-determination.

Conversations following that LASA session continued over email and at the 2018 meetings of the American Historical Association in Washington, DC, where all three of us participated on a panel with historian Micah Oelze.

A lunchtime discussion with Oelze and Sebastián Gil-Riaño ultimately inspired us to organize a workshop at the University of Washington titled "Ethics, Settler Colonialism, and Indigeneity in the History of the Human Sciences," which took place in November 2018. There, we deliberately brought together historians of science with Indigenous Studies scholars and scholars of race and empire to consider anew the relationship between knowledge production, scientific research, and ethics, and specifically how we might write about past ethical breaches in the history of the human sciences. We asked Indigenous Studies scholars to serve alongside historians of science as discussants and theorists at this workshop, rather than relegating them to the role of presenting case studies from their communities or speaking as representatives of those communities. Their feedback (together with others') generated a rich discussion, helped us to more clearly identify and articulate the stakes of the project, and improved immeasurably many of the papers presented that day, several of which are included in this book.

Following the workshop, we sought to deepen our engagement with Indigenous Studies literature through greater attention to work in Indigenous Science and Technology Studies, work on Indigenous epistemologies and knowledge making, and work focused on museum collections and relations with Indigenous communities.[1] We also identified models of the kind of Indigenous scholarship we sought to engage in the book, such as that of Abenaki anthropologist Margaret Bruchac in *Savage Kin: Indigenous Informants and American Anthropologists* (2018), and various chapters of Western Shoshone historian Ned Blackhawk and Isaiah Wilner's edited

---

[1] For Indigenous Science and Technology Studies, see Kim TallBear, *Native American DNA: Tribal Belonging and the False Promise of Genetic Science* (Minneapolis: University of Minnesota Press, 2013); Maile Arvin, *Possessing Polynesians: The Science of Settler Colonial Whiteness in Hawai'i and Oceania* (Durham, NC: Duke University Press, 2019); Eli Nelson, "Walking to the Future in the Steps of Our Ancestors: Haudenosaunee Traditional Ecological Knowledge and Queer Time in the Climate Change Era," *New Geographies* 09: Posthuman (2017): 133–138; Jessica Kolopenuk, "'Red Rivers,' No More Potlucks," http://nomorepotlucks.org/site/red-rivers-jessica-kolopenuk/. For Indigenous epistemologies and knowledge making, see Marisol de la Cadena, *Earth Beings: Ecologies of Practice across Andean Worlds* (Durham, NC: Duke University Press, 2015); Marisol de la Cadena, "Indigenous Cosmopolitics in the Andes: Conceptual Relations beyond 'Politics'," *Cultural Anthropology* 25, no. 2 (2010): 334–370; Marisol de la Cadena and Mario Blaser, eds., *A World of Many Worlds* (Durham, NC: Duke University Press, 2018); Helen Verran, "Reimagining Land Ownership in Australia," *Postcolonial Studies* 1, no. 2 (1988): 237–254; Helen Verran, "A Postcolonial Moment in Science Studies: Alternative Firing Regimes of Environmental Scientists and Aboriginal Landowners," *Social Studies of Science* 32, no. 5–6 (2002): 729–762. For relations between Indigenous communities and museums, see Amy Lonetree, *Decolonizing Museums: Representing Native America in National and Tribal Museums* (Chapel Hill: University of North Carolina Press, 2012). For theory in Indigenous Studies, see Audra Simpson and Andrea Smith, eds., *Theorizing Native Studies* (Durham, NC: Duke University Press, 2014).

volume, *Indigenous Visions: Rediscovering the World of Franz Boas* (2018).[2] Bruchac brilliantly complicates the process of knowledge production in the human sciences and decenters the authority of the ethnographer by emphasizing both the contributions of Indigenous informants to knowledge production and the practices ethnographers used to erase Indigenous roles.[3] Her work centers questions of relationality, encounter, and affect between researchers, research subjects, and intermediaries in ways that fundamentally question the frameworks historians of the human sciences and intellectual historians have long trusted and employed, some of which were on display in Lima, Peru. Likewise, various chapters in Blackhawk and Wilner's edited volume effectively demonstrate how Pacific Northwest and other Indigenous peoples contributed to key forms of globalized knowledge that we associate with modernity through their encounters and influence on Franz Boas and others. As they describe it, their collection "discloses the global sources of modern thought, bringing focus to the dissemination of knowledge from those supposedly under study to those who supposedly carry the study out – a binary that imposes false assumptions about who is acting and who is reacting, and that therefore requires rethinking and revision."[4]

We found it helpful to compare these works to the work of Warwick Anderson, who graciously participated in the 2018 workshop at the University of Washington. We found especially useful his work on scientific research and reciprocity among the Fore in *The Collector of Lost Souls*. In this study, Anderson provocatively situated the practices of the Fore as essential to shaping the novel research program of the mercurial Carleton Gajdusek, and in this important way pushed the history of science toward a paradigm of accepting the centrality and agency of Indigenous people in their encounters with Western biomedicine. At the same time, when contrasted with Bruchac's and Wilner and Blackhawk's studies, it is apparent that *The Collector of Lost Souls* provides but one approach to identifying the range of reciprocity on offer by granting agency to all knowledge creators present in the moment of relational exchanges. It prompts questions, furthermore, of what a decolonial approach to such projects might look like, especially one that is informed by Indigenous Studies, and whether it would offer something different from what Anderson's postcolonial method enables us to see.[5]

---

[2] Margaret Bruchac, *Savage Kin: Indigenous Informants and American Anthropologists* (Tucson: University of Arizona Press, 2018); Ned Blackhawk and Isaiah Lorado Wilner, eds., *Indigenous Visions: Rediscovering the World of Franz Boas* (New Haven, CT: Yale University Press, 2018).

[3] Bruchac, *Savage Kin*, 19.

[4] Blackhawk and Wilner, *Indigenous Visions*, xviii.

[5] Anderson classifies *The Collector of Lost Souls* as an example of postcolonial historical method and contrasts it to another of his works that he sees as decolonial, his 2012 article "Asia as Method in Science and Technology Studies." See Warwick Anderson, "Finding

As scholars who acknowledge their subject positions as settlers trained primarily as Latin American historians and historians of science and medicine, and as scholars who have only recently embraced decolonial methods themselves, the editors wish to make clear that they do not see this book as a major intervention in Indigenous Studies and Indigenous history. Admittedly, we have perhaps more to say about the structures of human sciences research and the history of human sciences knowledge creation in contexts defined by empire and colonialism – settler, internal, and otherwise – than about the rich and complex histories of the Indigenous peoples' lives that appear within its pages. These categories are often not easily separable, of course, and engagement of both varies across the chapters.

That said, the work of Indigenous and Indigenous Studies scholars has helped us to think through how we might historicize and explore the experiences of Indigenous communities and other communities in encounters with researchers in the human sciences. We recognize that, like Anderson, our ability to reconstruct encounters with scientists from the perspective of Indigenous historical actors is constrained not only by the limitations of the archive's fragmented records, which were often created by scientists themselves and do little to shed light on the experiences of those "researched," but also in many cases by our own subject positions and by differences of epistemology, worldview, experience, and desire. In acknowledging these limitations, we aim to acknowledge not only the challenges of accessing past experience that are common to all historical research, but also the specific responsibilities that reconstructing Indigenous histories entails. We believe engaging these concerns should form a key component of writing an ethical form of history.

Taking these concerns into account while asserting that accessing the historical experiences of Indigenous, Black, and mixed communities is not completely beyond reach, the chapters in this book engage both the limits and possibilities of what we might be able to say. In initial instructions to authors, we stressed the rich intellectual work that several key theoretical concepts in Indigenous Studies make possible for historians. These include Kahnawake Mohawk anthropologist Audra Simpson's theorizing of "ethnographic refusal," which Unangax̂ scholar Eve Tuck and her collaborator, K. Wayne Yang, engage and build upon in their own work, as one productive way to make sense of past Indigenous encounters with researchers.[6] For historians,

---

Decolonial Metaphors in Postcolonial Histories," *History and Theory* 59, no. 3 (2020): 430–438.

[6] Simpson develops ethnographic refusal as a concept to characterize behavior among Indigenous subjects who mediate and place limits on the efforts of anthropologists conducting this kind of work in settler colonial contexts and beyond. For Simpson, ethnographic refusal is an expression of sovereignty by those studied, who when speaking

their work proves valuable for thinking about silences in the archives of human sciences research, and it does so in ways that dovetail nicely with anthropologist Michel-Rolph Trouillot's work on archival silences and the making of history in *Silencing the Past: Power and the Production of History* (1995).[7] Remaining silent in the face of ethnographic questioning and engaging in other acts of refusal challenge the narrow, ritualistic, procedural, and unethical ways in which Indigenous people have been described and knowledge about them has been configured within anthropology as well as other human sciences. It is a way to reshape human scientific knowledge through withholding or resetting engagement, challenging moral and ethical frameworks, and placing limits on what the researcher can say, limits that are arrived at "when the representation would bite all of us [Indigenous people] and compromise the representational territory that we have gained for ourselves in the past 100 years."[8]

Other authors find especially valuable the way Eve Tuck and K. Wayne Yang have built on Simpson's work. Tuck and Yang argue that "refusal to do research and refusal within research" can also be a means of developing an ethics and a way of "humanizing researchers," particularly among communities who are over-studied in damage-centered research. In this sense, refusal constitutes an important collective strategy and response among a broad range of forms of engagement, and one carried out in the interest of establishing good (or better) relations. Tuck and Yang's critiques of damage-centered narratives and their call for a focus on Indigenous desire in academic research on Indigenous communities also resonated with contributors; the latter concept informs several chapters.[9] While not aimed at historians specifically, this

---

for themselves or refusing to speak, "interrupt anthropological portraits of timelessness, procedure and function that dominate representations of their past and, sometimes, their present"; Audra Simpson, "On Ethnographic Refusal: Indigeneity, 'Voice' and Colonial Citizenship," *Junctures* 9 (2007): 67–80; Simpson, *Mohawk Interruptus: Political Life across the Borders of Settler States* (Durham, NC: Duke University Press, 2014); Eve Tuck and K. Wayne Yang, "R-Words: Refusing Research," in *Humanizing Research: Decolonizing Qualitative Inquiry with Youth and Communities*, eds. Django Paris and Maisha T. Winn (Los Angeles, CA: Sage Publications, 2014), 223–248.

[7] Michel-Rolph Trouillot, *Silencing the Past: Power and the Production of History* (Boston: Beacon Press, 1995).

[8] Simpson, "On Ethnographic Refusal," 78.

[9] Eve Tuck, "Suspending Damage: A Letter to Communities," *Harvard Educational Review* 79, no. 3 (2009): 409–427; Tuck and Yang, "R-Words," 223–248. There are productive links to be made here between Tuck and Yang's scholarship and works outside of Indigenous Studies. In many respects, damage-centered narratives further processes of Othering, or Orientalism in the words of Edward Said, which is a cannibalistic stereotyping project that exaggerates difference and feeds on the confidence and sophistication of colonized peoples to normalize and have them internalize their "inferiority" vis-à-vis the "superior" modern West; see Edward Said, *Orientalism* (New York: Vintage Books, 1978).

work also serves as a useful provocation for rethinking whether, why, and how certain histories should be written. Combining their theoretical contributions with Anishinaabe scholar and writer Gerald Vizenor's concept of survivance and the related idea of thrivance, they can form part of an effort to more closely examine the archive for traces of how Indigenous communities and other communities aimed to relate to researchers through various kinds of encounters, and in some cases sought to transform them.[10]

As editors, we hope scholars, especially Indigenous Studies scholars, will see this book's engagement of theory from Indigenous Studies and other fields as a reflection of our recognition that writing about the human sciences demands an ethical choice to be inclusive, respectful, humble, and thoughtful in drawing upon and learning from interlocutors. Moreover, we hope this book invites further conversation about how Indigenous Studies theories and methods might take center stage in postcolonial and decolonial histories of the human sciences. We make no pretenses to having figured out the answers to these questions, or even to having done so as effectively as we could have, and our work certainly does not solve the structural inequalities, inequities, and violence that privilege certain voices over others in the history of science and the academy at large.[11] Finally, we acknowledge that actual theorizing within Indigenous Studies takes place from specific subject positions and is rooted in particular relationships and lived experiences. That said, we hope our work serves as a call for historians of science to read, take seriously, and think alongside the work of Indigenous and Indigenous Studies scholars from across the Americas and beyond and recognize the ethical necessity of doing so.

---

[10] Gerald Vizenor, ed., *Survivance: Narratives of Native Presence* (Lincoln: University of Nebraska Press, 2008). See also Lonetree, *Decolonizing Museums*.

[11] These structural inequalities, inequities, and violence marginalize not just Indigenous voices, but also those of scholars of many different backgrounds across Latin America and the Pacific, whose work receives insufficient attention in an academic world that privileges Anglophone scholarship. Admittedly, the process and circumstances under which this book came about unintentionally reproduced some of that marginalization. All contributing authors are based at institutions in the United States, though many have deep ties to Latin America and the Pacific. We hope that scholars in those parts of the world will see this book as an imperfect invitation to engage in further conversation.

ACKNOWLEDGMENTS

This book is the written manifestation of a long conversation among friends. We envisioned it, wrote it, and edited it digitally on lands now long occupied and often unceded. In both the meetings leading up to its publication and the book now in front of you, we strove to create a new approach to the history of the human sciences combined with a nuanced discussion of the ethical dimensions of historical methodologies. We acknowledge that the editors developed and brought the book to fruition at institutions that stand on the homelands and adjacent to the waters of the Coast Salish peoples (Duwamish, Puyallup, Suquamish, Tulalip, and Muckleshoot) as well as the Pennacook, Abenaki, Wabanaki, Akwesasne Mohawk, Haudenosaunee, Algonquin, and Huron-Wendat peoples.

Our book also owes its existence to the support and generous contributions of countless individuals and institutions. Their efforts have allowed us to collectively weave together threads of knowledge, insight, and passion, elevating the project and drawing out layers of meaning. We are indebted to every person who has contributed, directly or indirectly, in the numerous stages of its development. The collective efforts of the people listed below have enriched the content and depth of this book.

The first of two conferences related to the book's development took place in November 2018 at the University of Washington (UW), graciously hosted by the Walter Chapin Simpson Center for the Humanities. Special thanks to Simpson Center Director Kathleen Woodward and Associate Director Rachel Arteaga, along with the amazing staff at the center. University of Washington colleagues María Elena García (Quechua ancestry), José Antonio Lucero, Josh Reid (Snohomish), Jean Dennison (Osage), Dian Million (Tanana Athabascan), Sara Gonzalez, Jonathan Warren, and Ileana Rodriguez-Silva enriched the discussions with questions and insights that reflected their expertise in Indigenous Studies, Latin American and Caribbean Studies, and studies of empire. We are also grateful to other participants from outside the UW, among them Warwick Anderson, Hans Pols, Gabriela Soto Laveaga, Laura Stark, Ben Silverstein, Micah Oelze, and Sarah Walsh. Support for the conference was provided by the Simpson Center for the Humanities as well as UW's Department of History, the Latin

American and Caribbean Studies Program, and the Comparative History of Ideas Department. Many thanks to the chairs of those academic units: Anand Yang, José Antonio Lucero, and María Elena García. We also express gratitude for support in the form of a small seed grant from the Clarkson University Office of the Provost.

A follow-up conference to finalize the book was postponed by the COVID-19 pandemic until 2023. With the generous support from the University of New Hampshire (UNH) Department of History, the UNH Center for the Humanities, the Dunfey Family Gift Fund, and the Sidore Family Gift Fund, we held a large public event on "The Ethics of Encounter" at UNH, which brought all the book authors together to discuss and collaborate on the final draft of the book. Big thanks to UNH History Chair Kurk Dorsey; Administrative Assistant Jenna Scholefield; Meghan Howey and Katie Umans at the UNH Center for the Humanities; Rachel Nott and Lily Pudlo at UNH; and Paul and Denise Pouliot of the Cowasuck Band of the Pennacook, who shared their knowledge of Abenaki wisdom and language. The UNH College of Liberal Arts Dean Michele Dillon, Provost Wayne Jones, and President James Dean all made valuable contributions as well.

Finally, we extend gratitude to Lucy Rhymer at Cambridge University Press, who showed early interest in the project and ushered it through the review and production process, and to other staff at the press. Thanks as well to the two anonymous readers; their contributions made the book much stronger.

From the outset of this project, we were committed to making the book as widely available as possible. Thanks to Cambridge University Press for facilitating the book in Gold Open Access format. Funding for Open Access was generously provided by our three institutions: The Simpson Center for the Humanities at UW; the Dunfey Gift Fund, UNH Department of History; and Clarkson University's Department of Humanities and Social Sciences and Clarkson University's School of Arts and Sciences.

We dedicate this book to the spirit of friendship and collaboration.

A NOTE ABOUT THE COVER IMAGE

The cover of this volume includes an image of the entrance to the National Museum of Brazil in Rio de Janeiro, after it was destroyed by fire in September 2018. The image is one of profound tragedy and loss for many communities in Brazil, but especially for those Indigenous and Afro-Brazilian communities whose treasured possessions were housed, in some cases without their permission, within the museum's walls. Given that our decision to include it on the cover is likely to provoke discussion, we wish to clarify the reasoning behind this choice.

In the months following the fire, Brazilians took to the street to protest the conditions of government neglect and the budget cuts that led to this disaster. They rightly expressed outrage at the government's failure to value and adequately protect the museum's collection, which featured many items considered to be part of the nation's patrimony, including artifacts, fossils, works of art, and even a meteorite. Around the world, the press articulated similar claims about the causes of the museum's fire and its larger significance. The fire became a national tragedy and a tragedy for the world.

Missing from much of the national and global cosmopolitan debate about the disaster, however, was discussion of the museum's place within a larger settler colonial framework that has long structured the Brazilian state's relations with Indigenous peoples and other minoritized populations. While the fire was without question a tragedy, there is room to trouble what it meant in the first place for the nation to possess and exhibit the artifacts of Indigenous peoples, who continue to engage in survivance and assert their own forms of autonomy and sovereignty. Similarly, there is room to consider the role of the human sciences in this predicament while also acknowledging that many among the museum's staff went to heroic lengths to rescue important items during and following the fire. In many respects, the museum has revealed how alienated settlers are from the reality of their colonial values, both before and after the fire.

Our hope is that this volume can invite further consideration not just of the history of the human sciences and their relations to empire and colonialism, but also of paths forward that focus on repair. The image on the cover includes a depiction of destruction, but the beautiful, open door also gestures toward the possibility of alternative futures.

Introduction

ADAM WARREN, JULIA E. RODRIGUEZ,
AND STEPHEN T. CASPER

What is troubling about encounters in the history of the human sciences? Allow for one example among many that illustrate the challenges they present historians. In the mid-1940s, Carlos Gutiérrez-Noriega, a Peruvian psychiatrist and pharmacologist, criticized Carlos Monge, a high-altitude physiologist, who argued that coca leaf consumption had supposedly degenerative effects on human physiology and mental capacity.[1] Having conducted research experiments with coca and cocaine on medical students and subjects confined in Lima's custodial institutions for many years, Gutiérrez-Noriega left the capital for the Andean highlands, where coca chewing was widespread and where Monge had already experimented on mineworkers. There, he recreated his own laboratory in the field in the province of Huancayo. Armed with diagnostic and measuring equipment, he conducted interviews about coca use with 100 Indigenous and Mestizo peasants, shepherds, and mineworkers. In addition, he performed physiological and psychological examinations on subjects under the effects of coca. He aimed to observe the mental alterations during the "high" produced by the stimulant and also its long-term effects.[2]

While Gutiérrez-Noriega's publications offered elaborate conclusions about the effects of long-term coca use, they said little about the nature of his encounters with Indigenous populations in the Andean highlands. Much like Monge and other members of Lima's scientific community, nowhere does Gutiérrez-Noriega acknowledge the potentially troubling dimensions of his interactions with Indigenous research subjects, dimensions that appear

---

[1] Monge's research led to the concept of "Andean man," a racial variant of the human species that had evolved to be uniquely adapted to life at high altitude. Coca chewing was harmless as part of that adaptation. On Monge and Gutiérrez-Noriega, see Adam Warren, "Collaboration and Discord in International Debates about Coca Chewing, 1949–1950," *Medicine Anthropology Theory* 5, no. 2 (2018): 35–51, https://doi.org/10.17157/mat.5.2 .536. Also Marcos Cueto, *Excelencia científica en la periferia: Actividades científicas e investigación biomédica en el Perú, 1890–1950* (Lima: GRADE, 1989); Jorge Lossio Chávez, *El peruano y su entorno: Aclimatándose a las alturas andinas* (Lima: IEP, 2012).

[2] Carlos Gutiérrez-Noriega, "Alteraciones mentales producidas por la coca," *Revista Neuro-Psiquiátrica* 10, no. 2 (1947): 422–468.

problematic to us today as we historicize the making of research ethics in science. Nor does he acknowledge how Indigenous populations shaped inter-actions and knowledge making with him. Instead, he emphasized Andean peoples' alterity and embraced stereotypes. He positioned himself as the bearer of expertise, one who could bring about their redemption even as he used them to elevate his reputation.[3]

Gutiérrez-Noriega's studies occasionally involved attempts at relationality, motivated by desires to understand mental health, personality traits, and differences of behavior among different Indigenous peoples. He had little ability, however, to value them as interlocutors and collaborators. Racial thinking, Peru's internal colonialism, and concerns about ranking, categoriz-ing, and distinguishing among Indigenous and non-Indigenous populations conditioned his efforts.[4] His story reflects what historian Ann Zulawski calls the "tortured thinking" of physicians and scientists in her research on the history of public health in Bolivia.[5]

Gutiérrez-Noriega's encounters with his research subjects speak to the challenges of reconstructing such interactions in the history of the human sciences now. The one-sided descriptions of his interactions make clear that the organizing logic of many human sciences has often been outward from the scientists/agents toward objects/passive recipients. In other words, Gutiérrez-Noriega's encounters, methods, and knowledge are exemplary of the implicit logic of much human science. It is also a reflection of the human scientific archive and the structural conditions of its creation, which limit what the historian can access about the past.[6] This book, a collection of provocative case studies that rethink the ethical and material facets of human science in diverse settings, seeks to trouble all such encounters.

Clearly, scientists are important, and assessing them and their work in the past should not be understood as an anti-science stance. We are not investi-gating scientists as quasi-villains in a broader history of subordination. We see scientists as complex actors who are part of the story of encounters. Human subjects, nonhuman objects (many of which were once human or related to

---

[3] Ibid.
[4] This inability is common to other Latin American contexts shaped by racial thinking and internal colonialism. See Severo Martínez Peláez, *La Patria del Criollo: An Interpretation of Colonial Guatemala* (Durham: Duke University Press, 2009).
[5] Ann Zulawski, *Unequal Cures: Public Health and Political Change in Bolivia, 1900–1950* (Durham, NC: Duke University Press, 2007), especially chapter 5. Also Alberto Ortiz Díaz, "Pathologizing the Jíbaro: Mental and Social Health in Puerto Rico's Oso Blanco (1930s–1950s)," *The Americas* 77, no. 3 (2020): 409–441.
[6] For the UN Commission for the Study of the Coca Leaf's assessment of Gutiérrez-Noriega's work as well as that of Carlos Monge, see United Nations Economic and Social Council, "Report of the Commission of Enquiry on the Coca Leaf, May 1950," United Nations, Lake Success, NY, 1950.

human populations), and contexts explored by human scientists inevitably shaped knowledge and its applications. Our stance is thus one of knowledge constituted through encounters conditioned by the structures and dynamics of colonialism.[7]

The word "encounters" is ours, but we also understand it as a primary methodological premise of the human sciences. Thus, while the emphasis of the word encounter is toward the unexpected or the memorable, within the human sciences there was an anticipation of encounters, without which much of the work of those disciplines would have been impossible. Several features of the construction of such encounters are thus salient because they reflect at once this banal expectation and simultaneously its historical evolution through contexts that were themselves rapidly changing, both because of encounters and also as a predicate of settler, nationalist, or imperial struggles. The word encounter also speaks to an important, if uneven, reciprocity, one that may not have always been visible, but which allows for a balancing of scales and stories when taken as a constituent element in the history of the human sciences. Doing so does not naively elide power inequities or romanticize the objects of inquiries. Despite efforts to construe them as such, encounters are not politically neutral.[8] Histories that center encounters as an analytic create subjects and objects in both directions, thereby allowing for a fuller expression of the multiplicity of contexts and meanings evidently present within these interactions.

As an example of these encounters, consider Franz Boas. His name is synonymous with an American school of anthropology while his most significant Indigenous interlocutor, the Tlingit/British cultural broker George Hunt, became known as a mere source of material. Hunt accessed privileged

---

[7] This book acknowledges the important role of Indigenous Studies and Indigenous epistemologies in critiquing the Western division between the human and the nonhuman, and its consideration of more-than-human beings. See Zoe Todd, "Fish Pluralities: Human-Animal Relations and Sites of Engagement in Paulatuuq, Arctic Canada," *Études/Inuit/Studies* 38, nos. 1&2 (2014): 217–238; Todd, "Fish, Kin, and Hope: Tending to Water Violations in *amiskwaciwâskahikan* and Treaty Six Territory," *Afterall: A Journal of Art, Context, and Inquiry* 43, no. 1 (2017): 102–107; Kim TallBear, "Why Interspecies Thinking Needs Indigenous Standpoints," *Cultural Anthropology*, November 18, 2011, https://culanth.org/fieldsights/why-interspecies-thinking-needs-indigenous-standpoints; TallBear, "An Indigenous Reflection on Working beyond the Human/Not Human," in "Dossier: Theorizing Queer Inhumanisms," *GLQ: A Journal of Lesbian and Gay Studies* 21, nos. 2–3 (2015): 230–235; Robin Wall Kimmerer, *Braiding Sweetgrass: Indigenous Wisdom, Scientific Knowledge, and the Teaching of Plants* (Minneapolis, MN: Milkweed, 2013).

[8] An example of this phenomenon would be the Spanish government's 1992 framing of the 500th anniversary of Columbus's voyage to the Americas as an "encuentro de dos mundos," or encounter of two worlds, which aimed to elide acknowledgment of the violence, dispossession, and forms of enslavement that characterized Spanish invasion and colonization of Indigenous societies in the Americas and elsewhere.

knowledge through his Indigenous mother, sisters, and Kwakwaka'wakw wives, acting as a researcher prioritizing Boas's needs in ways that marginalized Indigenous women's voices in the resulting anthropological texts.[9] Recent scholarship has challenged us to ask why would it be strange to think of Hunt (or his mother, sisters, and wives, for that matter) as an agent cultivating and claiming ownership over his own anthropology, or seeing possibilities of discovery in Boas?[10]

Such an illustration shows that the historical picture of scientific work in the human sciences should foreground the ways Indigenous peoples, Mestizos, or other subalterns claimed scientists through their own dialogues, goals, and forms of collaboration, accommodation, and resistance. An encounter that only one person anticipates, claims, and records is likely more of a projection than an encounter.

This book troubles human sciences encounters in the nineteenth-, twentieth-, and twenty-first-century Western hemisphere's colonial, imperial, and national domains. In doing so, it seeks to undo the flat, one-sided narratives of these encounters that long characterized research in the history of the human sciences. Rather than privilege the voices of scientists themselves, contributing authors emphasize the tangled, rarely unproblematic claims different actors and groups made upon each other. Noting the transnational dimensions of such research, individual chapters explore the often internally divisive structural conditions in which such human science research took place as well as the importance of local, national, and global socio-political contexts. This book also navigates colonialisms – settler, internal, or otherwise – and nationalisms as its chapters explore the logics of human science and the practices of relationship-making and unmaking involving human subjects.

We recognize that this intervention in the history of the human sciences comes with the limitations and possibilities of the archive and oral histories, which provide fragmented and narrow accounts of the past and thus require reading "against the grain" or "along the archival grain," or thinking critically with the dominant narrative.[11] Unsurprisingly, postcolonial and decolonial

---

[9] Margaret Bruchac, "My Sisters Will Not Speak: Boas, Hunt, and the Ethnographic Silencing of First Nations Women," *Curator* 57, no. 2 (2014): 153–171.

[10] For studies of George Hunt relationship with and influence on Franz Boas, see Margaret Bruchac, *Savage Kin: Indigenous Informants and American Anthropologists* (Tucson: University of Arizona Press, 2018), especially chapter 2; Isaiah Lorado Wilner, "Transformation Masks: Recollecting the Indigenous Origins of Global Consciousness," in *Indigenous Visions, Rediscovering the World of Franz Boas*, eds. Ned Blackhawk and Isaiah Lorado Wilner (New Haven, CT: Yale University Press, 2018), 3–41.

[11] Numerous works have demonstrated the potential of top-down primary sources to form the basis of insightful, critical histories. See James Sweet, *Domingos Álvares, African Healing, and the Intellectual History of the Atlantic World* (Chapel Hill: University of North Carolina Press, 2011); Ann Laura Stoler, *Along the Archival Grain: Epistemic*

theories and methods, which will be discussed at length in the section on Indigenous Studies and postcolonial and decolonial histories in this introduction, provide all of the authors in this book with tools to grapple with the archive and reimagine these encounters. In addition, the authors are united in grounding their arguments in the methods, concepts, and claims of Indigenous Studies scholars globally, who have pushed for more ethical, restorative, and transformative forms of research while moving away from "victim-centered" or "damage-centered narratives." When combined with simplistic ethics, these narratives deny agency and desire, and assume homogeneity and the ubiquity of past and differentiated experiences of loss.[12]

The Scale and Legacy of Encounters

Empire, Colonialism, and the Human Sciences uses the phrase "troubling encounters" to signal a new direction in the history of the human sciences by expanding its critical lens both theoretically and geographically. It undertakes this work through engagement with global Indigenous Studies theories and methods and Latin American Studies scholarship. While it recognizes that included in the latter is a long tradition of research on and by Indigenous peoples that enriches and constitutes a part of the former, it also acknowledges that differences and tensions exist between Latin American Studies and global Indigenous Studies as fields. The centrality of indigenismo within the intellectual genealogies of research concerning Indigenous peoples in Latin America, which several chapters in this book historicize, led to approaches within Latin American Studies that historically centered the role of Indigenous peoples within national frameworks. In addition, much scholarship in Latin American Studies folds studies of Indigenous peoples into studies of race and ethnicity. These differ from approaches within global Indigenous Studies that start from Indigenous frameworks and center Indigenous epistemologies, and that treat Indigeneity as, among other things, a political concept. At the same time, important exceptions exist in Latin America and within Latin American Studies, including Indigenous working groups like the Taller de Historia Andina (Andean History Workshop) and the Comunidad de Historia Mapuche (Mapuche History Community).

Anxieties and Colonial Common Sense (Princeton: Princeton University Press, 2009); Marisa Fuentes, *Dispossessed Lives: Enslaved Women, Violence, and the Archive* (Philadelphia: University of Pennsylvania Press, 2016); and various works by Saidiya Hartman, including "Venus in Two Acts," *Small Axe: A Journal of Criticism* 26 (2008): 1–14.

[12] See Eve Tuck, "Suspending Damage: A Letter to Communities," *Harvard Educational Review* 79, no. 3 (2009): 409–427; Eve Tuck and K. Wayne Yang, "R-Words: Refusing Research," in *Humanizing Research: Decolonizing Qualitative Inquiry with Youth and Communities*, eds. Django Paris and Maisha T. Winn (Los Angeles: Sage, 2018), 223–248.

Scholarship produced in these Latin American contexts proves crucial to this book's work.[13]

Our chapters collectively ask about the nature of research encounters: their ontologies and epistemologies, their scales, their affective dimensions, their power imbalances and other social dynamics, and their ethics. Often the forms that human scientific knowledge took possessed a tangible existence, but the importance of such tangibility for scholars now derives from its legacies, artifacts, and archives, which have often been read by scholars for their alterity.[14] Chapter authors decenter scientists as the sole creators of such knowledge while exploring how affective dimensions, structural conditions, and colonial contexts – settler, internal, or otherwise – conditioned what would be taken as legitimate and become knowable.

Beyond these goals, this book records an uneven history of ethical action in the human sciences. Several chapters in this book historicize those ethics while also interrogating their centrality and efficacy in colonialism as an ideology, project, lived experience, and way of knowing and being in the Americas and the Pacific.[15] While accounting for the complex historical genesis of ethics, individual chapters show that scientists' ethical frameworks became muddied by the density of human interactions they experienced in these political and economic contexts, which intermingled with scientific beliefs about race,

---

[13] On Indigenous Studies critiques of Latin America as a concept, see Emil Kem'e, "For Abiayala to Live, the Americas Must Die: Toward a Transhemispheric Indigeneity," translated by Adam Koon, *Native American and Indigenous Studies* 5, no. 1 (2018): 42–68. For a comparative analysis of academic knowledge about Indigenous peoples in the Americas and its institutionalization, see Claudia Salomon Tarquini, "Academic Knowledge about Indigenous Peoples in the Americas: A Comparative Approach about the Conditions of Its International Circulation," *Tapuya: Latin American Science, Technology, and Society* 2, no. 1 (2019): 269–294. For scholarship on settler colonialism in Latin America, see M. B. Castellanos, "Introduction: Settler Colonialism in Latin America," *American Quarterly* 69, no. 4 (2017): 777–781; and Shannon Speed, "Structures of Settler Capitalism in Abya Yala," *American Quarterly* 69, no. 4 (2017): 783–790. For scholarship on Latin America conceptualized from an Indigenous Studies approach, see José Antonio Lucero, "'To Articulate Ourselves': Trans-Indigenous Reflections on Film and Politics in Amazonia," *Native American and Indigenous Studies* 7, no. 2 (2020): 1–28. The editors are grateful to José Antonio Lucero and María Elena García for these suggestions.

[14] Dipesh Chakrabarty, *Provincializing Europe: Postcolonial Thought and Historical Difference* (Princeton: Princeton University Press, 2008).

[15] Talal Asad, "Afterward: From the History of Colonial Anthropology to the Anthropology of Western Hegemony," in *Colonial Situations: Essays on the Contextualization of Ethnographic Knowledge*, ed. George Stocking (Madison: University of Wisconsin Press, 1991): 314–324; Asad's 1973 classic study *Anthropology and the Colonial Encounter* foregrounded these approaches. See Talal Asad, *Anthropology and the Colonial Encounter* (New York: Humanity Books, 2011).

Indigeneity, gender, place, and home. Historians of the human sciences have been too slow to center this contextualization.

By adopting a geographic sweep that includes Latin America, the Caribbean, the United States, and the Pacific, one of our goals is to break with frameworks in traditional histories of the human sciences that have circumscribed Europe and North America and separated them from the rest of the world. It will come as little surprise to Latin American and Indigenous scholars that the character of the United States as an imperial project haunted this geography with forms of colonialism and imperialism familiar across the Americas. These include both settler and internal colonialisms within the territorial United States, and near-hegemonic power in trade, global politics, and military interventions further afield. Our hemispheric and transnational approach, mindful of this reality, explores the ethical imperative of centering our research in theories, methods, and approaches developed in global Indigenous Studies and Latin American Studies, all of which contend with legacies of colonialism and the realities of empire.[16]

This book also takes ethics as central to the work of historians, and in contemplating multiple levels of agency in historical sources it unapologetically, even polemically, invites consideration of whether desires for restorative, procedural, and distributive justice are really so anathema to the

---

[16] For Indigenous Studies, see particularly Bruchac, *Savage Kin* and Blackhawk and Lorado, *Indigenous Visions*; on Indigenous Studies critiques of damage narratives and emphasis on Indigenous desire, see Eve Tuck, "Suspending Damage: A Letter to Communities," *Harvard Educational Review* 79, no. 3 (2009): 409–428; on respect and reciprocity as method, see Aileen Moreton-Robinson, *White Possessive: Property, Power, and Indigenous Sovereignty* (Minneapolis: University of Minnesota Press, 2015); on the theorizing of settler colonialism as a structure and the related concept of Indigeneity, see J. Kēhaulani Kauanui, "'A Structure, Not an Event': Settler Colonialism and Enduring Indigeneity," *Lateral* 5, no. 1 (2016), https://csalateral.org/issue/5-1/forum-alt-human ities-settler-colonialism-enduring-indigeneity-kauanui/; on Indigenous forms of relationality as an alternative to the logics and myths of settler colonialism, see Kim Tallbear, "Caretaking Relations, Not American Dreaming," *Kalfou* 6, no. 1 (2019): 24–41; on Indigenous STS, see Kim TallBear, *Native American DNA: Tribal Belonging and the False Promise of Genetic Science* (Minnesota, 2013); among others; on settler colonialism and STS, see Maile Arvin, *Possessing Polynesians: The Science of Settler Colonial Whiteness in Hawai'i and Oceania* (Duke, 2019), among others. For Latin American Studies, the work of early modernists on knowledge, technology, and healing in the Atlantic World proves influential. See Pablo Gómez, *The Experiential Caribbean* (Chapel Hill: University of North Carolina Press, 2017); Marcy Norton, "Subaltern Technologies and Early Modernity in the Atlantic World," *Colonial Latin American Review* 26, no. 1 (2017): 18–38. See also Sylvia Wynter, "No Humans Involved: An Open Letter to My Colleagues," in *Forum NHI: Knowledge for the 21st Century*, vol. I, no. 1 (Stanford: Stanford University Press, 1994): 42–71.

anti-presentism that is common in histories of science and medicine.[17] Make no mistake about our critique: we are not calling for a politically correct, anachronistic, or tendentious history in lieu of some alleged traditional approach, nor are we arguing that all human scientific research only caused harm, and never good. Instead, we invite readers to join us in a productive discomfort with an objectivity that demarcates speaking about *what happened* in narratives that reference only the affective experience of the powerful and dominant, while avoiding calls for restorative, presentist judgment on the part of the groups marginalized by the official frame and their allies. The contributors to this book thus explore how to decolonize the history of the human sciences and enact a critical and ethical form of historical research and analysis.

Reframing the History of the Human Sciences

This book positions itself as a historical contemplation of the human sciences aligned with much literature outside of traditional history of science scholarship.[18] While our approaches build on an earlier generation of scholars' critical insights, among them prominent cultural theorists such as Michel Foucault,[19] our turn toward global Indigenous Studies and Latin American Studies scholarship, including critical theory produced within those partially overlapping fields, signals our rejection of a Eurocentric framing of science, society and power. By engaging the work of Eve Tuck (Unangax̂), Sylvia Wynter, Silvia Rivera Cusicanqui, and others, we rethink the history of the human sciences vis-à-vis attention to multiple contexts, forms of agency and interaction, and intellectual and epistemological traditions. Recent scholarship on science in Latin America, moreover, focuses attention on power and resistance, asking how the social sciences have justified systems of inequality and motivated projects toward civil rights and the decolonization of knowledge.[20] Indigenous histories and settler colonial studies, on the other hand, have called attention

---

[17] For discussions of presentism in the history of science, see Naomi Oreskes, "Why I am a Presentist," *Science in Context* 26, no. 4 (2013): 595–609. On the political commitment of Latin American and Caribbean historians to the society and time period in which they live, see Manuel Moreno Fraginals, *La historia como arma y otros estudios sobre esclavos, ingenios, y plantaciones* (Barcelona: Crítica, 1999).

[18] The classic is Roger Smith, *The Norton History of the Human Sciences* (New York: W.W. Norton, 1997).

[19] See Michel Foucault, *The Order of Things: An Archaeology of the Human Sciences* (London: Routledge, 2010).

[20] For studies of the social sciences in relation to race and gender in Latin America, see Nancy Leys Stepan, *"The Hour of Eugenics": Race, Gender, and Nation in Latin America* (Ithaca, NY: Cornell University Press, 1991); Nancy P. Appelbaum et al., *Race and Nation in Modern Latin America* (Chapel Hill: University of North Carolina Press, 2003); Laura Briggs, *Reproducing Empire: Race, Sex, and U.S. Imperialism in Puerto Rico* (Berkeley: University of California Press, 2002); Julia Rodriguez, *Civilizing Argentina: Science,*

to the ways in which the human sciences are implicated in the dispossession of Indigenous peoples and inform ongoing criticism of the human sciences in the present.[21]

While we consider applying these perspectives to histories of science as self-evidently beneficial, we are struck by the fact that such approaches remain revisionist in a historiography that still primarily treats the human sciences as part of a North Atlantic intellectual tradition. Dorothy Ross's 1994 edited volume *Modernist Impulses in the Human Sciences* is a key example.[22] At the time a powerful contribution, the volume's authors treated the human sciences as unidirectional and shaped by European and North American internal patterns, cultural vogues, traditions, and religious, moral, and ethical frames. They elided the instrumental role Indigenous peoples, colonized populations, and women played in codifying those systems of knowledge. Contributors portrayed human scientific knowledge as only expanding outward to other groups through the hegemonic exercise of literacy, power, trade, imperialism, field study, and hedonistic engagement. While critiques of these methodological and framing concepts have emerged in the literature on the history of the human sciences, they nevertheless persist. To that end, the authors of *Troubling Encounters* take as a starting point that the human sciences themselves have been dependent upon troubling encounters with the very people and material objects marginalized in this canon.[23]

Medicine, and the Modern State (Chapel Hill: University of North Carolina Press, 2006); Alejandra Bronfman, *Measure of Equality: Social Science, Citizenship, and Race in Cuba, 1902–1940* (Chapel Hill: University of North Carolina Press, 2004); Ashley Elizabeth Kerr, *Sex, Skulls, and Citizens: Gender and Racial Science in Argentina (1860–1910)* (Nashville, TN: Vanderbilt University Press, 2020). For studies of the social sciences in relation to civil rights projects and the decolonization of knowledge, see Ella Shohat and Robert Stam, *Race in Translation: Culture Wars around the Postcolonial Atlantic* (New York: New York University Press, 2012).

[21] For studies of the social sciences and Indigenous dispossession, see Patrick Wolfe, *Settler Colonialism and the Transformation of Anthropology: The Politics and Poetics of an Ethnographic Event* (London: Cassell, 1999); Audra Simpson, *Mohawk Interruptus: Political Life Across the Borders of Settler States* (Durham, NC: Duke University Press, 2014); Tuck and Yang, "R-Words," 223–248; Linda Tuhiwai Smith, *Decolonizing Methodologies: Research and Indigenous Peoples* (London: Zed Books, 2012 [1999]); Arvin, *Possessing Polynesians*; Vine Deloria, *Custer Died for Your Sins: An Indian Manifesto* (New York: Macmillan, 1969); Joanna Barker, "The Specters of Recognition," in *Formations of United States Colonialism*, ed. Alyosha Goldstein (Durham, NC: Duke University Press, 2014), 33–56; Ailton Krenak, *Ideas to Postpone the End of the World* (Toronto: Anansi International, 2020).

[22] Dorothy Ross, ed., *Modernist Impulses in the Human Sciences* (Baltimore: Johns Hopkins University Press, 1994).

[23] The classics remain the massive volumes by George Stocking, *Victorian Anthropology* (New York: Macmillan Free, 1987) and *After Tylor: British Social Anthropology, 1888–1951* (Madison: University of Wisconsin Press, 1992).

An example of how this North Atlantic approach has persisted can be found in the recent work of historian of medicine Jan Goldstein. In her 2015 AHA Presidential address, later published in *The American Historical Review*, Goldstein grappled with whether historians should hesitate before passing judgment on the behavior of past historical actors, who often lived according to notions of right and wrong that differed from our own. She suggested that rather than engage in presentism or reject it altogether, historians should situate people in the context of their own historical "moral fields."[24] In her case study of French racial science, however, Goldstein situated moral thinking as ultimately shaped by and within a French culture of ideas, even as that culture itself was obviously contemplating human difference in a context where struggles over empire, exclusion, and belonging were quite tangible parts of its flourishing and reality. In doing so, she, like Ross, employed an approach that erased the way these sciences abetted domination and served central roles in imperial and elite claims to possessing others.[25] Goldstein's approach generated much discussion following the publication of her address, including among our authors, but of greater interest to us is the question of how imperial and colonial subjects exercised counter-agency and survivance and shaped those moral fields.[26]

Beginning in the early 2000s, historians of science challenged the preoccupation with the human sciences' internally evolving epistemologies or genealogies of norms and truth claims. Among them, historian Warwick Anderson called for a hermeneutics of context and encounter in his landmark book *The Collector of Lost Souls: Turning Kuru Scientists into Whitemen*.[27] Applying Anderson's strategy to the history of the human sciences entails studying collaboration, power, and the struggles over power that collaborations engendered. Like Anderson, we conceive of science as a transnational praxis and narrative that crossed borders but was itself also shaped by long histories of different kinds of imperialism and colonialism (and responses to each), both at the sites where research took place and within the societies and institutions where those sciences came into being and developed. Such an approach allows

---

[24] Jan Goldstein, "Toward an Empirical History of Moral Thinking: The Case of Racial Theory in Mid-Nineteenth-Century France," *American Historical Review* 120, no. 1 (2015): 5.

[25] See Henrika Kuklick, *The Savage Within: The Social History of British Anthropology, 1885–1945* (New York: Cambridge University Press, 1996).

[26] Early in our project, we found it useful to engage Goldstein's address. However, the "moral field" as defined by Goldstein became less central to our work as our perspective on the project's political stakes shifted; readers, however, will still note some authors' critical assessments of the concept.

[27] Warwick Anderson, *The Collector of Lost Souls: Turning Kuru Scientists into Whitemen* (Baltimore: Johns Hopkins University Press, 2008).

us to continue to trace the broader global movement of scientific ideas but also permits focused examinations of ethics, structural conditions, and encounters that reveal a spectrum of human scientific knowledge-making practices within and beyond societies.

As many studies have shown – especially studies by Indigenous scholars and scholars from the Global South – racial science engendered multiple forms of violence, dehumanization, and racism. Empirical histories of research structures and practices like those in this book probe the specific power relations imbued in knowledge production. Many histories of human science are histories of violence. Evidence gathering, photography, and other forms of documentation are themselves often violent acts, in addition to being motivated by curiosity, approximation, and desires for preservation. Anthropology, in particular, has facilitated cultural and physical disappearance and "salvages," with scientists a party to colonization, settlement, and modernization processes. As "civilizing projects," anthropology, and later economics and development studies also participated in constructing or attempting to impose social norms and expectations on those scientists researched.

The authors in this book have sought to account for these histories of violence and dispossession, so inextricable from practices in the human sciences, while also providing narratives of connection, survivance, and thrivance.[28] Thus, by drawing on recent theories from global Indigenous Studies, they seek to balance a reckoning of trauma and harm with celebration, futurities, and desire.[29] To that end, many of the chapters in this book draw on Indigenous theories of reciprocity, repair, and right relations to frame complex case studies from across the Americas and the Pacific that, taken together,

---

[28] Recent trends in the history profession reflect this new thinking about Indigenous history; see, for example, David Treuer, *The Heartbeat of Wounded Knee: Native America from 1890 to the Present* (New York: Riverhead, 2019); Susan Sleeper-Smith, et al., eds., *Why You Cannot Teach United States History without American Indians* (Chapel Hill: University of North Carolina, 2019). See also "In Future Issues: Decolonizing the AHR," *American Historical Review* 123, no. 1 (February 2018): xiv–xvii, and follow-up essay in the January 2019 issue. On Indigenous agency, voice, and authority, see Joshua L. Reid, "Introduction: Indigenous Agency and Colonial Law" (AHR Forum), *American Historical Review* 124, no. 1 (February 2019): 20–27. Recent works in Black diaspora studies also offer valuable models for this approach. For the history of medicine specifically in Black diaspora studies, see Gómez, *The Experiential Caribbean*; Londa Schiebinger, *Secret Cures of Slaves: People, Plants, and Medicine in the Eighteenth-Century Atlantic World* (Stanford: Stanford University Press, 2017); Sasha Turner, *Contested Bodies: Pregnancy, Childrearing, and Slavery in Jamaica* (Philadelphia: University of Pennsylvania Press, 2017).

[29] On survivance, see Gerald Vizenor, ed., *Survivance: Narratives of Native Presence* (Lincoln: University of Nebraska Press, 2008); and Treuer, *The Heartbeat of Wounded Knee*.

decenter standard North Atlantic histories of science and the narratives of interaction and encounter they traditionally emphasize.[30]

Building on scholarship by Jorge Cañizares-Esguerra, Mariola Espinosa, Gabriela Soto Laveaga, and others who have called for globalizing the history of science and medicine and centering Latin America's (and other regions') contributions to knowledge production, our approach is also transnational, though local and national scientific traditions often form part of the story.[31] We trace not only practices related to research on the ground at the field site, but also the movement of information and knowledge to research centers and their eventual publication, circulation, and reception among broader audiences. The chapters emphasize relations: between scientists and material objects, places, and human subjects, including intermediaries and local assistants. They explore how power and resistance have been understood and performed in the context of the researchers' gaze, how human science projects related to larger forces and structures of inequality and violence, and how they were (and are) critiqued and subverted from the perspective of the research subject. In this sense, we trace how moral justifications and frameworks came to be constructed over time and space in the human sciences.

Science, Empire, and Colonialism

What would it mean to collectively rethink the history of the human sciences in ways that center the role of such sciences as technologies, successful or failed, of imperialism and colonialism? Our endeavor builds on a robust literature on colonial/imperial science that has appeared in the last fifty years or so. Early influential works largely followed a center–periphery model of knowledge transfer and diffusion.[32] Later, circulationist, local, "bottom-up,"

---

[30] See Bruchac, *Savage Kin* on respect and reciprocity as method; Moreton-Robinson, *White Possessive*, on merging Indigenous theory with traditional academic disciplines to advance knowledge production about identity categories; Kauanui, "'A Structure, Not an Event'" on cooperation between Indigenous Studies and settler colonial studies; TallBear, "Caretaking Relations, Not American Dreaming."

[31] Jorge Cañizares-Esguerra, "Iberian Science: Ignored How Much Longer?" *Perspectives on Science* 12, no. 1 (2004): 86–125; Mariola Espinosa, "Globalizing the History of Disease, Medicine, and Public Health in Latin America," *Isis* 104, no. 4 (2013): 798–806; Gabriela Soto Laveaga, "Largo Discolare: Connecting Microhistories to Remap and Reconnect Histories of Science," *History and Technology* 34, no. 1 (2018): 21–30. On human science in the Pacific, see Arvin, *Possessing Polynesians*.

[32] See Alfred Crosby, *The Columbian Exchange: Biological and Cultural Consequences of 1492* (Westport, CT: Greenwood Publishing Group, 1972); Antonello Gerbi, *La naturaleza de las indias nuevas: Cristóbal Colón a Gonzalo Fernández de Oviedo* (Mexico: Fondo de Cultura Económica, 1978); Daniel Headrick, *Tools of Empire* (London and New York: Oxford University Press, 1981); David Arnold, *Colonizing the Body: State*

and subaltern studies of imperial science augmented and complicated these approaches.[33] There also developed a group of thoughtful theoretical analyses of colonial science as glimpsed through the lenses of postcolonial theory. In the last ten years, historians from and of the Global South have broadened these conversations significantly, building on the foundational work of scholars such as Frantz Fanon and Sylvia Wynter.[34]

At the same time, historians have reconsidered various cases of human science research and its imperialist dimensions in Latin America in the modern period, in some cases reinforcing the center–periphery model while deemphasizing the role of local interlocutors and flattening their desires. Ricardo D. Salvatore proposed in his 2016 book *Disciplinary Conquest: U.S. Scholars in South America, 1900–1945* that US-led projects in the first half of the twentieth century were prime examples of imperial knowledge production, reflecting expansionist tendencies of capital, technology, and culture.[35] Salvatore emphasized the unequal power relations, extraterritoriality, extraction, and exploitation implicit within the actual processes of knowledge production, which involved materials "constantly flowing toward centers of knowledge" in the Global North. Peripheries thus functioned "as great repositories of evidence to the center," as treasure troves of facts that researchers could gather and transmit for processing in the center without any possibility of a reciprocal relationship or concern about ethics.[36] In this way, Salvatore ensured that the center–periphery model endured, repackaging it as one in

---

Medicine and Epidemic Disease in Nineteenth-Century India (Berkeley: University of California Press, 1993).

[33] This literature is rapidly expanding and here we mention just a few examples: Martha Few, *For All of Humanity: Mesoamerican and Colonial Medicine in Enlightenment Guatemala* (Tucson: University of Arizona Press, 2015); Jorge Cañizares-Esguerra, *How to Write the History of the New World* (Stanford: Stanford University Press, 2001); Gómez, *The Experiential Caribbean*; Espinosa, "Globalizing the History of Disease," 798–806; Gabriela Soto Laveaga, "Largo Discolare."

[34] Sujit Sivasundaram, "Focus: Global Histories of Science," *Isis* 101, no. 1 (March 2010): 95–97; Stuart McCook, "Global Currents in National Histories of Science: The 'Global Turn' and the History of Science in Latin America," *Isis* 104, no. 4 (December 2013): 773–776; Frantz Fanon, *Black Skin, White Masks* (New York: Grove Press, 1967); Fanon, *The Wretched of the Earth* (New York: Grove Press, 1963); Sylvia Wynter, "No Humans Involved"; Sylvia Wynter, "Unsettling the Coloniality of Being/Power/Truth/Freedom: Towards the Human, after Man, Its Overrepresentation – An Argument," *CR: The New Centennial Review* 3, no. 3 (2003): 257–337; also Sylvia Wynter, "The Re-enchantment of Humanism: An Interview with Sylvia Wynter" (D. Scott, Interviewer), *Small Axe* 8 (2000): 119–207.

[35] Ricardo D. Salvatore, *Disciplinary Conquest: U.S. Scholars in South America, 1900–1945* (Durham, NC: Duke University Press, 2016).

[36] Ibid., 55.

which academic knowledge from the center was contingent on the extraction and accumulation of data from the periphery.

Other scholars took an alternative approach that showcased Latin America as a central site of transnational science. Histories of anthropology are indicative of this trend, as Karin Rosemblatt has recently shown.[37] In her 2018 book *The Science and Politics of Race in Mexico and the United States, 1910–1950,* Rosemblatt closely examined social scientists on both sides of the border. There were important distinctions between the two countries' intellectual trajectories, scientific thinking around race, and aims for policy outcomes, even as scientific communities in each nation freely shared ideas with each other. In the United States, the overarching goal was "sensitive" acculturation of Indigenous people matched by the assumption that Indigenous cultural expressions would "disappear." Mexican scientists' energies, in contrast, were largely devoted to documenting in great detail the cultural diversity in their nation, and they resisted generalizing theories. An article by Sebastián Gil-Riaño offers yet another transnational perspective. Carefully examining anthropologists' and social scientists' roles in the 1950 UNESCO Statement on Race, Gil-Riaño found that "in [the] transition to an economic-development paradigm, 'race' did not vanish so much as fragment into a series of finely tuned and ostensibly antiracist conceptions that offered a moral incentive for scientific elites to intervene in the ways of life of those deemed primitive." Even mid-century antiracist human scientists worked comfortably across borders within colonial, postcolonial, and modernization frameworks. As a result, "the retreat of scientific racism did not signify an end but rather an amplification of racial politics."[38]

Gil-Riaño's point is well-taken. Since its inception, anthropology has engaged many subjects, but most of its work has been with and on Indigenous peoples. Anthropologists' writings reveal that they were often handmaidens to colonial states and capital expansion, not to mention racial and gendered violence. Anthropological encounters were also almost always transactional, but there were numerous cases of ambiguity, opening to other types of relationships, and even a desire for right relations. Anthropologists, however, were not alone. As is evident in this book, psychologists, sociologists, criminologists, demographers, biomedical researchers, and geographers also navigated these situations. Even as they analyzed groups and subcultures in ways that accorded priority to the scientists' normative preferences (e.g., internal attributes, evolutionary history, proper behavior, propensity for

---

[37] Karin Alejandra Rosemblatt, *The Science and Politics of Race in Mexico and the United States, 1910–1950* (Chapel Hill: University of North Carolina Press, 2018).

[38] Sebastián Gil-Riaño, "Relocating Anti-Racist Science: The 1950 UNESCO Statement on Race and Economic Development in the Global South," *BJHS* 51, no. 2 (June 2018): 303, 281.

mental disease, gender and sexuality, and relationship to the natural environment), they also grappled with the dynamics of encounters.[39]

Race and racial science are a central focus in our book, because they prompted human scientists' appropriation of colonial logics that now trouble us. In the Americas, the race concept and racism have histories that extend back to the early Spanish colonial period.[40] As Peruvian sociologist Aníbal Quijano wrote in 2000, in the moment when the Americas emerged within the world market and created the opportunity for a world capitalism, a new mental category emerged codifying "the relations between conquering and conquered populations" and producing an "idea of 'race' as biologically structural and hierarchical differences between the dominant and dominated" as natural. In reflecting on these matters and articulating his notion of the "coloniality of power," in which race is central, he observed:

> New social historical identities were established: "Spanish" or "Portuguese" ("Whites" and "Europeans" came much later), "Indians," "Negroes" and "Mestizos." So "race" (biology and culture or, in our present terms, "race" and "ethnicity") was placed as one of the basic criteria to classify the population in the power structure of the new society, associated with the nature of roles and places in the division of labor and in the control of resources of production.[41]

For Quijano, this new mental category extended beyond mere matters of "external or physiognomic differences" and into considerations of "mental and cultural differences," hierarchized in terms of a superior point of reference or center and its others. These concepts of difference constituted classification schema with material consequences, against which colonized groups contended.

---

[39] For examples of human sciences histories that focus on biomedicine and demography, see Gabriela Soto-Laveaga, *Jungle Laboratories: Mexican Peasants, National Projects, and the Making of the Pill* (Durham, NC: Duke University Press, 2009); Raúl Necochea López, *A History of Family Planning in Twentieth-Century Peru* (Chapel Hill: University of North Carolina Press, 2014). On criminology in Latin America, see Robert Buffington, *Criminal and Citizen in Modern Mexico* (Lincoln: University of Nebraska Press, 2000); Carlos Aguirre, *The Criminals of Lima and Their Worlds: The Prison Experience, 1850–1935* (Durham, NC: Duke University Press, 2005); Rodriguez, *Civilizing Argentina*.

[40] We draw on George Stocking, ed., *Colonial Situations: Essays on the Contextualization of Ethnographic Knowledge.* History of Anthropology, vol. 7. (Madison: University of Wisconsin Press, 1991). For the history of race in the Americas and its links to empire, see James Sweet, "The Iberian Roots of American Racist Thought," *The William and Mary Quarterly* 54, no. 1 (1997): 143–166; Jorge Cañizares-Esguerra, "New World, New Stars: Patriotic Astrology and the Creation of Indian and Creole Bodies in Colonial Spanish America, 1600–1650," *The American Historical Review* 104, no. 1 (1999): 33–68.

[41] Aníbal Quijano, "Coloniality of Power and Eurocentrism in Latin America," *International Sociology* 15, no. 2 (2000): 216.

In 2003, Afro-Caribbean scholar Sylvia Wynter expanded on Quijano's "coloniality of power" and Walter Mignolo's related concept of "colonial difference" in her essay "Unsettling the Coloniality of Being/Power/Truth/ Freedom."[42] Wynter argued that "Man," the Western bourgeois notion of the human "which overrepresents itself as if it were the human itself," in practice defines itself against non-white, especially Black and dark-skinned others. Furthermore, "Man" must be examined both in relation to its development in the Enlightenment era and in relation to earlier Judeo-Christian notions of what it meant to be "human."[43] Positioning the Copernican Revolution as a turning point, Wynter suggested that these differing concepts of the human correspond to contexts of knowledge making distinguished by the shifting influence of religion and science. Arguing that the notion of "Man" as a rational subject is in fact predicated on its opposite, the nonrational subject who is excluded from the category of human, Wynter links the development of these ideas to Iberian colonization of the Americas and European expansionism into Africa. By focusing on "'the rise of Europe' and its construction of the 'world civilization' on the one hand, and, on the other, African enslavement, Latin American conquest, and Asian subjugation," her work firmly links the very subject of the human sciences, "Man," to empire.[44]

Scholarship in settler colonial theory, which has long been used to interrogate structures of colonialism in former British colonies such as Canada, the United States, Australia, and New Zealand, adds further nuance to these discussions of classification schema with relevance across the hemisphere. Drawing Latin America into discussions of settler colonialism and settler colonial science in US territories throws into relief the complexities, diversity, and overlapping histories of colonialism and imperialism in these different locations. In fact, the chapters in our book ultimately required engagement with various theories of colonialism and neocolonialism that have long been treated as applicable to separate contexts in North America, Latin America, and the Pacific. To that end, chapters focus not just on postcolonial nations like Argentina, Mexico, Peru, and Brazil, but also territories of the formal US Empire such as Hawai'i, Puerto Rico, and the Akimel O'odham nation, which also variously claim sovereignty.

This hemispheric scope complicates the common sense of what we mean by colonial encounters and requires that we acknowledge how much there is to

---

[42] Sylvia Wynter, "Unsettling the Coloniality of Being/Power/Truth/Freedom: Towards the Human, after Man, Its Overrepresentation – An Argument," *CR: The New Centennial Review* 3, no. 3 (2003): 257–337. For the concept of "colonial difference," see Walter Mignolo, *Local Histories, Global Designs: Coloniality, Subaltern Knowledge, and Border Thinking* (Princeton: Princeton University Press, 2000).

[43] Wynter, "Unsettling," 260.

[44] Ibid., 263.

learn from Latin America, in particular, when held up to histories of British and British-derived settler colonialism.[45] With its distinct local and regional histories and its large and in some places dense Indigenous populations, the region and its many nations bring distinct cases and conditions of domination and contestation to the table. Latin America has long been home to processes of colonialism and internal colonialism and its peoples have experienced the legacies of multiple empires, including pre-Columbian Indigenous empires as well as settler ones. These histories result in distinct experiences regarding, for example, the impact of Iberian versus British legal regimes on Indigenous rights.[46]

Recently, scholars have also made compelling arguments about the value of settler colonial theory for understanding Latin America, especially in the contexts of modern nation building, US imperialism, and global capitalism.[47] Similarly, insights from Latin American and Latin Americanist theorists inform our relational framework for studying the region and the US Empire. Such scholarship pushes back against the flattening effects of settler colonial theory, which sometimes reduces historical actors to settlers and non-settlers with little nuance. Many of the chapters engage an expanded corpus of scholarship on colonialism and Indigeneity, one that places theories such as settler colonialism in productive dialogue with theories of internal colonialism and coloniality, among others. Some also touch on Blackness in the Americas to better incorporate and engage case studies of human sciences research involving populations of African descent. In this way, the chapters herein

---

[45] A significant body of scholarship shows that settler colonialism has taken on different forms in different local settings. For contemporary Latin America, see Speed, "Structures of Settler Capitalism in Abya Yala," 783–790. Other essays in that issue of *American Quarterly* are equally valuable for theorizing settler colonialism in a Latin American context. For Mexico specifically, see Natasha Varner, *La Raza Cosmética: Beauty, Identity, and Settler Colonialism in Postrevolutionary Mexico* (Tucson: University of Arizona Press, 2020). For a comparative analysis of settler colonialism across the US–Mexico border, see María Josefina Saldaña Portillo, *Indian Given: Racial Geographies across Mexico and the United States* (Durham, NC: Duke University Press, 2016). For Hawai'i, see Arvin, *Possessing Polynesians*.

[46] On colonial legal regimes, see Lauren Benton, *Law and Colonial Cultures: Legal Regimes in World History, 1400–1900* (New York: Cambridge University Press, 2001); Patricia Seed, *American Pentimento: The Invention of Indians and the Pursuit of Riches* (Minneapolis: University of Minnesota Press, 2001); Bianca Premo, *Enlightenment on Trial: Ordinary Litigants and Colonialism in the Spanish Empire* (New York: Oxford University Press, 2017); Yanna Yannakakis, *The Art of Being In-Between: Native Intermediaries, Indian Identity, and Local Rule in Colonial Oaxaca* (Durham, NC: Duke University Press, 2008); Adrian Masters, "The Two, the One, the Many, the None: Rethinking the Republics of Spaniards and Indians in the Sixteenth-Century Spanish Indies," *The Americas* 78, no. 1 (2021): 3–36.

[47] Speed, "Structures of Settler Capitalism in Abya Yala," 783–790.

focus on more than elimination or salvage and include themes of race, gender, labor, Mestizaje, acculturation, and nationalism.[48]

We highlight these overlapping histories of imperialism and colonialism with an eye to unique dynamics in local and national contexts. A notable consequence of earlier North Atlantic and settler colonial frameworks is that they have hindered the development of hemispheric discussions and theorizations of Indigeneity and Blackness in relation to multiple forms of colonialism. Moreover, different societies have varying timelines in terms of their colonial and imperial experiences. The end of European colonial rule in the "Latin" part of the Americas unfolded by and large in the nineteenth century, but the entire region eventually fell under US purview. Some societies have yet to break free from the yoke of colonialism and imperialism. This is especially true of US territories, all of which are at different junctures in their relationship with the United States.[49] Long a subject of Latin American scholarly and political critique and more recently a focus of much Anglophone scholarship, the US Empire was forged in slavery and dispossession and shifted over time toward capital extraction and concentrated positions of military power, which by the twentieth century became global in their extent. Like other nations in the Western hemisphere, the territorial United States also had and has its own colonial subjects, including Indigenous peoples, emancipated slaves, and their descendants.[50]

While practically unknown among historians of the human sciences in the North Atlantic context, Latin American theorists of empire recognize Quijano's critiques of the human sciences and their links to ongoing manifestations of colonialism. They also engage Wynter's discussions of science in the construction of "Man" and the notion of the human to understand racism in the present. By analyzing encounters between human scientists and Black,

---

[48] Our hope is that this book may begin a productive conversation with scholars examining histories of Blackness, settler and internal colonialisms, and the human sciences. On Blackness and settler colonialism, see Tyla Miles, "Beyond a Boundary: Black Lives and the Settler–Native Divide," *William and Mary Quarterly* 76, no. 3 (2019): 417–426; Stephanie Smallwood, "Reflections on Settler Colonialism, the Hemispheric Americas, and Chattel Slavery," *William and Mary Quarterly* 76, no. 3 (2019): 407–416.

[49] Recent reviews of Daniel Immerwahr's *How to Hide an Empire: A History of the Greater United States* (New York: Farrar, Straus, and Giroux, 2019) shed light on how to conceptualize the United States's territorial holdings. See Anne S. MacPherson, "A Caribbean Historian Extends Scholarly Critiques of "How to Hide an Empire: A History of the Greater United States," *Society for US Intellectual History*, March 15, 2020, https://s-usih.org/2020/03/a-caribbean-historian-extends-scholarly-critiques-of-how-to-hide-an-empire-a-history-of-the-greater-united-states/; Daniel Immerwahr, "Puerto Rico in the U.S. Empire: A Reply to Anne MacPherson," *Society for US Intellectual History*, March 22, 2020, https://s-usih.org/2020/03/puerto-rico-in-the-u-s-empire-a-reply-to-anne-macpherson/.

[50] Immerwahr, *How to Hide an Empire*, Introduction.

Indigenous, and mixed communities, our book dialogues with these scholars and acknowledges that colonialism and coloniality are familiar frames for exploring uneven and violent encounters. For example, Quijano pointed out the contributions of Eurocentric knowledge production alongside shifting relations of production in the establishment and expansion of Iberian colonies. Natural philosophy and eventually science in the nineteenth century codified "the relations between conquering and conquered populations" and naturalized and ascribed to biological theories of race "hierarchical differences between the dominant and dominated," ideas that persisted into the national period in Latin America.[51]

Silvia Rivera Cusicanqui has taken these critiques yet further, proposing that an ethical coherence be applied in a practical decolonization of knowledge production in Latin American contexts, and that it be based on the Andean concept of *ch'ixi* (roughly: motley or mixed).[52] By critiquing the notion of hybridity and centering ch'ixi, Rivera Cusicanqui, writing as a person of double ancestry in a historically colonial context, refers to the experience of living with unreconciled and coexisting cultural differences that inform one's identity and "antagonize and complement each other." In her words, ch'ixi shapes epistemologies by constituting "a double and contentious ancestry, one that is denied by the process of acculturation and the 'colonization of the imaginary' but one that is also potentially harmonious and free if we liberate our half-Indian ancestry and develop dialogical forms for the construction of knowledge."[53] While Rivera Cusicanqui does not endorse a decolonial approach to knowledge, her scholarship informs how we interrogate and historicize different actors' ways of perceiving and acting upon research in the human sciences when positioned through relationships understood as colonial.

Indigenous Studies and Postcolonial and Decolonial Histories

Mohegan writer, historian, and storyteller Melissa Tantaquidgeon Zobel has eloquently and forcefully articulated how, through various unethical research projects involving Indigenous communities in North America, "outside researchers have posed a real and constant threat." She advises that "if we are ever to recover from these issues, the ethics of engagement call for serious

---

[51] Aníbal Quijano, "Coloniality of Power and Eurocentrism in Latin America," *International Sociology* 15, no. 2 (2000): 216.

[52] Silvia Rivera Cusicanqui, "Ch'ixinakax utxiwa: A Reflection on the Practices and Discourses of Decolonization," *South Atlantic Quarterly* 1, no. 111 (2012): 95–109. See also Kauanui, "'A Structure, Not an Event.'"

[53] Silvia Rivera Cusicanqui, *Ch'ixinakax utxiwa: On Practices and Discourses of Decolonization* (Cambridge: Polity Press, 2020), 66–67.

consideration."[54] In conversation with Indigenous Studies scholars at a 2018 University of Washington workshop, we further expanded our consideration of decolonizing methodologies and recognized the need to diversify our list of contributing authors and our approaches. Among other things, we encouraged contributors to consider the questions at the heart of Māori education scholar Linda Tuhiwai Smith's work about who owns research, who designs it and carries it out, where research findings are disseminated, whose interests research serves, and who benefits from it.[55]

Since not all the chapters in this book directly address questions of Indigeneity to the same degree, we have encouraged a multiplicity of approaches to navigating the ethical, relational, political, epistemic, and reparative questions that decolonizing methodologies rightly center. We invited our authors, when possible, to ground their research in relationships with the communities whose contemporaries or ancestors appear in the pages of this book. When seeking authorization from communities has been feasible and authors have raised this possibility, we have encouraged it along with other forms of collaboration. However, we have also made the deliberate decision that this should not be a requirement of the book. In multi-sited studies that trace the travels of scientists across broad regions in the past, questions of scale and the sheer number of communities involved may make such practices impossible. In other chapters, authors focus on scientists who researched and discussed Indigenous communities in the abstract through data and statistical information, rather than engaging in direct relations with such populations.

Decolonizing methodologies, of course, entail far more than just reliance on authorizing strategies and are wider ranging than Smith's guidelines. In her recent book on photography and the leper colony in Molokai, for example, Adria Imada advocates for an "ethics of restraint," a strategy that reflects deeply on the subject position and responsibility of the researcher to accommodate the desire of some communities not to be contacted or subjected to an outsider's gaze. Imada's example does not avoid questions of ethics, accountability, and obligations, but rather makes them central to her research.[56]

---

[54] Melissa Tantaquidgeon Zobel, "Foreword," in *Savage Kin*, Margaret Bruchac (Tucson: The University of Arizona Press 2018), xi.

[55] Smith, *Decolonizing Methodologies*, 10. For decolonizing methodologies in Latin America and the Pacific, see Florencia E. Mallon, ed., *Decolonizing Native Histories: Collaboration, Knowledge, and Language in the Americas* (Durham, NC: Duke University Press, 2012); Florencia E. Mallon, *Courage Tastes of Blood: The Mapuche Community of Nicolás Ailío and the Chilean State, 1906–2001* (Durham, NC: Duke University Press, 2005).

[56] Imada engages with "surrogates who work with and have developed deep relationships with patients and people connected to the settlement" while adopting "active forms of distancing and restraint" with residents of the settlement itself; Adria Imada, *An Archive of Skin, An Archive of Kin: Disability and Life-Making during Medical Incarceration*

Miranda Johnson's recent article in *History and Theory*, likewise, questions to what degree Smith's methods are uniformly applicable to different contexts. Johnson reads Smith's work in the context of its production to understand its limitations and possibilities. She troubles tensions within Smith's framework "between objectivity and intersubjectivity, on the one hand, and between essentialist identity and hybridity, on the other," and asks why Smith's methodology "hinges on dichotomizing nonindigenous and indigenous researchers, who are by turn constrained in a colonial present."[57] By historicizing Smith's framing of decolonizing methodologies in relation to the characteristics and genealogies of Indigenous politics and engagement of history and anthropology in Aotearoa New Zealand in the late twentieth century and earlier, and by revisiting these ideas in relation to contemporary problems and historical justice movements, Johnson asks what other categories of historical actors and formulations of relationality might be possible within a decolonizing framework.[58]

More generally, historians of science have reflected productively on the differing uses and understandings of decolonizing methodologies in postcolonial and decolonial historical research. Sharing Warwick Anderson's caution that attempting to cordon off postcolonial approaches from their decolonial counterparts "might limit the power, range, and agility of both," this book's contributions fall along a spectrum between postcolonial and decolonial historical analyses. They offer readers multiple models for centering and writing about, or refusing to write about, Indigenous peoples and other subaltern or colonized actors.[59]

Settler Australian scholars Timothy Neale's and Emma Kowal's recent discussion of decolonizing methodologies is especially helpful. Neale and Kowal identify postcolonial approaches as engaging in "epistemic decolonizing," a practice in which historians and other scholars, having identified the origins of social inequalities between groups "in the domination of one episteme, or way of knowing, over others," seek "more 'horizontal' relations between histories and knowers, or foster 'a pluriverse' of onto-epistemes." They continue, "whatever the particular pathway, and many have been mapped, the ambition is to halt the domestication of othered subjects by first

(Berkeley: University of California Press, 2022). Julie Livingston wrestles with similar questions. See Julie Livingston, "Figuring the Tumor in Botswana," *Raritan* 34, no. 1 (2014): 10–24.

[57] Miranda Johnson, "Toward a Genealogy of the Researcher as Subject in Post/Decolonial Pacific Histories," *History and Theory* 59, no. 3 (2020): 429.

[58] Ibid., 429.

[59] Warwick Anderson, "Finding Decolonial Metaphors in Postcolonial Histories," *History and Theory* 59, no. 3 (2020): 431.

resisting their conceptual domination."[60] Such observations have also been made elsewhere, as in, for example, Achille Mbembe's work on colonial vernaculars in African contexts.[61] These methods within postcolonial historical analysis have taken various forms,[62] but they have generally served to overturn and eliminate grand narratives of the rise of science by "provincializing science into one 'indigenous knowledge tradition' among others or assiduously drawing attention to hybridity and 'contact zones' in order to undercut the past, present or future supremacy of supposedly universal knowledge practices."[63] They thus reinforce what Anderson describes as the "'unstable economy' of science's shifting spatialities as knowledge is transacted, translated, and transformed across the globe."[64]

Neale and Kowal contrast these epistemic decolonizing practices with what they call "reparative decolonizing," a decolonial approach that works "toward an explicitly material end: returning to Indigenous peoples the power and resources taken from them through (ongoing) colonialism."[65] In this approach, "the analysis of the West moves from being favored to becoming

---

[60] Timothy Neale and Emma Kowal, "'Related' Histories: On Epistemic and Reparative Decolonization," *History and Theory* 593 (2020): 404.

[61] Mbembe has analogized colonial vernaculars to "illicit cohabitation" created by subjects sharing "the same living space." He encouraged intellectuals thinking about the legacies of colonialism to consider the power relations among multiple epistemes. Achille Mbembe, "Provisional Notes on the Postcolony," *Africa* 62, no. 1 (1992): 4. In an interview published in *Esprit* in 2006 Mbembe observed that postcolonial thought "is not an anti-European thought. On the contrary, it's the product of the encounter between Europe and the worlds it once made into its distant possessions. In showing how the colonial and imperial experience has been codified in representations, divisions between disciplines, their methodologies and their objects, it invites us to undertake an alternative reading of our common modernity." Achille Mbembe, Olivier Mongin, Nathalie Lempereur, Jean-Louis Schlegel, and John Fletcher, "What Is Postcolonial Thinking?" *Esprit* 12 (2006): 117–133.

[62] See, for example, Saul Dubow, *Scientific Racism in Modern South Africa* (Cambridge: Cambridge University Press, 1995) for a discussion of the social sciences in Africa, especially chapters 2 and 6.

[63] Neale and Kowal, "'Related' Histories," 406. For the concept and use of "contact zones" in Latin American history, see Gilbert Joseph, Catherine LeGrand, and Ricardo Salvatore, *Close Encounters of Empire: Writing the Cultural History of U.S.–Latin American Relations* (Durham, NC: Duke University Press, 1998). In their volume, "contact zones" are described as "sites where ideologies, technologies, capital flows, state forms, social identities, and material cultures meet, and where multiple messages are conveyed; a series of communicative exchanges in which insiders and outsiders engage, act on, and represent each other" (15); "areas of intense interaction between two or more cultures in contexts of unequal power and resources" (336); and "social spaces where disparate cultures meet, clash, and grapple with each other, often in highly asymmetrical relations of domination and subordination" (403).

[64] Anderson quoted in Neale and Kowal, "'Related' Histories," 406.

[65] Ibid., 405.

the central analytical strategy," and the analysis itself serves a liberatory purpose.[66] The works of Tuck and Yang, Sisseton-Wahpeton Oyate scholar Kim TallBear, and Métis scholar Michelle Murphy are examples of this method. TallBear's deliberate refusal to analyze Native Americans' own views of DNA serves as "an *explicitly* ethical move from an *explicitly* situated place," one that supports Indigenous governance of knowledge production by studying "the under-studied non-Indigenous actors who currently dominate that sphere."[67] TallBear provides a valuable provocation for troubling how we center ethics in the study of human science encounters.

Michelle Murphy's intersectional interventions are also relevant here. Murphy views the main objective of her research as "the 'dismantlement' of colonial power," and proposes "world-building" as an active complement to this process.[68] She argues for

> going from *being* to *doing*. I want to start with creating alter-embodiments, alter-objects of care – even if only conjecturally – that call in our complicities, that require less-violent practices, that require different worlds. I want to think with you about tactically moving from being to doing – calling forth alter-embodiments, alter-being, or what I am calling here alter-life.[69]

Murphy invokes the work of Frantz Fanon "as a starting point for a decolonial STS," highlighting

> how he navigated, and refused to disavow, the contradictions making up embodiment, how Fanon theorized within a set of tensions or toggles: for example, the toggle between the hopeful care for embodied difference, and the pessimistic ways bodies are already materialized in colonial and racist worlds, the toggle between medicine's racist apprehension of pathological bodies, and the ways that bodies also exceeded those materializations.[70]

Given that this book's authors engage postcolonial and decolonial methods in different ways and to different ends, we believe the chapters provide valuable material for discussing how decolonizing methods should be conceptualized and incorporated into research on the history of encounters, ethics, and affect within the human sciences. Moreover, many chapters provide an opportunity to consider Neale and Kowal's provocative, and from our vantage point problematic, assertion that "a reparative approach that centers Indigenous peoples' politics and analytics while refusing to analyze Indigenous people

---

[66] Ibid., 406.
[67] TallBear, *Native American DNA*, 9, quoted in Neale and Kowal, "'Related' Histories," 406.
[68] Murphy cited in Neale and Kowal, "'Related' Histories," 407, 412.
[69] Michelle Murphy, "What Cannot a Body Do?" *Catalyst: Feminism, Theory, Technoscience* 3, no. 1 (October 18, 2017): 7.
[70] Ibid., 4.

and communities generates its own consequences and problems."[71] Such an assertion would certainly generate discussion and potentially disagreement among our authors.

Overview of the Book

The book is divided into three groups of historical studies, in addition to this introduction, the conclusion, and epilogues. In the first section, three chapters reexamine interactions between people and subjectivities in the field, with special attention to the affective dynamics of encounters on the part of both scientists and Indigenous human subjects and informants. The first chapter in this section, by Julia E. Rodriguez, rethinks Indigenous bodies and remains as unstable sources of scientific knowledge during a period of great violence and settler expansion in the late nineteenth-century Indigenous lands of Southern Argentina. Rodriguez reads scientists' reports of their own emotive states as well as their interpretation of Indigenous peoples against the grain, revealing that underneath the authoritative scientific conclusions lay uncertainty and unease.

Next, Adam Warren takes the reader to witness micro encounters engendered by the Yale Peruvian Expedition, exploring via textual and photographic evidence the racial scientific research that formed the relationality that shaped encounters in Peru between expedition members and Indigenous and Mestizo peoples, some of whom served as the expedition's workers and assistants. Warren questions how different groups imagined and contested the moral and ethical dimensions of such work. Drawing on the concept of ethnographic refusal in Indigenous Studies while also identifying other forms of engagement, Warren criticizes the univocal conception of moral fields as the possession of imperial researchers but not of Indigenous and Mestizo people subjected to their gaze.

Like Rodriguez and Warren, Sebastián Gil-Riaño's chapter turns to unexplored dimensions of contact in the field between distinct actors, interrogating a little-known story of Indigenous child abduction in twentieth-century Paraguay. Gil-Riaño uses his case to explore the underside of antiracist, redemptive stories of cultural assimilation that circulated in cosmopolitan institutions like UNESCO, arguing that these liberal ideas were in fact congruent with other instances of material and cultural decimation. As an examination of the scientific effacement of settler colonial violence, Gil-Riaño's study of forced child removal and its role in the human sciences captures the affective complexities that accompanied these violent encounters.

The second section turns toward institutions located within the United States Empire and the settler colonial logics that shaped encounters between

---

[71] Neale and Kowal, "'Related' Histories," 407.

researchers and research subjects within their walls. In many respects, what these three chapters interrogate is how settler colonialism operates within science. Moreover, they demonstrate how the normative practices of settler colonial science sought to deny and discipline Indigenous and racial differences and thereby created patterns of agency, endurance, and refusal.

In her study of settler colonial social science in action in the territory of Hawai'i, Maile Arvin calls attention explicitly to the way American social sciences shored up white supremacist imperial domination of the islands. What colonial authorities of the United States Empire construed as rehabilitative and objective truths about and applied to Native Hawaiians and immigrants of color, Arvin, focusing on the fertile ground of settler training schools, shows to be tools of domination that abetted countering and enduring patterns of resistance to the structure of Hawaiian settler colonialism.

Alberto Ortiz Díaz, meanwhile, shows how similar logics played out in mid twentieth-century Puerto Rico. Ortiz Díaz argues that human sciences researchers faced pressure to abandon earlier traditions and embrace the methods and biomedical enterprise of the US Empire's scientific modernity. Drawing on the history of mental testing and inmate assessment as well as designs for a new penitentiary, Ortiz Díaz contends that while mid twentieth-century American social science engaged in intense processes of othering that aligned with imperial expansion, Puerto Rican social scientists combined American psychometrics with older Spanish ethnographic traditions that powerfully resurfaced in the 1940s. This resulted in a blended, "creole" nationalist science with decolonial aspirations, but one that was colonial-populist in practice.

In both cases, Arvin and Ortiz Díaz show that while settler colonial science informed and exemplified attitudes about what "civilizing," disciplining, and reforming dominated subjects required, it bore signs of local inflections. Laura Stark adds to these deliberations by looking at boundary making in the Americas – scientific, bioethical, and racial. She leverages the provocative case of the National Institutes of Health and the surprising trajectory of one so-called normal research subject, Carolyn Matthews. Troubling Matthews's vernacular archive, Stark shows that over the course of a life spent participating in scientific research as an experimental subject and a technician gathering human subjects' data, Matthews acquired a new bioethical awareness. Yet, Mathews's time spent as an x-ray technician in a field study involving the Akimel O'odham tribe in the Sonoran desert, which was predicated on past and present US racist imperialism, did not form part of her reflections and criticisms. Stark asks whether Matthews's example shows state settler power operating through and with science, and in a compelling twist, emphasizes in her essay that bioethics was itself a product of the settler state project.

Themes of affective relationality and the colonial logics of science – settler, internal, or otherwise – bring issues of governance, politics, and self-determination to the foreground. The three studies in this book's third section show that self-determination needs to be interrogated beyond the work and experiences of history's subjects to include political contexts and desires for sovereignty. Analyzing the public controversy in Mexico over "Cuauhtémoc's Bones," a set of human remains that were discovered alongside other objects under the floor of a church in rural Ixcateopan, Guerrero, in 1949 and that villagers and some scientists attributed to the last Aztec emperor, Cuauhtémoc, Karin Rosemblatt problematizes how locals came to engage the human sciences and how they, alongside members of the scientific community sent to examine the remains and political officials, debated the discovery's significance at the local and national levels. Rosemblatt argues that the rich tensions in her story resulted from conflicts over gender, sexuality, and scientific authority and a specific vision of Mexico at the national level, which favored "a whitened, cosmopolitan, masculine identity and was unconcerned with the needs or histories of villages like Ixcateopan." Locals, however, articulated alternative forms of nationalism.

Navigating a transnational framework rather than a national one, Eve Buckley engages the local and the global to look at how formative years spent living alongside impoverished Black and mixed-race populations in the northeast of Brazil shaped Brazilian scientist Josué de Castro's critiques of overpopulation discourse and the eugenicist, neo-Malthusian arguments advanced by North American conservationist William Vogt. Through his book, *The Geography of Hunger*, and subsequent writings and correspondence, de Castro articulated a radically different vision of resource scarcity, hunger, population growth, and development that was grounded in relationality and a critique of economic systems. Buckley observes that these decisions were inherently political ones in a context where wealthy nations deployed the tools and expertise of the human sciences, often through a racial lens, to inform ideas about global security, modernization, development, and governance. De Castro proved more than able to turn the tools of human science around to challenge naturalistic arguments about scarcity.

Rosanna Dent's chapter examines not only how the Brazilian government has regulated researchers' access to Indigenous peoples through webs of bureaucratic oversight, but also how a particular Indigenous community in Brazil, A'uwẽ (Xavante) of the Pimentel Barbosa Indigenous Territory, has constructed its own frameworks and protocols for engaging outsiders. Dent argues that Brazilian state regulatory systems, which were intended to protect Indigenous interests, resulted in new forms of risk for them. As some contemporary geneticists seek ethical workarounds to avoid Brazil's bureaucracy, they use old biosamples created under prior ethical regimes. Indigenous peoples, in turn, find themselves in what she describes as a bureaucratic double bind, one in which non-Indigenous experts are inevitably "called on to justify and validate their claims in the eyes of the [Brazilian] state." Dent shows that by establishing their own relationship-based practices for engaging outsiders,

A'uwẽ navigate the possessive logics of both the state and researchers in ways that further their sovereignty.

Conclusion

Reflecting on encounters in the ways outlined in this chapter should in no way abet reactionary anti-scientism or challenge claims to expertise in the United States and elsewhere. Our concern is to gain insight into the colonial structures and tangled, complex encounters that have shaped research in the human sciences – for indeed they did. That concern, ethical in its frame of reference, does not end with the book's historical case studies. We argue that its subject matter presents an ongoing challenge for historians of the human sciences now and in the future as they write about encounters. The difficulty of doing so stems not just from guiding logics that have informed approaches within the history of the human sciences, but also from the way research scientists have customarily written about encounters, their adherence to protocols, and their own understanding of ethics and what ought to be said.

This challenge can be seen clearly in a 2020 article in *Nature* and its related *New York Times* op-ed, both of which discussed new findings on the genetic history and population size of pre-contact Caribbean peoples and their relations to contemporary Caribbean islanders. In the op-ed, geneticist David Reich and sociologist Orlando Patterson characterized research for the project as having been carried out "in collaboration with Caribbean scholars, with permission from Caribbean governments and institutions and in consultation with Caribbean people of Indigenous descent."[72] Likewise, the research article in *Nature* directs readers to a broader discussion of ethics in a supplementary information section, having noted that results of DNA analysis performed on ancient skeletal remains "were discussed before submission with members of Indigenous communities who trace their legacy to the pre-contact Caribbean and their feedback was incorporated."[73] The supplementary section, however, mentions not a single community or individual specifically identified as claiming such a legacy. Instead, it lists the museum officials and representatives of other institutions who granted permission for analysis of skeletal remains in their custody. How was the feedback of communities of Indigenous descent sought, heard, and incorporated? What did this collaboration look like? Who was considered a legitimate interlocutor? The statements caution that genetic data "are a form of knowledge that contributes to

[72] David Reich and Orlando Patterson, "Ancient DNA Is Changing How We Think about the Caribbean," *The New York Times*, December 23, 2020, www.nytimes.com/2020/12/23/opinion/dna-caribbean-genocide.html?searchResultPosition=1.

[73] Daniel M. Fernandes et al., "A Genetic History of the Pre-Contact Caribbean," *Nature*, December 23, 2020, www.nature.com/articles/s41586-020-03053-2.

understanding the past; they co-exist with oral traditions and other Indigenous knowledge," and that genetic ancestry "should not be conflated with perceptions of identity, which cannot be defined by genetics alone." In this sense, the researchers may have provided clues for answering these questions and suggest their respect for Indigenous epistemologies.[74]

While the editors of *Empire, Colonialism, and the Human Sciences* have no doubt that the authors of the study went to significant lengths to engage Indigenous communities, the form by which this protocol is communicated may leave the more critical reader to wonder about that which is always left unsaid, regardless of who does the narrating. It is time for historians to interrogate these encounters in depth, and in ways that attend to multiple voices, as we reconstruct the broader trajectory of the human sciences.

[74] Fernandes et al., "A Genetic History."

PART I

Relationality in Field and Expedition Science

2

Skull Hunters on the Pampa

Anthropology as Uncanny Encounter in Argentina's "Last Massacre"

JULIA E. RODRIGUEZ

Rudolf Virchow, the pioneering German pathologist and one of the Atlantic world's most renowned scientific figures in the late nineteenth century, was excited. It was 1879, and his colleague, Carl Hagenbeck, had traveled from South America with three visitors – a man, woman, and child named Pikjiojie, Batzinka, and Luis. The three hailed from the southernmost tip of Chile, a territory also claimed by Argentina. They were Indigenous people from Patagonia, on display for the German scientific audience. Virchow presented his expert assessment of their cranial types, facial measurements, and body features. He recorded his remarks and published them in the top anthropology journal of his day.[1]

Anthropometry was but one of Virchow's scientific interests. He was one of the most famous scientists in late nineteenth-century Europe, a brilliant doctor with a broad range of influence, best known for his pioneering work in cell pathology. Virchow had by then also become a major political figure in German liberal reform movements.

Meanwhile, just north of Punta Arenas, Argentina, the young Argentine naturalist Francisco Pascasio Moreno was getting ready for his third major expedition in Patagonia. Moreno, a top scientist in Argentina and member of a powerful family dynasty, published lengthy accounts of his trips, including descriptions of interactions with local Indigenous peoples. On these journeys, Moreno filled multiple roles: geographer, diplomat, adventurer, and scientist. But his main goal was to gather specimens – including human remains – for his growing collection, destined to be housed in Argentina's first Natural History Museum.[2] The most valuable bounty was the skulls and skeletons he

---

[1] Rudolf Virchow, "Drei Patagonier," *Zeitschrift für Ethnologie* XI (1879): 198–204. Hagenbeck was an animal importer who also trafficked in humans for zoos and living exhibits; see Pascal Blanchard et al., eds., *Human Zoos: Science and Spectacle in the Age of Empire* (Liverpool: University of Liverpool Press, 2008). Unless noted otherwise, all translations from Spanish, French, and German are my own.

[2] On the history of the Museo de La Plata, see Máximo Farro, *La formación del Museo de la Plata: Coleccionistas, comerciantes, estudiosos y naturalistas viajeros a fines del siglo XIX*

came across on his journeys. Some were from recently deceased Aònikenk (Tehuelche) and Genniken people; others he hoped were remnants of ancient "Man."[3]

Moreno and Virchow never met in person, yet their work in physical anthropology was intertwined. The intellectual and professional context of their work with human skulls was Americanism, an interdisciplinary scientific project that emerged in near-parallel with anthropology in the late nineteenth century. Initially born in the 1860s of the efforts of naturalists, linguists, and archaeologists primarily from France, Germany, and Britain, within a few decades the movement expanded from a handful of national institutions such as museums and scientific groups to a transnational network of scientists. In 1875, men (and a few women) from scientific backgrounds ranging from medicine to classics joined together to establish a more broad-based group specifically focused on the prehistory, culture, and racial traits of New World populations. Both Moreno and Virchow participated in the meetings of this group, the International Congress of Americanists (ICA). They also crossed paths in other ways, such as exchanging material and correspondence.

Americanist anthropologists dipped into novel scientific methodologies, including the physical analysis of human bodies, or forensics. Anthropology – the "science of man" – itself emerged from a medicalized and body-based approach to human difference, and its practitioners, many of whom were trained in medicine, applied new techniques from biology and anthropometry to their study of cultures and civilizations, human origins, and heritage. These endeavors garnered new levels of state support from nations in Europe and the Americas, as governments funded scientific expeditions as part of larger colonial or postcolonial stratagems.

The emergence of scientific anthropology coincided with the opening of new areas for exploration in the Americas and a corresponding tidal wave of material evidence. Sites in postcolonial Latin America, finally stable after

---

(Rosario, Argentina: Prohistoria Rosario, 2009); Jens Andermann, *The Optic of the State: Visuality and Power in Argentina and Brazil* (Pittsburgh: University of Pittsburgh Press, 2007); Carolyne R. Larson, *Our Indigenous Ancestors: A Cultural History of Museums, Science, and Identity in Argentina, 1877–1943* (University Park: Penn State University Press, 2015); and I. Podgorny and M. Lopes, *El desierto en una vitrina: museos e historia natural en la Argentina* (Mexico City: UNAM, 2008). On the international circulation of Argentine objects, see Ashley Kerr, "From Savagery to Sovereignty: Identity, Politics, and International Expositions of Argentine Anthropology (1878–1892)," *Isis* 108, no. 1 (March 2017): 62–81.

[3] Another dimension to consider is the role that Patagonian artifacts played in the history of origins theory; see Irina Podgorny, "Bones and Devices in the Constitution of Paleontology in Argentina at the End of the Nineteenth Century," *Science in Context* 18, no. 2 (2005): 249–283; Irina Podgorny, "Human Origins in the New World? Florentino Ameghino and the Emergence of Prehistoric Archaeology in the Americas, 1875–1912," *Paleoamerica* 1, no. 1 (2016): 68–80.

decades of civil war, emerged as promising new sources of scientific material. Argentina, lacking the architecture of Aztec and Maya civilizations, nonetheless constituted a promising site for paleontologists and physical anthropologists. As early as the 1840s, British, French, and Spanish scientific explorers took to the systematic exhumation and analysis of animal fossils in Argentina, and nowhere more than in Patagonia, which remains a major site of fossil finds to this day.[4] Americanists practiced what we now call "salvage" anthropology – the imperative to preserve and record human and cultural products just as they were "disappearing" – as a key component of settler colonialism.

Recognizing cultural production as an ingredient to understanding settler colonialism as "a structure, not an event" extends to the reconstruction of human emotions, or in Ann Laura Stoler's words, the "*distribution of sentiments* within and between empire's subjects and citizens as part of imperial statecraft."[5] Indeed, historians have turned to the psychodynamic experiences that scientists and their subjects navigated in the context of inherited power dynamics, recognizing affective relations as important and revealing facets of colonial and postcolonial encounters.[6] In any such moment, a range of behaviors is possible; there can be subtle nuances in even the most brutal situations. On the individual level, each scientist who ventured into a foreign zone – just as each person who found themself confronted by an invader –

[4] Irina Podgorny, "De ángeles, gigantes y megaterios. El intercambio de fósiles de las provincias del Plata en la primera mitad del siglo XIX," in Salvatore, ed., *Los lugares del saber. Contextos locales y redes transnacionales en la formación del conocimiento modern* (Buenos Aires: Beatriz Viterbo, 2007), 125–157. On scientific expeditions to Argentina, see also Adriana Novoa and Alex Levine, *From Man to Ape: Darwinism in Argentina* (Chicago: University of Chicago Press, 2010), chapter 1.

[5] Ann Laura Stoler, "Intimidations of Empire: Predicaments of the Tactile and Unseen," in *Haunted by Empire: Geographies of Intimacy in North American History*, ed. Ann Laura Stoler (Durham, NC: Duke University Press, 2010), 4, emphasis in original; for more on the "affects" of empire and refusal primarily in the North American context, see essays in Stoler, *Haunted by Empire*; and Carole McGranahan and John F. Collins, eds., *Ethnographies of U.S. Empire* (Durham, NC: Duke University Press, 2018). See also Mark Rifkin, *Settler Common Sense* (Minneapolis: University of Minnesota Press, 2014); J. Kēhaulani Kauanui, "'A Structure, Not an Event': Settler Colonialism and Enduring Indigeneity," *Lateral* 5, no. 1 (2016), https://csalateral.org/issue/5-1/forum-alt-humanities-settler-colonialism-enduring-indigeneity-kauanui/.

[6] In Warwick Anderson, Deborah Jenson, and Richard C. Keller, eds., *Unconscious Dominions: Psychoanalysis, Colonial Trauma, and Global Sovereignties* (Durham, NC: Duke University Press, 2011), the authors highlight the "globalization of the unconscious as a mediating discourse of modern civilization, its discontents, and its others." (1) See also Christopher Heaney, "A Peru of Their Own: English Grave Opening and Indian Sovereignty in Early America," *William and Mary Quarterly* 73, no. 4 (October 2016): 609–646; and Fenneke Sysling, "Racial Science and Colonial Practice in the Netherlands East Indies, ca. 1890–1960," unpublished paper presented at "Phrenologies" workshop, Clarkson University, August 2015.

experienced some form of psychological drama.[7] We can characterize many of these moments as triggering the Freudian Uncanny, that is, something or someone who is strangely familiar, always disconcerting. The feeling of unease could result in a specter-like image of Indigenous peoples, simultaneously seen as alive and dead, passive and active, weak and dangerous.[8]

The two men shared a passion for scientific discovery; in particular, they sought answers for large and pressing questions about human evolution and racial classification.[9] Comparing Moreno and Virchow's experiences with Indigenous Argentine individuals expands our understanding of the emotive aspects of relationality; the impact of historical context on these affective relations, in particular the varieties of colonial and postcolonial science in settler societies; the construction of material and spiritual meaning in early anthropology; and like other works in this book, encourages us to consider the boundaries – and the limits – of the moral field concept. Finally, insofar as the behavior and emotional responses of Indigenous individuals were noted, the encounters revisited here provoke us to reflect on the concepts of reciprocity and relationality in moments of contact in the human sciences.

The intertwined stories of the two skull scientists discussed here illustrate, however, that despite significant variations in their affective relations with their subjects, similar psychological and professional goals overruled their humanism. In instances of skull science with Indigenous bodies in South America, our attention to affect, intention, and agency highlights the joint significance of material context and power dynamics on the one hand, and personal experiences of actors on the other.[10] Whether colonialism is driven by external or internal forces, both are harmful, albeit in different ways. In this sense, a comparative view of the dynamics between anthropologist and human

---

[7] Stephen Greenblatt has described the first emotion at the moment of encounter as that of wonder; almost at the speed of light, the individual must decide what to do with that emotion. He or she has a choice: constructively engage the Other or annihilate it. Stephen Greenblatt, *Marvelous Possessions: The Wonder of the New World* (Chicago: University of Chicago Press, 1991).

[8] Renee Bergland observes a "literary Indian removal" that includes the "ghosting" of Indians as a "technique of removal"; Renee Bergland, *The National Uncanny: Indian Ghosts and American Subjects* (Dartmouth, NH: University Press of New England, 2000), 3–4. See also Jesse Alemán, "The Other Country, Mexico, the United States, and the Gothic History of Conquest," in *The Spectralities Reader: Ghosts and Haunting in Contemporary Cultural Theory*, eds. María Pilar del Blanco and Esther Peereen (London: Bloomsbury, 2013), 507–526.

[9] See Robert E. Bieder, *Science Encounters the Indian, 1820–1880* (Norman: University of Oklahoma Press, 1986); L. Stephen Jacyna, "Medicine in Transformation, 1800–1849," in *The Western Medical Tradition, 1800–2000*, eds. W. F. Bynum et al. (New York: Cambridge University Press, 2006), 92–99.

[10] Ruth Leys, "The Turn to Affect: A Critique," *Critical Inquiry* 37, no. 3 (2011): 434–472.

subjects allows us to trace key and intersecting aspects of the material and psychological dimensions of colonial violence and their implications for the construction of ethical norms for encounters in the human sciences.

Skulls as Uncanny Objects

That prominent scientists like Moreno and Virchow accelerated the collection, analysis, and display of human skulls reflected a growing interest in, if not an obsession with, human parts as the centerpiece of Americanist investigation in the late nineteenth century.[11] Anthropological expeditions and displays of artifacts in museums in all major cities on both sides of the Atlantic reflected the growing fetishization of human body measurement as the centerpiece of Americanist investigation. Craniology began in the 1830s, established by figures such as US scientist Samuel George Morton, who assembled a vast collection of human skulls and used them not just to provide evidence for polygenic theories of human biological descent, but also to justify superiority of the Nordic race.[12] While Morton had his detractors (such as Rudolf Virchow, who believed in the unity of humankind), the fascination with crania cut across all sides of the evolutionary debate.

While later generations would thoroughly debunk craniology's scientific and moral failures as well as anthropology's symbiosis with colonial regimes, late nineteenth-century anthropologists saw skulls at the centerpiece of their field's most pressing questions. Even when handled in volume, skulls are not just any type of scientific evidence.[13] Cara Krmpotich and coauthors remind us that bones are not mere lumps of matter but are "constantly constituted and negotiated as persons or things, subjects or objects, meanings or matter . . . The materiality [of bones] engages those they encounter."[14] They draw our

---

[11] Benoit Massin, "From Virchow to Fischer: Physical Anthropology and 'Modern Race Theories' in Wilhelmine Germany," in *History of Anthropology: Volksgeist as Method and Ethic*, ed. George Stocking (Madison: University of Wisconsin Press, 1998), 79–154, 83. On bone collecting in US museums, see Samuel J. Redman, *Bone Rooms: From Scientific Racism to Human Prehistory in Museums* (Chicago: University of Chicago Press, 2010).

[12] Ann Fabian, *The Skull Collectors: Race, Science, and America's Unburied Dead* (Cambridge, MA: Harvard University Press, 2016).

[13] The dead "do things the living could not do on their own." See Thomas W. Laqueur, *The Work of the Dead: A Cultural History of Mortal Remains* (Princeton: Princeton University Press, 2015), 18.

[14] Krmpotich et al. go so far as to say the bones themselves have agency, flipping the direction of the typical question to "what do bones do to people?" Cara Krmpotich, Joost Fontein, and John Harries, "The Substance of Bones: The Emotive Materiality and Affective Presence of Human Remains," *Journal of Material Culture* 15 (2010): 371–384 at 372–373, 381. For an illustration of this approach, see Kim Wagner, "Confessions of a Skull: Phrenology and Colonial Knowledge in Early Nineteenth-Century India," *History Workshop Journal* 69, no. 1 (March 2010): 27–51.

attention to bones as "uneasy, ambivalent subject/objects" with the ability to "make present that which is absent."[15] Bones can remind the living of the deceased, creating a haunting effect as scientists imagine who lived in those bones, speculating about the deceased's spirit, soul, or personality. By the same token, the materiality of human bones, their hard substance lasting longer than flesh, is a reminder of past historical events.[16] This uniquely spiritual materiality of human remains impacts our understanding of the person's place in past events. In the context of skull science in Argentina, the materiality of human remains expands the historian of science's toolkit.

Moreno and Virchow, embedded as they were in a larger network of Americanist anthropology, placed human skulls at the center of their scientific work. To that end, they both repeatedly engaged with their main sources: dead and living Indigenous bodies. While both scientists directly engaged with Indigenous individuals, one had extensive, close, even life-long contact with his human subjects, and the other was removed from the field. So while there are important differences in the dynamics of these encounters, in the end both were active participants in violent acts of Indigenous erasures, enhanced as they were by their own ambivalence. In this sense, the story of skulls on the Pampa fits Amy Lonetree's framing of history, which calls us to pay attention to both agency and harm, trauma and resistance. "Hard truths" are recognized just as productive refusal and other forms of Indigenous agency and desire emerge in the narrative.[17] Moreover, the story reveals how personal dynamics between people created multiple, and sometimes unpredictable, outcomes.

The quest for Indigenous peoples' skulls on the fertile plains of the Argentine Pampa after 1870 was set in motion by a tangle of scientific, economic, and military agendas. The episode implicated actors from local Indigenous people to wealthy elites in Buenos Aires, to intellectuals in North Atlantic capitals. The two scientists seeking skulls in a contested territory were of course shaped by their specific contexts, including differences in Argentine and German political, intellectual, and institutional milieus. For example, Moreno worked for an expansionist Argentine state that had specific goals and assumptions about comparative human worth framing its mission.[18] Argentina sits uneasily between models of Northern/British settler colonialism and Latin American internal colonialism. A crowded urban center with a

---

[15] Krmpotich et al., "The Substance of Bones," 372, 378. For broader discussion of the ambiguities of human remains, see Karin Sanders, Bodies in the Bog and the Archaeological Imagination (Chicago: University of Chicago Press, 2009).

[16] Krmpotich et al., "The Substance of Bones," 375.

[17] Lonetree, Decolonizing Museums.

[18] For a recent intersectional analysis of the frontier sciences in Argentina, along with a discussion of sexuality and desire in scientific excursions, see Ashley Elizabeth Kerr, Sex, Skulls, and Citizens: Gender and Racial Science in Argentina (1860–1910) (Nashville, TN: Vanderbilt University Press, 2020).

European-descended majority, a relatively low-density population with large "unconquered" tracts until the 1890s, and a late-stage scramble for land were the factors that led to Argentina's distinct hybrid postcolonial relations. Moreno's participation in salvage anthropology must be viewed in this unique context.[19]

Virchow, on the other hand, was shaped by German liberal cosmopolitanism, a philosophy that would later inspire radical egalitarianism in some of his students, most famously Franz Boas. Virchow, a German scientist of great stature, was a political progressive who "saw no contradiction between [his elite status]" and his liberal views.[20] A close reading of his and Moreno's own words about their observations and interactions with Indigenous Americans, along with a reading between the lines to extrapolate the experiences of their human subjects, reveals intricate and layered encounters – occurrences that had real consequences for European science, Argentine national goals, and Indigenous peoples' lives.[21] Their interactions with skulls demonstrate how body parts, especially crania, are key to understanding historical acts of dehumanization, including colonial scientific encounters and state campaigns to physically and culturally annihilate Indigenous peoples in postcolonial settler societies, even as they reveal elements of ambiguity.

The scientific encounter between anthropologist and subject, then, can highlight some of the nuances of what we have come to call colonial science, and the difficulty of navigating damage narratives. The scientists, for their

[19] See Fernanda Peñaloza, "On Skulls, Orgies, Virgins and the Making of Patagonia as a National Territory: Francisco Pascasio Moreno's Representation of Indigenous Tribes," *The Bulletin of Hispanic Studies* 87, no. 4 (2010): 455–472; Ricardo D. Salvatore, "Live Indians in the Museum: Connecting Evolutionary Anthropology with the Conquest of the Desert," in *The Conquest of the Desert: Argentina's Indigenous Peoples and the Battle for History*, ed. Carolyne Larson (Albuquerque: University of New Mexico Press, 2020), 97–121.

[20] Jacyna L. Stephen, "Medicine in Transformation, 1800–1849," in *The Western Medical Tradition, 1800–2000*, eds. W. F. Bynum et al. (New York: Cambridge University Press), 100. For a darker historical perspective on German anthropology, see Andrew Zimmerman, *Anthropology and Antihumanism in Imperial Germany* (Chicago: University of Chicago Press, 2001). See also Stocking, *Victorian Anthropology*.

[21] See Kerr, *Sex, Skulls, and Citizens*; also William Y. Adams, *The Boasians: Founding Fathers and Mothers of American Anthropology* (New York: Hamilton Books, 2016); H. Glenn Penny, "The Politics of Anthropology in the Age of Empire: German Anthropologists, Brazilian Indians, and the Case of Alberto Vojtech Frič," *Comparative Studies in Society and History* 45, no. 2 (April 2003): 249–280; see also Warwick Anderson, *The Collectors of Lost Souls: Turning Kuru Scientists into Whitemen* (Baltimore: Johns Hopkins University Press, 2019) on interpersonal exchange between anthropologists and subjects. For an explanation of another dimension of German cultural activity in Latin America, see H. Glenn Penny, "Material Connections: German Schools, Things, and Soft Power in Argentina and Chile from the 1880s through the Interwar Period," *Comparative Studies in Society and History* 59, no. 3 (July 2017): 519–549.

part, performed ambivalence about the human relations they initiated. Indigenous perspectives are legible in these encounters in terms of acts of both reciprocity and refusal. Affective reflections revealed that all parties existed in uneasy tension with each other, characterized by occasional reciprocity and more frequently, heartbreaking acts of annihilation. Not surprisingly, individuals like Moreno and Virchow did not regard themselves as abusive monsters; quite the contrary. They believed they were humanistic champions of progress, imbued with genuine affection for Indigenous people. Philosopher of dehumanization David Livingstone Smith has pointed out that the mere fact of considering the Other as less than human provides a rationalization for violence, but not in a straightforward way: "Dehumanizers often *behave* towards their victims in a manner that implicitly acknowledges their humanity... In dehumanizing others, we categorize them *simultaneously* as human and subhuman."[22] In colonial and postcolonial settings, uncanny emotions arise when we recognize our savage past in the Other, a concept of direct relevance to anthropologists.[23]

As these actors ran up against each other on the Argentine Pampa, the skull scientists left a record not just of their findings, but also their personal and emotive reactions to the skulls. At the same time, they noted to varying degrees the reactions, behavior, and emotional states of the Indigenous people they engaged. Moreno and Virchow's overlapping but distinct encounters examined here dramatically reveal the contours of the uncanny encounter between liberal anthropologist and subject. The affective dimensions of their skull work seamlessly merged with the material realities of frontier violence and the scramble for objects of economic and scientific value. These two scientists, with their close contact with Indigenous people, provide unusually direct examples of these intimate exchanges. How did Aònikenk and other Indigenous peoples' active roles in the process contribute to the emotive and intellectual dynamics of encounters? Were Moreno and Virchow swayed by

---

[22] David Livingstone Smith, "Paradoxes of Dehumanization," *Social Theory and Practice* 42, no. 2 (April 2016): 417–418. Emphasis in original. The uncanny, or a destabilizing feeling of strange familiarity, derives from Sigmund Freud's concept of *Unheimlich*. Freud argued it is necessary to repress uncanny emotions to create a civilized society (although we usually fail).

[23] Priscilla Wald saw this dynamic in nativism against immigrants in the nineteenth-century United States, observing that "anxiety [or estrangement] ... grows out of the transmutation of something 'known of old and long familiar' into something frightening ... [One] recognizes the stranger, whose appearance he dislikes, as himself." Priscilla Wald, *Constituting Americans: Cultural Anxiety and Narrative Form* (Durham, NC: Duke University Press, 1995), 5–7. For a broader discussion of othering of the stranger, see Julia Kristeva, *Strangers to Ourselves* (New York: Columbia University Press, 1991).

their personal relationships with Indigenous individuals, enough to alter the momentum of postcolonial violence? This had been an implicit choice facing centuries of interlopers – and was also true for anthropologists pushing south from Buenos Aires into Argentina's interior.

Moreno: "A Sacrilege Committed for the Sake of Osteological Study"

Until the nineteenth century, a good portion of the southern half of the nation was largely independent territory, with an Indigenous majority loosely controlled by scattered state representatives in a handful of European settlements.[24] The official image of Patagonia was one of vast, "empty" space. Spaniards and Argentines imagined the South and its inhabitants as "at the edges of the world" and culturally marginal.[25] British explorer Julius Beerbohm, author of *Wanderings in Patagonia*, expressed a typical European attitude toward Southern Argentina in 1881, declaring that Patagonia would remain uninhabited forever.[26] Another trope about Southern Argentina was its supposed wildness, and in the modern expression, barbarism. This barbarism in turn linked Indigenous peoples with primitivism, brutal or animalistic violence, and racial inferiority, creating a dichotomy with "civilization" and legitimating increasingly violent measures to claim the land and tame its inhabitants. In the 1870s, fifty years after Argentina's independence from Spain, the Pampas were still inhabited by semi-sedentary communities with diverse identities and speaking different languages, and a shared history of intertribal relations for millennia before European incursions.[27] By 1870, an

---

[24] On the historical context of the racial concept in Argentina, see Paula L. Alberto and Eduardo Elena, eds., *Rethinking Race in Modern Argentina* (New York: Cambridge University Press, 2016).

[25] Claudia Briones and José Luis Lanata, "Living on the Edge," in *Archaeological and Anthropological Perspectives on the Native Peoples of Pampa, Patagonia, and Tierra del Fuego to the Nineteenth Century*, eds. Briones and Lanata (Westport, CT: Bergin and Garvey, 2002), 1–12; 1. See also Walter Delrio, *En el país de nomeacuerdo: Archivos y memorias del genocidio del estado argentine sobre los pueblos originarios, 1870–1950* (Viedma, Argentina: Editorial Universidad Nacional de Río Negro, 2018); Pilar Pérez, *Archivos del silencio: Estado, indígenas y violencia en Patagonia Central, 1878–1941* (Buenos Aires: Prometeo Libros, 2016).

[26] Alejandra Pero, "The Tehuelche of Patagonia as Chronicled by Travelers and Explorers in the Nineteenth Century," in *Archaeological and Anthropological Perspective*, eds. C. Briones and J. L. Lanata (Westport, CT: Bergin and Garveys), 120.

[27] In the northern area of the Pampas region, the Rankulche people lived; in the southern Pampas the Puelche and Mapuche. Most or all "Pampas" Indians may have called themselves Tehuelche (in Araucano, *che* means people and *tehuel* means south; *puel* means east). The Selk'nam live further south, in Tierra del Fuego. See Ana Ramos and Claudia Briones, eds., *Parentesco y política: Topologías indígenas en Patagonia* (Viedma, Argentina: Editorial Universidad Nacional de Río Negro, 2016).

estimated 40,000 Indigenous people lived in Southern Argentina, including the Pampas and Patagonian regions.

Centuries of intermittent violence between Indigenous peoples and Argentines peaked in April and May of 1879 in a short but intense military campaign the government dubbed the "Conquest of the Desert."[28] With this war, the Argentine state intended to seize the Pampas and achieve three goals: to prevent the Chileans from claiming Southern Argentina, to create settlements and Europeanize the province, and to exploit its fertile land for the production of cattle, wheat, and other crops. European Argentina lauded the Desert Campaign as a strategic success, yet the Mapuche called it the "Last Massacre." In a two-month period the Argentine military killed about 1,300 Rankulche, Puelche, and Mapuche people and captured or displaced 15,000 more.[29] The campaign also contributed to the developing idea of Argentine national identity as racially white.[30]

While the basic outlines of "the last massacre" and other postcolonial wars in late nineteenth-century Argentina are well known, less appreciated is that along with the generals and soldiers, there rode geographers, naturalists, and anthropologists. These scientists were without exception wealthy, upper-class men, who sought to build careers in arts and sciences through the study of their own largely unexplored country and its inhabitants. They were scientific pioneers and enjoyed celebrity as such. Elite men of science in countries like Argentina found themselves in a liminal state in transatlantic scientific power structures. Lords of their realms at home, they were looked down upon in North Atlantic (especially European) scientific institutions. Seeking to transcend centuries-long-repeated theories of Latin American inferiority, scientists like Moreno translated their desire for scientific recognition for themselves and for Argentina into an aggressive push inland. The anthropologists embedded in the military expeditions were primarily looking for artifacts and living

---

[28] See Larson, *The Conquest of the Desert.* On the political economy of frontier violence in Latin America, see Silvio R. Duncan Baretta and John Markoff, "Civilization and Barbarism: Cattle Frontiers in Latin America," in *States of Violence*, eds. F. Coronil and J. Skurski (Ann Arbor: University of Michigan Press, 2005), 587–620; orig. in *CSSH* 20: 587-620. On violence and "whitening" in historical and contemporary Argentina, see Gastón Gordillo, "The Savage outside of White Argentina," in *Rethinking Race in Modern Argentina*, eds. P. Alberto and E. Elena (New York: Cambridge University Press, 2016), 241–267; see also Susana Rotker, *Captive Women: Oblivion and Memory in Argentina* (Minneapolis: University of Minnesota, 2002).

[29] Larson, *Our Indigenous Ancestors*, 14–15.

[30] See María E. Argeri, *De Guerreros a Delincuentes. La desarticulación de las jefaturas indígenas y el poder judicial; Norpatagonia, 1880-1930* (Madrid: CSIC, 2005); Rodriguez, *Civilizing Argentina*; Mariela Eva Rodríguez, "'Invisible Indians,' 'Degenerate Descendants': Idiosyncrasies of *mestizaje* in Southern Patagonia," in *Rethinking Race*, eds. Paula L. Alberto and Eduardo Elena (New York: Cambridge University Press), 126–154.

subjects, to build collections in new (or planned) national museums. At their most ambitious, based on their immediate access to relics and living people, Argentine anthropologists hoped to contribute to new understandings of American populations, and provide insight on the origin and meaning of humanity itself.

Francisco P. Moreno was one of his nation's prominent scientists; born in 1852, he founded Argentina's first natural history museum, the Museo de La Plata in 1888, and eventually the first national park, Nahuel Hapi in 1903. One of the first Argentine anthropology textbooks described Moreno's groundbreaking role in the field: "Moreno's... long view, [and] his deep knowledge of the country, led him to launch a great initiative and to be forever known abroad as the authoritative spokesperson of this incipient Argentine science [of anthropology]."[31] Later in life, Argentines often referred to him as "Perito [the expert] Moreno," an honorific recognizing his achievements. Moreno hailed from Argentina's landed aristocracy; as a teenager, the land on his family's estate provided the naturalist with his first area of exploration. Like his scientific compatriots, he was also invested in the national and cultural development of the nation. He would eventually serve in federal government, building on his reputation attained from founding the Museo de La Plata, and his role in negotiating the border with Chile. Moreno also looked beyond the borders of his homeland to exercise his talents. He was an active participant in transnational anthropology, including time in Europe in the early 1880s, where he met with prominent scientists, gave talks, and visited museums. (Extended European trips were common, if not expected, of wealthy and educated young Latin American men at the time.) Moreno's wider circle included local intellectuals and government officials such as Ernesto Zaballos and Eduardo Holmberg, as well as foreign ones like Paul Topinard, Secretary of the Societé d'Antropologie de Paris, and Rudolf Virchow.

Moreno pioneered the collection and examination of human skeletons culled from his nation's territory and was one of a small handful of Euroamericans in the 1870s engaging directly with Mapuche, Aònikenk, and Genniken culture. As such, he left a rich record of his encounters with Indigenous people, both dead and alive. Human remains were central to Moreno's plan to expand the Museo de la Plata, an institution that became one of the top natural history museums in Latin America. The Museo's anthropology exhibits would eventually feature displays of human remains – about 1,500 pieces to begin with – built on Moreno's personal collection. His larger goal was to put together a collection of crania representing all of Latin America and the Canary Islands, whose early inhabitants he believed to be

[31] Feliz Outes and Carlos Bruch, *Los aborígenes de la República Argentina* (Buenos Aires: Estrada y Cia., 1910), 25.

related to South American Natives.[32] In addition to skulls, Moreno displayed skeletons, bone fragments, clay and bronze busts of Indigenous individuals, masks, and photographs and daguerreotypes of racial types. He even forced living people on display: between 1885 and the early 1890s, a group of Indigenous people resided in the Museo de La Plata.[33]

Moreno began digging for fossils and artifacts as a young man, and after his first early excursion in 1871, he undertook five more between 1876 and 1880. In his first scientific journey, he collected animal fossils, ceramic shards, and carved rock objects. Within a year, Moreno was headed to Patagonia to look for human skulls. Two years later, he was granted permission to join a military campaign to the South, a position acquired thanks to his family connections in government. In 1873, Moreno gathered skulls of long-dead inhabitants of the southern region, and also examined his Indigenous contemporaries, taking measurements of the skulls, heights, torso, and feet; separating women and men, he attempted to distinguish the individuals by bloodline (i.e., between tribes).[34]

Moreno's published reports on these journeys follow a typical pattern: first he would describe the landscape and expedition party members (including Indigenous guides); next he would recount the meetings with people in great detail. Finally, he would mention whether he had met his goal of gathering enough relics for the Museum. Once he had acquired objects to his satisfaction, he recorded that it was time to embark on his next excursion. This pattern underscored his personal prioritization of the collection of materials, with skulls and bones being the most coveted, over other aspects in the context of these expeditions. Moreno's narrative of his trip to Patagonia in 1875–1876 contained multiple descriptions of his personal and extensive meetings with Indigenous peoples, but repeatedly stressed that his main reason for the expedition was to gather objects for his collection. In the process, however, he also documented the living conditions, customs, habits, and beliefs of the people he met in rich detail. Moreno's second expedition brought him further into the southern region. Reflecting on this journey later in life, Moreno revealed these other goals:

> My objective was not only to study the regions along the way and cross the Cordillera to Chile, but also to see the Indians [indios] in their surroundings, far from civilization, by living in an Indian hut. I wanted

---

[32] Farro, La formación del Museo de La Plata, 139.

[33] There is little documentation of exhibits of living people in the Museo de la Plata, however, such exhibits are referred to in a number of publications. See Mónica Quijada, "Ancestros, ciudadanos, piezas de museo. Francisco P. Moreno y la articulación del indígena en la construcción nacional argentina," EIAL 9, no. 2 (1998): 21–46; Salvatore, "Live Indians in the Museum."

[34] Farro, La formación del Museo de la Plata, 59.

to gather information from among these tribes facing extinction. I wanted to document what I simply knew from hearsay since that method fell short of my goals.[35]

In many instances, Moreno stayed for days, weeks, or months in Aònikenk communities, and sometimes referred to his hosts as his friends or *compadres*. Moreno mentioned that in some situations they brought him into kinship circles; he claimed that "the Chief Chacayal, my supposed father in law" called him Tapayo, "the name that some Indians gave me."[36] Later on in the narrative, Moreno reported that "as it is necessary [to have] a title equal to the chief, I take the name Comandante."[37] Years later, in his memoir, Moreno reflected on the rapport he felt with the tribal people. He described his "friendship," for example with chief Quinchahuala, who helped him acquire safe passage to Nahuel Huapi. "Quinchahuala took a liking to me since I accepted a plate of food from him consisting of cornmeal with blood and raw tripe, and I ate it without a visible display of revulsion. That was proof of my outpouring friendship ... These foods were eaten as a matter of course in the wilderness. Suffice it to say, the stomach adapts to the circumstances far beyond one's expectations."[38]

Moreno's descriptions of these get-togethers read like diplomatic meetings. He emphasized the intimate, personal, and emotive exchanges between himself and Indigenous informants, as well as the bonds he believed were formed. And yet the differing intents among the actors surface in his recollections as well. They reflect his pattern of recording communication successes and failures, his observations of Indigenous appearance and behavior, participation or observation of rituals and trade, and finally his attempts to acquire scientific objects and body measurements. While meticulously describing the distinctive landscape, living or communal structures, clothing, and customs, he often advanced a larger perspective that modern civilization, in particular, science, was the antidote and inevitable corrective to tribal life patterns: "Only science can give us the conviction that everything stops after our departure from the earthly realm, but science is unknown in the uncultured primitive mind."[39]

Moreno's narratives also revealed the opportunities for real or potential reciprocity. For example, while in the field he relied on Indigenous hosts for his needs: food and shelter. In an 1878 description of his earlier Patagonian

---

[35] Francisco Moreno, *Perito Moreno's Travel Journal: A Personal Reminiscence*, compiled by Eduardo V. Moreno and translated by Victoria Barcelona (Buenos Aires: Elephante blanco, 2002), 31. All quotes from Moreno's *Travel journal* translated by V. Barcelona.

[36] Francisco Moreno, *Viaje a la Patagonia Austral* (Buenos Aires: La Nacion, 1879), 110, 112.

[37] Moreno, *Viaje*, 220.

[38] Moreno, *Travel Journal*, 39.

[39] Moreno, *Viaje*, 119.

expedition, he recorded a moment that recognized their status and power. In an encounter between his group and the Aònikenk (whom he called Tehuelche):

> Our provisions were extremely scarce, and consisted only in a few sand-wiches, a gift of the Aònikenk Rosa, wife of Manuel Coronel, another good gaucho countryman who had accompanied Monsieur Pertuíset [a French explorer] to Tierra del Fuego, and who pretends to appear [*muy farzante hace aparecer*] like the Peruvian Yupanqui, [and] with the same formality later assured us that Rosa was a princess of the Imperial race of the Incas; to the sandwiches she added meat for a day and two boxes of paté de foie gras.[40]

One can almost imagine the camp site encounter, with Manuel Coronel and Rosa insisting on being seen, in demanding recognition of their ancestry. Even as they offered food to the soldiers and expedition members, perhaps in exchange for money, the Aònikenk also insisted on their presence, moreover in terms of their unique cultural heritage. Through these acts they declared themselves as alive in the present as well as connected to the past.

Moreno reacted to offers of reciprocity with conflicted emotions. In detailing his interactions, Moreno's memoirs contained more nostalgia than did his careerist field notes as an aspiring scientist. He vividly reminisced about his first encounters on his 1875 journey, reporting that he relied on four Indigenous guides to help him search for abandoned Indigenous settlements and burial grounds, as he attempted to "[develop] better relations with the Tehuelche, Gennaken, and Mapuche tribes."[41] At other times, Moreno recounted, the two parties relaxed together: "Every now and then, bands of friendly Indians pierced the silence and cheered us up. About a hundred of them traveled with us to Chichinal, now called General Roca. They made the days go by faster as they enthusiastically hunted ostriches."[42]

What these moments meant to the Indigenous travelers is difficult to perceive, as Moreno largely described them as backdrop to his adventures. Yet the living subjects of science, of course, had their own complex belief systems. At the time of the Pampas wars, Indigenous peoples held their own long-standing ideas and practices around death and the body.[43] Moreno's writing reveals that he was aware of their worldviews; he knew he was violating

---

[40] Francisco Moreno, "Apuntes sobre las tierras patagónicas," *Anales de la Sociedad Científica Argentina* V (1878): 1–19, 4.

[41] Moreno, *Travel Journal*, 28.

[42] Ibid.

[43] Pero in Briones and Lanata, *Archaeological and Anthropological Perspectives*, 117; Celia N. Priegue, "Mortuary Rituals Among the Southern Tehuelche," in *Archaeological and Anthropological Perspectives on the Native Peoples of Pampa, Patagonia, and Tierra del Fuego to the Nineteenth Century*, eds. Claudia Briones and José Luis Lanata (Westport, CT: Bergin and Garvey, 2002), 54.

Indigenous peoples' bodies *and* their belief systems. For example, reflecting on a visit in 1874 to a burial site called "Indian Pascual's ranch," Moreno mentioned that "one of [Pascual's] sons died there, and as the Indians believe that death takes over the place where one person died and all the other members of the family perished if they remain there, Pascual moved by setting fire to the cursed dwelling [*toldo*; tent made of leather]."[44]

Years later, Moreno would describe in his memoirs an incident in which he approached a burial heap near Chocón-Geyú, which, according to his local guides, held "nine burial mounds made of loose stones and dry branches [covering] the skeletal remains of an entire Indian family."[45] The family had died in a sudden snowstorm. Moreno elsewhere discussed the Aònikenk beliefs about the afterlife, or as he described it, their "fetish," which he described as such: "The Indigenes believe in the persistence of the spirit and in the voyage that takes them to another world after having abandoned, by death, the body that generated it."[46] He expressed distaste for their views, which he saw as primitive superstition: "How much better would it be if they recognized [death] as the work of nature! But let us not blame the savage. We ourselves, the civilized, are full of superstitions, some worthy of the Southernmost people [*australianos*], and we are generally the same. We deny the tangible, to believe in the intangible."[47] Moreno saw himself as an enlightened, forward-thinking scientist. Metaphysical beliefs, whether rooted in Indigenous or European, Christian worldviews, were trumped by science. And science demanded skulls.

Moreno's single-minded drive for skulls, in fact, ultimately overrode his ambivalence about Indigenous peoples' humanity. Moreno saw people as obstacles in his path for human *materiel* in three ways. His scientific and national drives led Moreno to reduce Indigenous people to body parts, to overlook their individual identities and define them primarily as members of a group, to in effect "kill" and dismember them metaphorically before their actual demise. Writing to his father from Fort Mercedes in October 1875, Moreno reassured him that "I could not be in better health. I just had a minor headache on the day I arrived but finding the Indian bones cured it completely."[48] On an earlier journey, in April 1873, he celebrated his accomplishments in terms of the wealth of human remains: "I conclude by providing a list of the principal *objects obtained* during my short trip; [I am] happy if this result can demonstrate the anthropological riches contained in the valley of

---

[44] F. Moreno, "Cimetiéres et paraderos préhistoriques de Patagonie," *Revue d'anthropologie* (1874): 72–90, 84.

[45] Moreno, *Travel Journal*, 30.

[46] Moreno, *Viaje*, 94.

[47] Ibid., 107.

[48] Moreno, *Travel Journal*, 64.

the Rio Negro."[49] He listed sixty skulls "of both sexes," along with two skeletons, tools, pottery, and wood items. The skulls had a variety of characteristics, Moreno noted, including signs of cranial deformation, a topic of great fascination to anthropologists at the time. Despite the regrets he expressed toward the end of his life, Moreno's scientific agenda in the field led him to a pattern of deception (and self-deception) that included violating his informants' trust, bodily integrity, and belief systems. It also precluded, or at least delayed, an expansion of options for ethical norms of interaction in the emerging field of Argentine anthropology.

Moreno described an encounter in 1874 that reveals the subterfuge and duplicity required to acquire skulls and human remains from people and communities with whom he sought (or claimed to seek) emotional kinship.[50] Any moral calculation went out the window when presented with prime objects: human skulls. He frequently noted Indigenous peoples' agitation at and refusal of invasive requests. Moreno's awareness of these refusals was clear, as he wrote in a top French anthropology journal:

> I was able to get six of these painted skulls, but I only kept two complete ones; they were exhumed very quickly, as the Indians opposed it. While I was busy collecting anything that could be of interest to my studies, a few Indians from the family of the former owner of the place *approached to observe me and to ask what I was doing.* My answer that I was only concerned with the stones did not satisfy them, [so] they called their leader, Pascual, to drive me away. *This Indian forbade me* to touch anything; he then told Mr. Real, who accompanied me, that he was a fool to allow me to extract these bones, which belonged to the Tehuelche Indians and were red because they had died of an epidemic of small-pox a thousand years ago. The Indian believed this, indicating that he did not know which race of men the rest of them were, and fearing smallpox like the *galichu* (devil), he changed his mind and allowed us to extract the bones, which we did immediately by picking up all the objects that we could.[51]

Similarly, a few years later, during his 1879 expedition, Moreno confessed his manipulation of one of his Aònikenk "friends" (an affective term similarly deployed in the 1912 Yale Expedition, as described in Adam Warren's chapter in this book.) Describing the interaction, Moreno revealed that:

> He consented that we photograph him, *but by no means wanted that we measure his body and even more so, his head.* I do not know the source of his strange preoccupation, but later, upon returning to meet him in

---

[49] Moreno, "Cimetiéres," 88. Emphasis added.

[50] For a discussion of reciprocity and its breakdown in twentieth-century anthropology, see Anderson, *Collectors of Lost Souls.*

[51] Moreno, "Cimetiéres," 85–86. Emphasis added.

Patagones, although continuing to be friends, *he did not allow me to approach him* while he was drunk, and a year later, when I returned to this site to embark on my journey to Nahuel-Huapí, *I proposed that he accompany me and he refused, saying that I wanted his head.* This was his destiny. Days after my departure, he was treacherously taken to Chubut and there murdered by two other Indians during an orgiastic night.[52]

That was not the end of the story, however, at least for Moreno:

Upon my arrival, I learned of his disgrace, [and] figured out the place in which he had been buried and, in the moonlight, exhumed his cadaver, preserving the skeleton in the Anthropological Museum in Buenos Aires; *a sacrilege committed for the sake of the osteological study of the Tehuelches* ... I did the same with the Chief Sapo and his wife, who had died at this spot a few years earlier.[53]

We can only imagine the horror or dismay the Aònikenk experienced; the death and theft of body parts was a violation of their sensibilities.

Despite his multiple intense and personal encounters with the Mapuche and others, Moreno repeatedly rewrote reality, describing the Indigenous peoples of Argentina as either disappearing or disappeared. In 1874, Moreno remarked that in Carmen de Patagones and Chubut, in the Northern part of the province, even though:

civilization has barely penetrated there ... The nomadic tribes ... are marching ever faster towards their extinction, dragged [there] by deadly causes and absorbed through civilizing forces that will replace them through the peaceful possession of the land. And, these remote and extensive regions, until recently [seen as] mysterious and the subject of fables, will pass to the domain of science that studies everything, offering more appeal and utility.[54]

Thus, he constructed his Indigenous subjects as already dead, in the past – therefore available as *materiel* for his science, as well as labor power needed for Argentina's economic expansion.

Moreno, especially as time went by, not only romanticized the "disappearing Indian," but also presented himself as the champion and defender of Indigenous humanity. In his old age, Moreno would wax nostalgic, even mildly regretful, about the drastic decline in Argentina's Indigenous population after 1880, although he stopped short of acknowledging his role in those disastrous events. In his memoirs, Moreno wrote:

[52] Moreno, *Viaje*, 93. Emphasis added.
[53] Ibid.
[54] Moreno, "Apuntes," 5.

I hope I will have enough time to report on my impressions of the primitive environment in which these native tribes lived. Indeed, I was the last one to experience them before they were wiped out by those who never bothered to listen to opposing views. I lived among these self-reliant natives, masters of highlands and plains, followers of no laws other than those imposed by their limited needs.[55]

Viewing Indigenous peoples as part of the natural world was closely linked to a commonly held idea that they were on the brink of extinction. This, too, was a widely expressed belief among late nineteenth-century European and American anthropologists, and similar to those in the United States but not common in most other Latin American countries, which had large Indigenous and Mestizo populations. The act of retrospection had inserted nostalgia in Moreno's narrative.

Moreno's reminiscences were no doubt altered by Indigenous peoples' marked failure to disappear, that is, their continued presence, survival, and survivance. Recalling a moment of shared experience during his past adventures between himself and local people, Moreno recalled much later:

> During those hunts with elusive nomadic tribes, or when we'd take a moment to rest, I would often talk to my Indian guides about the future of these territories without stopping to think whether in my need to find an outlet for my aspirations I was exposing myself to harm. I would speak to them as I satiated my hunger with raw intestines from a worn-out mare or eagerly watched a tasty, skewered ostrich being barbecued over heated rocks, a cooking method that preceded the use of pottery. It gave me great pleasure to recall this scenario twenty-five years later when I revisited the same locales and saw that they had blossomed into towns. *Perhaps my former listeners' grandchildren were attending the local schools.*[56]

Here Moreno, astonishingly, projected a place for Indigenous peoples in Argentina, if not a peaceful coexistence. Reflecting in particular on the role he had played in military violence, Moreno lamented the loss of life on "both sides." Time had made Moreno charitable toward his former frenemies, even stating that the European Argentines had committed far greater atrocities. Indigenous peoples, he now judged, saw themselves as defending their own land, and "they also vividly recall the government's 'Desert Campaign,' dating back less than twenty years, in which executions were almost a daily occurrence." He concluded with a plea for assimilation, to heal the national body and make up for the slaughters: "Our beloved country thus lost thousands of her native sons, useful hands when properly overseen! Even as we speak, those

[55] Moreno, *Travel Journal*, 32.
[56] Ibid., 29. Emphasis added.

who view the natives without bias can see that the remaining few have more good than bad in them."[57]

Moreno's words revealed the emotional imprint made during his interactions with numerous Indigenous people, that he saw them, then and subsequently, as complex, living human beings. At the same time, he opted repeatedly to exploit their bodies in service to his professional and nationalistic goals. Far from a paradox, however, these two seemingly incompatible positions are causally related. Moreno's proximity, immediacy, and the personal nature of his encounters in fact heightened his sense of ambivalence toward his Indigenous neighbors, and, by creating discomfort and unease, ultimately bolstered his urgency to subordinate and erase them. Moreno, like most other Argentine anthropologists at the turn of the twentieth century, found annihilation the best response to the disquiet of his conscience.

Virchow and American Skulls: "These People ... Destined to Be Presented Here Today"

Across an ocean, Rudolf Virchow also prepared to engage with human remains plundered from Argentina. In 1871, he awaited a precious shipment of skulls from Peru and Argentina for what would be his first foray into an extensive study of American skulls. By then, Virchow was established as a pioneer of forensic science, and had helped initiate a whole new discipline: *Anthropologie*, the study of physical remnants of ancient and primitive societies. While secondary to his work in pathology, he pursued his interest in craniometry for the next thirty years, amassing thousands of skulls that he stored at the Pathology Institute at the University of Berlin.[58] Virchow's commitment to anthropology is reflected in his founding role in creating the first Berlin Anthropology Society [*Berliner Gesellschaft für Anthropologie, Ethnologie und Urgeschichte*, or BGAEU] in 1869.[59] Never to set foot in the Americas himself, Virchow had commissioned the skulls from adventurers in the field as early as the late 1860s, aided by local colleagues in Argentina and Central America.[60] Over the next decades, Virchow published numerous analyses of these skulls, including a significant volume on American craniometry in 1892.[61]

---

[57] Moreno, *Travel Journal*, 108.

[58] Massin, "From Virchow to Fischer," 85.

[59] See Patrick Schilling Dowd, "Rudolf Virchow and the Science of Humanity," PhD thesis, University of Pittsburgh, 1999; Rudolf Virchow, "Schädel von Chiriqui (Panama)," *Sämtliche Werke* 52 (1871): 9.

[60] Virchow, "Schädel von Chiriqui (Panama)," 9.

[61] Rudolf Virchow, *Crania ethnica americana. Sammlung Auserlesener amerikanischer Schädeltypen herausgegeben von Rudolf Virchow* (Berlin: A. Asher & Co., 1892). See also Byron A. Boyd, *Rudolf Virchow: The Scientist as Citizen* (New York: Garland, 1991); Gabriel Finkelstein, *Emil du Bois-Reymond: Neuroscience, Self, and Society in Ninteenth-Century Germany* (Cambridge, MA: MIT Press, 2014).

As Virchow pursued cutting-edge work in forensic anthropology, with a focus not just on German subjects but global ones, his international reputation grew. He was a key figure in building anthropological institutions at home, such as the Berlin Ethnological Museum. At the same time, as an active early Americanist he corresponded with, taught, or mentored nearly all the well-known scientists exploring the material traces of human societies in the Americas and around the globe. Virchow, like many European Americanists at this time, relied on collectors and informants to find and ship him evidence for analysis.[62] He was an armchair scientist, distant for the most part from any face-to-face experience with Americans. Virchow corresponded frequently about skulls with scientists in the field, including other German-speaking anthropologists working in the Americas.

One of Virchow's most important South American correspondents was Francisco P. Moreno. The Argentine exchanged publications with Virchow and procured human remains for the German to study. In July of 1878, Moreno wrote to Rudolf Virchow to thank him for the BGAEU's invitation as a corresponding member. In this letter (likely translated into German on Moreno's behalf), and in another written three days later in Spanish, to a German diplomat, Moreno acknowledged the honor and also expressed his hope that "there will arrive the opportunity to be useful."[63] Similarly, a form letter (in French) from Francisco Moreno on a Museo de La Plata letterhead to the BGAEU in 1890 reflected the intellectual and material exchange between Europeans and Latin Americans. Thanking the German organization for the gift of a brochure, the letter offered to continue the "exchange of objects and publications." Moreno's letter suggested the unique value of his collections: they contained "all the information [one] may require about the physical and moral history of the southernmost area of America."[64]

Virchow published his first Americanist study, a description of a single skull from Panama, in 1871. The skull, according to Virchow, had been "rescued from the old Indian burial ground" and brought to Europe by a former French consul in Panama.[65] According to Virchow, the burial site included various other artifacts, including jeweled objects and animal figures. Virchow, however, was only interested in the skull, even though it was badly damaged. He measured it carefully, reporting the dimensions. Just a few years later, Virchow was able to locate larger groups of skulls and carry out comparative

---

[62] For a discussion of the scientific division of labor in transatlantic archaeology see Podgorny, "Bones and Devices." See also David N. Livingstone and Charles W. J. Withers, eds., *Geographies of Nineteenth-Century Science* (Chicago: University of Chicago Press, 2011).

[63] Francisco Moreno to Baron de Holleben, July 23, 1878, BGAEU MIT260/1–3, bl. 2–3.

[64] Francisco Moreno to the BGAEU, September 25, 1890, BGAEU MUS 37.

[65] Virchow, "Schädel von Chiriqui (Panama)," 9.

studies. Then, in 1873, Virchow wrote a report on some skulls that had been sent to him from Argentina by Burmeister, and about which he corresponded with Moreno, who himself had an even larger stash. Virchow carefully compared these four skulls with two in his possession from another region of Argentina, measuring the brain capacity, horizontal and vertical diameters, and length of various parts and bones. He also calculated ratios and percentages of the variable parts of the skull, which he then arranged in a chart in order to compute relative dimensions.[66] In carrying out these measurements, Virchow, despite his humanistic philosophy and geographical and emotional distance from his human subjects, trod the same path as Moreno. When given the opportunity to examine Indigenous people or their remains, Virchow also reduced them to body parts, classified those bodies by group, and placed them in racial schema. He too expressed anxiety about Native American extinction but consoled himself with a role in salvaging their bones. In this sense, Virchow engaged in a common anthropological practice of simultaneously personalizing his subjects and presenting them as "frozen in time" by the assumed fact of impending extinction.[67]

Virchow, a European outsider only indirectly affected by Argentina's internal power struggles, could more easily distance himself from the darker aspects of anthropological collection practices. He had no land holdings or emotional attachment to Argentina; rather, he identified as a seeker of knowledge about humanity. He enjoyed the luxury of remove from the souls who had inhabited the skulls in his possession. His scientific distancing was also a product of his philosophical outlook. Virchow was engaged in the most politically progressive movements of his time. Philosophically and spiritually, Virchow saw himself as the intellectual heir of Alexander von Humboldt, sharing the great naturalist's underlying belief in the unity of humankind. Uncomfortable with German and British hierarchical classifications of races and peoples, Virchow argued for diversity and variety in human culture.[68]

---

[66] Rudolf Virchow, "Altpatagonische, altchilenische und moderne Pampas-Schädel," *Sämtliche Werke*, 52 (1873): 20.

[67] See Johannes Fabian, *Time and the Other: How Anthropology Makes Its Object* (New York: Columbia University Press, 1983).

[68] Massin, "From Virchow to Fischer," 87; William Y. Adams, *The Boasians: Founding Fathers and Mothers of American Anthropology* (New York: Hamilton Books, 2016). Humboldt's cultural status influenced many German scientists' interest in America as a location for acquiring knowledge about the world. According to H. Glenn Penny, most late nineteenth-century German anthropologists were inspired by cosmopolitanism, not colonialism, and "not all explorers, German or otherwise, sought territory or possessions." See H. Glenn Penny, *Kindred by Choice: Germans and American Indians Since 1800* (Chapel Hill: University of North Carolina Press, 2013), 33; see also Matti Bunzl and H. Glenn Penny, "Rethinking German Anthropology, Colonialism, and Race." In *Worldly Provincialism: German Anthropology in the Age of Empire*, eds. Matti Bunzl and H. Glenn Penny (Ann Arbor: University of Michigan Press, 2003).

In fact, he was part of a greater scientific enterprise in the human sciences and biomedicine, embraced by scientists across the Atlantic world, to seek universal truths about human nature and origins. Ironically, it was clear that they needed to look to the Global South for answers to the mysteries of humankind. As inhabitants of these areas exhibited, supposedly, various degrees of "savagery" and "barbarism," scientists experienced ambivalence about their subjects and themselves. They struggled to reconcile competing ideas about universal humanity and racial hierarchies.[69]

At the same time, and much like his Argentine counterpart, Virchow's ultimate ambition vis-à-vis America was the acquisition of Indigenous body parts for science.[70] In the 1873 article, Virchow recounted how he acquired the skulls of two "Pampas Indians," characterized not as humans who lived in community with others, but as members of a racial group. In his discussion of how he acquired the skulls, he repeated the account that an Argentine bureaucrat he was working with, Herr Oldendorff, wrote in the letter that accompanied the skull to Berlin: "I came to possess the Pampa-Indian skulls as per your wish through Herr Litzmann, and it was taken care of through arrangement by my friend General Rivas, commander of the southern border of this province ... It is not so easy to acquire full blood Pampa-Indian skulls, as there has already been much cross-breeding with the mixed races."[71] The remains uncovered by locals such as Olberdorff and Francisco Moreno were not enough, Virchow stated. "One can only hope ... that other areas of South America can be searched."[72]

As he racially parsed the skulls in terms of groups, Virchow also speculated about their individual identities. Despite no direct contact with Indigenous people at this point, Virchow described in the 1873 article what he imagined was their typical physical appearance, including a detailed description of infant board swaddling that might result in a particular formation of the skull. Virchow also mentioned that Oldendorff had noted that one of the old Pampas Indian skulls was that of the *"formidable Capitanejo,* known to and feared by our border patrol as *'Juan por Siempre'* [Juan Forever]; a horribly bloodthirsty bandit (his forehead is nearly two fingers wide) who carried out numerous murderous and harmful deeds."[73] Here he combined tropes of

---

[69] See Anderson et al., *Unconscious Dominions* for parallels observed in psychoanalysis in colonial contexts.

[70] Friendship networks among scientists from Europe and Latin America were crucial to the advancement of their work; see Patience Schell, *The Sociable Sciences: Darwin and His Contemporaries in Chile* (New York: Palgrave Macmillan, 2013).

[71] Virchow, "Altpatagonische," 20.

[72] Ibid., 27.

[73] Ibid., 20.

barbarism with anthropometric references to capture the danger of Capitanejo.

Virchow's persistent view of Indigenous peoples primarily as specimens was further reflected in his 1878 article on "American craniology," in which he sought to compile anthropometric data from available objects from North and South America. His main concern was the timing of different groups' appearance in the human record, based solely on skull shapes and measurement. Different skull types such as "brachycephalic" and "mesocephalic" would prove the relative antiquity of Indian ancestors in different parts of the continent.[74] He even attempted to create an atlas of "ethnic American skulls," organizing skulls by type.[75] Nevertheless, he avoided generalizing about a common origin:

> Today I confine myself to declaring that the physiognomic characters of the American heads show such a manifest divergence that the construction of a universal and common type of the American natives must be definitively abandoned. They are also mixtures of several native races, and the program of future research will find its final expression in the separation of different ethnic elements, which are included in the composition of the *various living and extinct tribes.*[76]

This last sentence reveals his ambivalent, uncanny sentiment toward the skulls: even at a remove, Virchow recognized the present-tense existence of his Indigenous subjects.

During the live demonstration with Pikjiojie, Batzinka, and Luis in 1879, Virchow seemed unsettled by their presence in his face-to-face meeting with the three Patagonians. (For an interesting parallel, see Sebastián Gil-Riaño's chapter in this book.) He announced to his audience, "These people [are] destined to be publicly presented today ... You will be astonished, as I was, to see these extraordinary phenomena before you." Repeatedly referring to them as "people" [*Personen*], Virchow could not avoid describing aspects of their humanity. He noted that "According to the man [Pikjiojie], his tribe consists of only 80 individuals; for this reason, perhaps, it is the group of the Patagonian tribes closest to civilization." Virchow regarded the three as informants about their land and culture. But, in a dualistic feat of simultaneous humanizing and objectifying, Virchow placed his guests in a classificatory schema of nine Patagonian tribes, "as [Pikjiojie] named them."[77]

Virchow characteristically stated that he was hesitant to draw "anthropological" conclusions, that is, to weigh in on these tribes' place in the pantheon

---

[74] Rudolf Virchow, "La craniologie américaine," *Sämtliche Werke,* 52 (1878): 127–134.
[75] Virchow, "La craniologie américaine," 127.
[76] Ibid., 134. Emphasis added.
[77] Virchow, "Drei Patagonier," 198.

of human evolution. What he could do, however, was examine, measure, and describe in detail the people before him. Thus the bulk of Virchow's presentation in 1879 consisted of the comparative analysis of measurements and physiognomic description of both the three living subjects, as well as the skulls in his own collection. He conducted precise measurements of the skull dimensions for his audience, along with typologies of skull shape; he also provided charts of body parts and their lengths. He even compared the living Patagonian in front of him to skulls he had previously classified, from South America and Europe, concluding for example, that their heads appeared similar to the Sami people of Scandinavia.[78]

At the same time that Virchow poked, prodded, and measured the Patagonians like the scientific specimens he took them to be, he also noted aspects of their individuality. From his reports we can also infer the mood and experience of the Indigenous people on display. He described their temperaments, for example, stating that:

> Piktschotsche or Pikjiojie, 43 years old, from the Haveniken [sic] tribe . . . usually exhibits a very serious, proud, and also melancholy appearance. He decides to speak, with difficulty; when it happens, the whole face suddenly comes alive, but he restricts himself to a brief, fast phrase. At rest, his face has a strict, almost hard expression: the fine lips are tightly closed, the lips around the mouth and nose are very prominent, the eyes look straight ahead.[79]

Virchow must have had to engage in extremely close proximity and communicated intimately with his guests to draw these conclusions, though, notably, he devoted significantly less detail to describing the woman and child.[80]

In theory, Virchow could retain a universalist view toward his human subjects, preserving a shred of their personhood, or at least withholding racial judgment until enough incontrovertible evidence emerged. He probably thought he was acting humanely, and compared to others, perhaps he was. Virchow held high moral ground for his era, based on his humanistic philosophy and clinical methods with Indigenous bodies. At the same time, when presented with opportunity to advance his science, he objectified dead and living Indigenous people, reducing them to specimens and body parts. Thus, the violent practices of colonialism, both foreign and internal, were to be tolerated, even subconsciously anticipated, in service to the scientific demand for human skulls.

---

[78] In this article, Virchow mentioned in passing that the skulls in question had been provided by Moreno. Virchow, "Drei Patagonier," 199.

[79] Virchow, "Drei Patagonier," 202.

[80] The next chapter in Virchow's larger investigation of anthropology and European cultural and racial studies was his *Crania ethnica americana*, published in 1892.

Skulls, Ambivalence, and Dehumanization

The encounters between these two skull scientists and their Patagonian con-temporaries mark a fulcrum, a point of concentrated ambiguity, disquiet, and incipient violence that shadowed the more obvious acts of Argentine internal colonialism and the expansion of international science. The history of the American continent, writ large, can be understood as a series of multiple encounters, including but not limited to enslavement, migration, trade, colon-ization, war, and scientific exploration. At the same time, these large-scale confrontations contain millions of individual conflicts or compromises, each of which has indeterminate outcomes. These micro encounters are subject to global forces and the inheritances of power dynamics but also to the individual acts and motivations of the participants. Arguably, this is especially true for scientists facing patients or human research subjects.

To wit, late nineteenth-century physical anthropologists and their Indigenous interlocutors carried with them their respective beliefs about land, material wealth, bodies, and identities. Anthropologists like Moreno and Virchow operated within the structures of power born of rich pre-contact societies, centuries of intertribal relations, of European invasions, and finally of postcolonial exploitation. Forensic studies of Pampas skulls occurred in the context of a new era of racial violence with hemispheric and local relevance, and the manner in which the two scientists carried out their work on skulls ultimately served the goal of taking possession of the frontier societies opening up in the Americas. For them, like so many others in the history of anthro-pology, the temptation to dehumanize outweighed the rewards of cultural exchange, relationality, and reciprocity as they both achieved their goal: turning Indigenous bodies into evidence for the growing corpus of scientific knowledge.

Moreno and Virchow operated within distinct national traditions and worldviews, but their approaches to the Indigenous Other ran parallel and were, ultimately, extractive. They both felt ambivalent about their human subjects – at times disparaging, at others quite positive or even affectionate – as they prioritized their material and professional agendas over human rela-tionships. For his part, Virchow's work on skulls contributed to the objectifi-cation and dehumanization of many others in the Americas in the following decades. He may have resisted hierarchical racial schemes but did not advocate for the full humanity of those he studied, or rail against racial violence, as field collector Alberto Frič, and later Boas, would.[81] When given the chance, Virchow chose to see Indigenous peoples as above all objects of science.

---

[81] The exceptional story of Frič, a Czech collector hired by the Germans to gather artifacts in the early 1900s, and who refused to participate further once he met real people in the

Virchow and Moreno, unable to convert their societies into polycultural utopias, and unwilling to fully drop out (as did Frič), remained agents of the larger system. In their personal interactions with their subjects, they may have sometimes held themselves in the balance, but in the end, both scientists participated in the objectification and dehumanization of the people who supplied their evidence. Ultimately, those bodies existed primarily as means to the ends of European science. The temptation of scientific progress was too great, and their status and access to political power made it easy for them to override whatever curiosity about, and personal frisson with, the Indigenous Other they might have had. Moreover, Moreno and Virchow's scientific wonder and care for Indigenous peoples did not lead to respect and coexistence (despite Moreno's later mild regrets), but rather facilitated participation in campaigns of violence and erasure, while convinced that they were on the right side of history.[82]

Moreno, as an Argentine, was more familiar with the Aònikenk, Mapuche, Genniken, and Selk'nam (Ona) peoples. The proximity of traditions created even more competition and ultimately appeared to short-circuit any generosity he might have felt. European-descended Argentines and Indigenous peoples had long had encounters on the land. This historical reality in fact raised the stakes, as the question of racial difference was a living and very real issue for Argentines. Moreno had more opportunity and proved more willing than Virchow to judge in detail his informants' characters and human value, to scrutinize them and their place in the pantheon of humanity. He was motivated in an immediate sense to objectify, classify, and distance himself from his human subjects. Primarily concerned with his professional standing and scientific zeal for collecting and reconstruction the national history of his country, he participated in the displacement of peoples and occupation of land carried out by his government. The annihilating impulse was heightened by the knowledge that Argentines of Indigenous and European descent could (and did) merge biologically and culturally. Moreover, he recognized, and recorded, in his many close encounters with Indigenous people their distress and refusal along with offers of reciprocity. Knowing that, Moreno's willful betrayal of his "amigos" was all the more brutal.

In the name of scientific exploration, Moreno and his colleagues sought to salvage the remains of a population whose attempted genocide they actively participated in. They focused great attention on Mapuche and Aònikenk peoples and sought to entomb them prematurely in the nation's past. As Argentines built their institutions, their nationalist science was indelibly marked by dynamics of internal colonialism and racism specific to their

field, demonstrates the range of the possible in the "moral field" of that moment. Penny, "The Politics of Anthropology."

[82] Smith, "Paradoxes of Dehumanization," 416–443.

particular settler society. Indeed, for anthropologists like Moreno, there is evidence that a psychological need to escape the discomfort of the uncanny led these men to distance themselves from the violent acts of erasure. In Argentina, especially, anthropologists paid a steep price for their scientific prestige: the denial of the Indigenous parts of the national body.[83]

More broadly, the story of skull science on the Pampa illustrates the importance of context on the formation of scientific and clinical method, and even anticipates the emergence of ethical conundrums in anthropological practice. Talented and successful scientists, then and now, are expected to overcome their repulsion for the strange. Anthropologists, specifically, must hold themselves in the tension of competing desires to understand and destroy. Like other human scientists, anthropologists have often been profoundly changed by their interaction with the objects of their research. Outcomes, however, are not predetermined. In the twentieth century, humanitarian, antiracist, and even egalitarian racial schema would emerge, and many anthropologists came to reject the destructive, colonializing aspects of science in this period. The incidents of skull science examined here, including the affective responses of scientists and human subjects alike, demonstrate both the utility and the limits of the moral field concept. Without the benefit of ethical norms and rewards and punishments for unethical behavior, however, it appears that the weight of science's material demands creates a strong bias toward dehumanization and away from reciprocity and respect.

[83] See Argeri, *De Guerreros a Delincuentes*; M. Rodríguez and Gordillo chapters in Alberto and Elena, eds., *Rethinking Race*. According to Walter Delrio, this tendency, which he argues should be defined as genocide, continues to the present day; see Delrio et al., "Discussing Indigenous Genocide in Argentina: Past, Present, and Consequences of Argentinean State Policies toward Native Peoples," *Genocide Studies and Prevention* 5, no. 2 (summer 2010): 138–159.

3

Subverting the Anthropometric Gaze

Racial Science in the 1912 Yale Peruvian Expedition

ADAM WARREN

In the notebooks used to record the anthropometric measurements and physical features of Peruvian highlanders during the Yale Peruvian Expedition's 1912 visit to the Cusco region, the entry for Justo Rodríguez stands out. Listed as approximately thirty-eight years old and hailing from Abancay, in the rural province of Apurímac, Rodríguez encountered American researchers on August 23, 1912 at the Huadquiña Hacienda, a sugar and sheep-raising estate in the neighboring province of La Convención. There in the semi-tropical mountain forest region known as the *ceja de selva*, or eyebrow of the jungle, on the eastern slopes of the Andes, the expedition's surgeon, Luther T. Nelson, took his measuring tools and camera and sought to examine and photograph Rodríguez, one of six men he would study that day and 145 people he would examine that year on haciendas, at the nearby ruins of Machu Picchu, and in the city of Cusco. Working as part of a larger expedition remembered mainly for excavating Machu Picchu, Nelson's research agenda focused almost exclusively on Andean bodies, race, and health. Using pre-printed forms, he recorded descriptions and measurements of Rodríguez's hair, nose, teeth, eyes, skin, malar bones, head, face, trunk, arms, legs, hands, feet, and height, and he wrote the number 6 under "No. Children in Family."[1] Taking fingerprints from Rodríguez's right hand and writing that he "has Spanish blood," Nelson then added a surprising comment. Since he lacked a designated place for such information on the form, he wrote sideways up the middle of the second page "Subject became impatient and would not ~~stay~~ submit to further measurement."[2]

Rodríguez is the only research subject at the Huadquiña Hacienda whom Nelson described in his anthropometric notebooks as refusing to be measured. Indeed, his is the only reference of its kind in the 1912 anthropometric notebooks. It seems to correspond, however, to a more vivid description

---

[1] It is unclear whether this meant Rodríguez was one of six children or had six children; records of other research subjects suggest the latter is more likely.

[2] Yale Peruvian Expedition Papers (hereafter YPEP), Series III, 1912, Box 20, Folder 28, Notebooks of Luther T. Nelson.

elsewhere of what is likely the same man's act of refusal. Although Rodríguez is not mentioned by name, Nelson and others included an account of insubordination in their reports and correspondence, now archived in Yale's Sterling Memorial Library.[3] The Yale Peruvian Expedition's leader, Hiram Bingham, moreover, published the account almost verbatim in magazine and book-length descriptions of the expedition's work. For example, in *The National Geographic Magazine's* April 1913 article "In the Wonderland of Peru," Bingham conveyed the following:

> At Huadquiña the Indians were ordered to a room to be measured. One subject objected strenuously and made it as difficult as he could for any measurements to be taken. He would not stand straight, nor sit straight, nor assume any position correctly. Finally, when the measurements were all taken, he was offered the usual *medio* for his trouble. This small coin, with which one could purchase a large drink of native beer, was usually gratefully accepted as a *quid pro quo*, but in this case the Indian decided he had been grievously insulted, and he threw the coin violently to the ground and strode off in high dudgeon.[4]

This account circulated worldwide not just in *National Geographic Magazine*, but also in publications that reprinted images and text from "In the Wonderland of Peru" in translation.[5]

What can these sources tell us about Rodríguez's motivations for insubordination and the experiences of others whom Nelson and his collaborators sought to measure and photograph between early July and mid November 1912? Was Rodríguez, the 102nd person measured, the only one who refused outright to cooperate? Did others engage or resist in ways not immediately recognizable to Yale scientists? Nelson's description of Rodríguez, hastily jotted down in pre-printed anthropometric notebooks, demonstrates how

---

[3] I cannot say conclusively that the account corresponds to Rodríguez. In the earliest version I have found, which appears in a handwritten letter Nelson wrote to accompany a draft of his report, the references to numbered photographic negatives differ by fourteen exposures from those listed for Rodríguez in Nelson's notebooks. This is likely an error by Nelson, who wrote the description aboard ship when returning to the United States, over three months after measuring Rodríguez. If Nelson's references to photographic negatives in his description are correct, then the account corresponds to Nasario Ortíz, a forty-year-old man from Talavera, in the province of Andahuaylas. Ortiz's anthropometric entry, however, is more complete than that for Rodríguez and mentions medical conditions, the nature and descriptions of which suggest he was more cooperative.

[4] Hiram Bingham, "In the Wonderland of Peru: The Work Accomplished by the Peruvian Expedition of 1912, under the Auspices of Yale University and the National Geographic Society," *National Geographic* 24, no. 4 (1913): 562.

[5] In Peru alone, "Wonderland" appeared in the periodicals *Peru To-Day, La Crónica*, and *Ilustración Peruana*; Amy Cox Hall, *Framing a Lost City, Science, Photography, and the Making of Machu Picchu* (Austin: University of Texas Press, 2017), 86.

record-keeping forms themselves, the categories that organize them, and scientists' use of them preclude making anthropometric research subjects' subjective experiences known. The notebooks included a category for "Expression of Emotions," which Nelson almost always left blank, and there was little room elsewhere for descriptions of behavior.[6] Subjective experience, in other words, did not have scientific value for the expedition; the form reduced the person to a set of measurements. By publishing the anecdotal description of resistance, on the other hand, Bingham transformed an Indigenous research subject's struggle into a whimsical account, one that helped frame the expedition's work as an adventure story designed to entertain a US and global reading public. Rather than providing insight into research subjects' moral thinking, the expedition's use of this anecdote mocked and stereotyped Indigenous behavior as irrational.

In many respects, reconstructing Justo Rodríguez's lived experience presents challenges similar to those Marisa Fuentes encountered in researching enslaved and formerly enslaved women's histories in Barbados. Fuentes found only fleeting references to such women in the colonial archive, and she argued popular depictions distorted understandings of who specific women really were and their positions in colonial society. The colonial archives' violence thus silences and prevents their histories from becoming fully knowable, and it challenges historians to find new ways to reconstruct what their lived experiences might have been like.[7] In the Yale Peruvian Expedition's archival records, traces of Indigenous and Mestizo research subjects' behavior and practices of engagement and refusal can be identified and recovered to some degree, however tentatively. That said, one must also take seriously our limited ability to fully access, via the archive, past Indigenous Andean forms of perceiving, knowing, and world-making, which Marisol de la Cadena describes in the present as both engaging and exceeding Western forms.[8]

Clues as to what the encounters themselves were like can nevertheless be found not only in the expedition's written records, but also in the hundreds of anthropometric photographs of those subjected to the researchers' measuring instruments and camera. These photographs were published as part of

---

[6] In perhaps the only use of the "Expression of Emotions" category, Nelson described the eighth person he measured, a twenty-seven-year-old man, as "sullen and stupid"; YPEP, Series III, 1912, Box 20, Folder 28, Notebooks of Luther T. Nelson.

[7] Fuentes, *Dispossessed Lives*. Fuentes builds on Rolph-Trouillot's *Silencing the Past*.

[8] de la Cadena, "Indigenous Cosmopolitics in the Andes: Conceptual Reflections beyond 'Politics'," 334–370; de la Cadena, *Earth Beings*. De la Cadena provides an earlier historical and ethnographic analysis of race and overlapping concepts of Indigenous and Mestizo identity in Cusco in *Indigenous Mestizos: The Politics of Race and Culture in Cuzco, Peru, 1919–1991* (Durham, NC: Duke University Press, 2000).

H. B. Ferris's "The Indians of Cuzco and Apurímac."[9] They shed light on how ordinary Indigenous and Mestizo people toiling on haciendas, working at Machu Picchu, and traversing Cusco's streets may have found more subtle ways than those Rodríguez employed to refuse or resist Nelson's requirements, and in some cases embrace and subvert them. Moreover, they along with written materials shed light not only on how US expedition members understood their work, but also on how local intermediaries and collaborators may have perceived the ethics and practice of racial science.

Building on Julia Rodriguez's study of expedition science's affective dimensions in this book, this chapter examines the history of moral thinking among Indigenous and Mestizo research subjects, foreign scientists, and local collaborators to reconstruct a fuller picture of the Yale Peruvian Expedition's field encounters and the forms of relationality they engendered and entailed. It argues that histories of the human sciences can and should combine Indigenous Studies methods and concepts with the moral field framework to understand more fully the nature of research encounters involving Indigenous peoples. Inspired by works by Helen Verran and de la Cadena on encounters between Indigenous and Western knowledge systems, this study posits "moral fields" in the plural to describe coexisting, separate, yet partially overlapping practices of perception, evaluation, and judgment among populations brought into contact with one another.[10] While the archival record has limits, it enables us to reconstruct behavior, identify the structures of expedition scientific research, and map the broader political and social world that shaped moral thinking and decisions among figures like Justo Rodríguez.

Focusing on moral thinking, however, should not require withholding judgment of historical actors or treating the past as separated off from the present. This chapter explores what kinds of decolonial work engaging the lived experiences and moral thinking of past research subjects can do in the present. It asks, in particular, how this history of engagement and challenges to the Yale Peruvian Expedition's racial science research might inform recent Indigenous encounters with transnational scientific projects rooted in similar settler colonial logics, such as that which Rosanna Dent describes later in this book and its counterpart in Peru. In this sense, it hopes to be of use to those

---

[9] H. B. Ferris, "The Indians of Cuzco and the Apurimac," Reprinted from the *Memoirs of the American Anthropological Association* 3, no. 2 (1916): 59–148, 60 unnumbered leaves of plates.

[10] De la Cadena, "Indigenous Cosmopolitics"; Verran, "A Postcolonial Moment in Science Studies: Alternative Firing Regimes of Environmental Scientists and Aboriginal Landowners," 729–762. For moral fields, see Goldstein, "Toward an Empirical History of Moral Thinking: The Case of Racial Theory in Mid-Nineteenth-Century France," 1–27.

engaged in moral thinking around the politics of human scientific research in the Andes today.

A Note on Method

In *Decolonizing Methodologies: Research and Indigenous Peoples*, Linda Tuhiwai Smith noted in the 1990s that within many Indigenous communities, vigorous debate centers on the practices of research and the roles of both researchers and those who are researched. These debates, she suggests, can be summarized by questions about the ownership and control of research, the methods employed and questions asked, the production and dissemination of results, and the interests served.[11] I have taken these questions to heart in thinking through my work and subject position as a privileged, non-Indigenous, non-Andean, white American scholar from a North American research university. I am still figuring out the answers and sitting with the discomfort these questions present.

Admittedly, this chapter deviates from the decolonizing methods Smith and others rightly and forcefully articulate and promote. It is not a collaborative project framed in consultation with contemporary Indigenous people in the Cusco region, nor does it draw on their participation in gathering and analyzing archival materials or drafting findings. In part this is due to the location of archives in the United States, but it is also because I did not initially envision this project through a decolonizing framework. That said, there are also practical issues involved in seeking the approval of communities, whose past residents Nelson and his assistants measured and photographed. The anthropometric notebooks from 1912 include records of Indigenous and Mestizo research subjects from sixty-three communities in sixteen provinces.[12] Seeking the approval of each community's *asamblea comunal*, or communal assembly, is unfeasible. At the same time, the only entities that claim to speak for these communities as a whole are provincial and departmental governments and the Peruvian national government. For various reasons, I resist organizing my work around seeking approval from governing bodies and institutions of the Peruvian state, a political formation that has its own troubled history of anti-Indigenous racism and internal colonialism, and that continues to enact violent and extractive policies that disadvantage Indigenous peoples to the benefit of others.

Beyond these matters, I share historian of Hawai'i Adria Imada's concern about intruding in communities that may have been "talked out." Indigenous peoples in the Cusco region have been subject to what Eve Tuck describes as

---

[11] Smith, *Decolonizing Methodologies*, 10.
[12] Ferris, "The Indians of Cuzco and the Apurimac," 62.

"ongoing colonization by research," and many communities are deeply entangled with foreign and domestic tourism, which require and entail particular ways of performing Indigeneity and engaging outsiders.[13] Taking these considerations into account, over time I hope to strengthen existing relationships in the Cusco region and explore new ones, while remaining mindful of Imada's modeling of an "ethics of restraint."[14] I also plan to make this research available in Quechua to Indigenous communities encountering a new generation of research scientists, who seek DNA samples in a renewed effort to study human variation.[15]

This project has also required that I give thought to the kinds of sources employed and how I describe and reproduce them. Working with anthropometric photographs and written records of the expedition's racial scientific work is a fraught exercise. As the published account of Rodríguez's refusal shows, the images and descriptions that Nelson and others created capture various subjective experiences for Indigenous and Mestizo research subjects, among them trauma, and the violence of transnational scientific projects rooted in US imperialism, its settler colonial logics, and its goals of capitalist expansion. To borrow from Eve Tuck and K. Wayne Yang, such sources serve as vestiges of "damage-centered research" in action and evidence of the harm it can cause.[16] That harm, moreover, extends beyond the invasive, ephemeral moments in which the images were taken and engages Peru's history of internal colonialism. Amy Cox Hall argues that by emphasizing themes of poverty and primitivism, the expedition's anthropometric images reflected and furthered damage, lending credence "to the notion of a glorious Incan past and a miserable indigenous present."[17] As a central feature of creole nationalism long before Bingham arrived,[18] such depictions of Indigeneity denied

---

[13] Tuck, "Suspending Damage," 415. Note that members of many communities view tourism positively, given the revenue and opportunities it provides. See *Earth Beings*, in which de la Cadena's main interlocutor, Nazario Turpo, builds a career as an Andean shaman who meets with tourists.

[14] Adria Imada, *An Archive of Skin, an Archive of Kin: Disability and Life-Making during Medical Incarceration* (Berkeley: University of California Press, 2022). For my project, COVID-19 interrupted progress toward these goals.

[15] I hope my work's discussion of engagement and resistance to earlier research efforts proves empowering, though I make no assumptions about specific communities' positions on research initiatives.

[16] Eve Tuck and K. Wayne Yang, "R-Words: Refusing Research," in *Humanizing Research: Decolonizing Qualitative Inquiry with Youth and Communities*, eds. Django Paris and Maisha T. Winn (Los Angeles: Sage Publications, 2018), 223–248. Here I also draw on Margaret Bruchac, whose work argues that practices of salvage anthropology facilitated and furthered efforts at Native populations' disappearance. See *Savage Kin*.

[17] Cox Hall, *Framing a Lost City*, 179.

[18] There is significant scholarship on creole nationalism's tenets and the inclusion or exclusion of Indigenous people. For Peru and the Andes, see Cecilia Méndez, "Incas Sí,

Indigenous people full rights of citizenship and correspond to what Aníbal Quijano theorizes as the centrality of race in the "coloniality of power."[19] In working with the 1912 images, I have chosen to reproduce anthropometric photographs only when their inclusion as historical evidence can be purposeful. Photographs appear when analysis of their composition or subjects' visible actions and behaviors is necessary for understanding arguments about encounters with researchers and corresponding forms of resistance, refusal, engagement, and desire. In this sense, my approach builds upon scholarship on the history of photography in Andean and Latin American Studies, which emphasizes the behavior and self-fashioning of photographed subjects.[20]

On Regional Contexts

What were social and political conditions like in the Cusco region in 1912, the year Nelson measured and photographed Justo Rodríguez and 144 other Indigenous and Mestizo research subjects? How might Rodríguez have experienced and navigated this world, and how did it inform the Yale Peruvian Expedition's work? In *Los sueños de la sierra: Cusco en el siglo XX*, José Luis

Indios No: Notes on Peruvian Creole Nationalism and Its Contemporary Crisis," *Journal of Latin American Studies* 28, no. 1 (1996): 197–225; Mark Thurner, *From Two Republics to One Divided: Contradictions of Postcolonial Nationmaking in Andean Peru* (Durham, NC: Duke University Press, 1997); Brooke Larson, *Trials of Nation Making: Liberalism, Race, and Ethnicity in the Andes, 1810–1910* (Cambridge: Cambridge University Press, 2004); Alberto Flores Galindo, *In Search of an Inca: Identity and Utopia in the Andes*, eds. and trans. Carlos Aguirre, Charles F. Walker, and Willie Hiatt (Cambridge: Cambridge University Press, 2010). For Latin America, see Rebecca Earle, *The Return of the Native: Indians and Myth-Making in Spanish America, 1810–1930* (Durham, NC: Duke University Press, 2007).
[19] See Aníbal Quijano, "Questioning 'Race'," *Socialism and Democracy* 21, no. 1 (2007): 45–53; Quijano, "Coloniality of Power, Eurocentrism, and Latin America," *Nepantla: Views from South* 1, no. 3 (2000): 533–580.
[20] For the Andes, see Deborah Poole, *Vision, Race, and Modernity: A Visual Economy of the Andean Image World* (Princeton: Princeton University Press, 1997); Jorge Coronado, *Portraits in the Andes: Photography and Agency, 1900–1950* (Pittsburgh: University of Pittsburgh Press, 2018); Yazmín López Lenci, *El Cusco, paqarina moderna: Cartografía de una modernidad e identidad en los Andes peruanos (1900–1935)* (Lima: Instituto Nacional de Cultura, Dirección Regional de Cultura de Cusco, 2007); Jason Pribilsky, "Developing Selves: Photography, Cold War Science and 'Backwards' People in the Peruvian Andes, 1951–1966," *Visual Studies* 30, no. 2 (2015): 131–150. For elsewhere in Latin America, see Kevin Coleman, *A Camera in the Garden of Eden: The Self-Forging of a Banana Republic* (Austin: University of Texas Press, 2016); Greg Grandin, "Can the Subaltern Be Seen? Photography and the Affects of Nationalism," *Hispanic American Historical Review* 84, no. 1 (2004): 83–111; Deborah Poole, "An Image of 'Our Indian': Type Photographs and Racial Sentiments in Oaxaca, 1920–1940," *Hispanic American Historical Review* 84, no. 1 (2004): 37–82.

Rénique describes the southern departments of Peru as a "land of Indians and lords, united by the bonds both violent and subtle of a paternalist culture woven over centuries."[21] Yet, he and others also shed light on specific features of the Cusco region at this time, noting that broader economic changes and aspirations among the region's elite in the previous decades had transformed conditions in the valleys to the north of Cusco, where Machu Picchu and the Huadquiña and Santa Ana haciendas are located. These changes did not put an end to "the inherited structures of colonial domination"[22] to which Indigenous people were subjected, but rather intensified them. Such intensification likely proved fundamental in shaping Indigenous peoples' perceptions of the Yale Peruvian Expedition.

Long seen as a backwater in Peru, the city of Cusco was experiencing a resurgence in 1912. While parts of the regional economy had begun to grow and diversify as early as 1895,[23] such processes accelerated with the construction of a railroad connecting Cusco to Arequipa, other southern departments, and the port of Mollendo on the Pacific coast. The railroad opened in September 1908, inserting Cusco into a broader commercial network fueled by Arequipa's expanding wool economy. In the department of Cusco the railroad transformed the power equation, resulting in importers-exporters from Arequipa joining with large hacendados of La Convención province, where Huadquiña Hacienda is located, and Lares province as the dominant sectors of the department.[24]

Located in the northern part of the department, La Convención and Lares were nowhere near the railroad line to Arequipa, They had been transformed, however, by efforts to modernize Cusco and expand its economy prior to the railroad's inauguration. As Mark Rice explains, Machu Picchu's environs were "not the uncharted wilderness described in Bingham's accounts but a key economic frontier of Cusco."[25] Beginning in 1897, members of the newly formed Centro Científico del Cusco (Cusco Scientific Center) argued that the eastern slopes of the Andes in these regions and elsewhere could become sites for increased production of lucrative goods; trade routes to the Atlantic, moreover, could be established via Amazon waterways. Motivated by the boom in rubber and gum exploitation already underway, their efforts to attract state investment increased the political and economic power of large hacendados from these provinces and the output of their estates. In the 1890s, a road

---

[21] José Luis Rénique, *Los sueños de la sierra: Cusco en el siglo XX* (Lima: CEPES, 1991), 29. Note: All translations from Spanish are mine.
[22] Rénique, *Sueños*, 32.
[23] Mark Rice, *Making Machu Picchu: The Politics of Tourism in Twentieth-Century Peru* (Chapel Hill: University of North Carolina Press, 2018), 18.
[24] Rice, *Making Machu Picchu*, 18.
[25] Ibid., 19.

was blasted through the Urubamba Valley toward Vilcabamba to expand commerce with the area's rubber producers and haciendas.[26]

In 1910 and 1911, these same hacendados campaigned to create and extend the Santa Ana railroad north from Cusco to their provinces. Local politics focused on this project, which Peru's president, Augusto B. Leguía, approved. His decision reflected the persistence of an entrenched power structure in the Cusco region, in which large hacendados and other local powerholders called the shots and exercised influence over government officials. This structure had consequences for people like Justo Rodríguez. According to Rénique, "a large part of the Indigenous population remained at the mercy of local powerholders, resulting from a combination of economic forces and authority, of despotism and paternalism, a system known as gamonalismo."[27] As figures wielding authority, hacendados "could demand that indigenous people comply with a series of personal services, which despite being legally eradicated, would continue in effect until well into the twentieth century."[28]

As will become clear, the Yale Peruvian Expedition exploited these relationships between hacendados, local officials, and the national government in the city of Cusco and the region around Machu Picchu. They also drew on military support to carry out their work and gain access to Indigenous and Mestizo bodies, both as sources of labor and as subjects for anthropometric experimentation. These entrenched power structures, however, should not be interpreted as preventing resistance. Rather, while much scholarship on peasant rebellions in early twentieth-century Cusco has focused on the period after 1915 and especially the 1920s, the early 1910s were also a time of struggle. As Christopher Heaney notes, when Bingham arrived in Peru in 1911, President Leguía warned him that the region he intended to visit, the valleys of the Urubamba and Vilcabamba rivers, had witnessed upheaval a few months before, when "Indian farmers and rubber collectors there had rebelled against the region's landowners."[29] Describing the lower Urubamba Valley, Heaney adds that "The place was a human tinderbox" because "The state was weak, and the landowners dominated the peasants with a mixture of paternalism and abuse."[30] Some villages that had rebelled remained unwelcoming to American visitors.[31] Rodríguez's refusal should thus be understood within this

---

[26] Christopher Heaney, *Cradle of Gold: The Story of Hiram Bingham, a Real-Life Indiana Jones, and the Search for Machu Picchu* (New York: Palgrave, 2010), 84.

[27] Rénique, *Sueños*, 34.

[28] Ibid., 33–34.

[29] Heaney, *Cradle of Gold*, 77.

[30] Ibid., 97.

[31] Chaullay's residents "glared from their doorways at the tall white stranger and his hated military escort"; Heaney, *Cradle of Gold*, 98.

broader picture of simmering tensions, as should the behavior of others Nelson measured at Huadquiña Hacienda and elsewhere.

Expedition Strategies, Racial Thinking, and Coercion

In this political, social, and economic context, members of the 1912 Yale Peruvian Expedition adapted their approach and reframed their own rationales to fit local conditions. That said, given that the expedition was the fourth of its kind that Hiram Bingham, a lecturer in history at Yale, led to South America, it also drew on his ingrained prejudices and learned strategies. Born in Hawai'i to American missionaries, Bingham trained as a historian at Yale, UC Berkeley, and Harvard. As Cox Hall notes, however, he "might best be characterized as an explorer and collector with a scientific purpose."[32] He was a generalist who ended up carrying out research on history, geography, and archeology before eventually entering politics. According to Ricardo Salvatore, he might also be considered a "'gentleman scholar' with no financial limitations on travel overseas," since he had married an heir to the Tiffany fortune.[33]

An interest in exploring South America's past inspired and shaped Bingham's various trips. In 1906–1907, Bingham and Dr. Hamilton Rice, a fellow of the Royal Geographical Society, traveled from Venezuela to Colombia, retracing Simón Bolívar's route nearly a century earlier.[34] In 1908, before attending the Pan-American Scientific Congress in Santiago, Chile, Bingham traveled from Buenos Aires to Potosí to map the old royal road. He then continued with Clarence Hay, the secretary for the US delegation to Chile, to Cusco in early 1909, where they established contacts with locals and traveled to the ruins of Choqquequirau. According to Salvatore, Bingham became fascinated with Inca civilization and history on this journey, most likely at a local museum of Incaica in Cusco, in a visit to the ruins of Sacsayhuaman, and through conversations with local informants.[35] This journey inspired a return trip with a team of researchers in 1911, in which Indigenous people brought them to Machu Picchu. Bingham's original goal was to find Vilcabamba, the site to which the Inca leadership had fled after the Spanish invasion, but Machu Picchu came to dominate his subsequent work.

Bingham's racism permeated all aspects of these journeys and was evident in his published accounts. According to Cox Hall, in his description of his 1906–1907 travels with Rice, Bingham acted "as the anointed translator and knowledge broker for future travelers" and explained the moral character of

[32] Cox Hall, *Framing a Lost City*, 5.
[33] Salvatore, *Disciplinary Conquest*, 40.
[34] Ibid., 41.
[35] Ibid.

the people they encountered through racial categories prevalent at that time.[36] In his description of his time in Potosí in 1908, moreover, Bingham's racism and arrogance led interactions with an Indigenous man to turn violent. He wrote that when an innkeeper rejected a bill that appeared to be from an untrustworthy bank, "The idea of having a servile [Quechua] decline to receive good money was irritating." A scuffle ensued, in which Bingham, on horseback, rode the man up against a wall. The explorer wrote, "I fully expected that he would follow us with stones or something else, but as he was only a [Quechua] he accepted the inevitable and we saw no more of him."[37]

Bingham viewed Indigenous peoples in the Cusco region through this same disparaging lens in 1909, including them in broad generalizations about South Americans that ultimately shaped the 1912 expedition's thinking and behavior in the field. Influenced by salvage anthropology in North America and by settler colonial logics within science and beyond it, he organized the expedition's research around the assumption that "pure" Indigenous people would disappear as modernization progressed across the continent. He understood South America through an evolutionary and industrialist paradigm, one in which the continent's challenges stemmed from its climate and "race history" while the United States constituted "the apex of modernity."[38] As a result, from his perspective little effort could or should be made to learn from contemporary Indigenous people. Notably, none of the expedition members in 1911 or 1912 could speak the Indigenous language Quechua; it was not until the third expedition in 1914–1915 that any were tasked with learning it.[39] While Bingham wrote of "securing from the natives (by the offering of rewards for certain highly desired information) what data they could give regarding the presence of ruins, the frequence [sic] of certain animals, the peculiarities of the climate, etc.," he employed disparaging terms like "stupid boy" to describe young assistants and guides.[40]

Bingham's previous expeditions also laid groundwork for how the 1912 expedition would navigate local power relations and inequalities to curry favor, engage different communities, and secure Indigenous labor in the field. In his 1908 trip from Buenos Aires to Potosí and his 1909 trip to Cusco, for example, Bingham initiated a practice of coordinating with government officials, the military, and powerful landowners to facilitate his work. He did so through the language of friendship. According to Salvatore, when Bingham

---

[36] Cox Hall, *Framing a Lost City*, 10.
[37] Quoted in Heaney, *Cradle of Gold*, 40.
[38] Cox Hall, *Framing a Lost City*, 11.
[39] Ibid., 15.
[40] YPEP, Series V, 1915, Box 28, Folder 19, "Methods Followed in the Field Work of the Peruvian Expeditions of Yale University and the National Geographic Society," 5; Series III, 1912, Box 19, Folder 15, Hiram Bingham, Journal, 6.

arrived in Potosí, "The local prefect received the U.S. party with red-carpet treatment. Celebrations in his honor lasted a week, including bullfights, dinners, balls, fireworks, and illuminations."[41] When he and Hay traveled to Abancay, the city listed as Justo Rodríguez's place of origin, some months later, they did so with a military lieutenant's assistance. They were greeted by twenty-four landowners and soldiers, who "cheered and escorted the American *científicos* into the small city, where they met Abancay's prefect."[42] Leaving by mule two days later, Bingham and Hay were accompanied by the lieutenant and a team of Indigenous men the lieutenant had conscripted, who "were paid a pittance and could be jailed if they refused to work."[43]

Bingham had no moral quandaries with these exploitative practices and returned to them in 1911, drawing on the support of the military and the Abancay prefect, who by then had become prefect of Cusco. Emphasizing the rhetoric of friendship and diplomacy in later published descriptions, Bingham wrote of first "establishing friendly relations with the foreign Government and securing of requisite permits from various governmental bureaus and introductions to large landowners; and second, in making local arrangements such as establishing connections with reliable business houses, purchasing equipment and supplies, and securing the most efficient native assistants."[44] At the same time, he lamented building connections with locals as time "wasted in diplomacy" and noted that "stupid officials, suspicious land owners, and ignorant natives" could interrupt or undermine the expedition's work.[45] The rights and wishes of Indigenous and Mestizo men, furthermore, mattered little to him. Two government officials accompanying the expedition traveled from farm to farm, greeting such men, and slipping silver dollars into their palms when they shook hands. In doing so, they effectively paid the men in advance for their work, "threatening them with imprisonment or worse if they refused." According to Heaney, "The farmers pleaded that they had to tend to their crops, that their families could not spare them, that they lacked the food for a week's march into the jungle. But the officials were implacable, and Bingham soon had a dozen porters."[46]

In 1912, such practices formed a routine part of the expedition's strategy, one consistent with its members' views but infrequently mentioned in published accounts. Indigenous men from the town of Ollantaytambo served as conscripted paid laborers, having been rounded up by the governor and

[41] Salvatore, *Disciplinary Conquest*, 41.
[42] Heaney, *Cradle of Gold*, 50.
[43] Ibid., 52.
[44] YPEP, Series V, 1915, Box 28, Folder 19, "Methods Followed in the Field Work of the Peruvian Expeditions of Yale University and the National Geographic Society," 2.
[45] Ibid., 7.
[46] Heaney, *Cradle of Gold*, 108–109.

imprisoned "to see that they did not run away" before joining the expedition.[47] Relations between these workers and American expedition members generated discord, with the reluctance of the former stemming from being uprooted from their communities and "forced into the abusive practices of the lower Urubamba, where tensions between Indians and whites ran high."[48] Particular events further undermined trust and pretenses of actual friendship. At Mandor, Pampa workers encountered the stabbed body of an Indigenous man. They likely also knew that an Indigenous child had died accompanying Bingham's expedition a year earlier, swept away by the Urubamba River under suspicious circumstances. At Machu Picchu workers set a fire that nearly took the lives of the Peruvian soldier accompanying the expedition and an expedition member, leading to suspicion of foul play.[49] While it is unclear if such stories reached Justo Rodríguez and informed his act of refusal the following month, they would have shaped the thinking and behavior of others conscripted to work for the expedition and measured by Nelson.

Tensions, resistance, and refusal of various kinds continued in the months that followed and during the 1914–1915 expedition. In analyzing Bingham's correspondence, Salvatore notes signs of unease, including "Indian laborers who abandoned the camp without reason, peasants who refused to sell mules to the expedition, Mestizo guides who kept Indian laborers away from the Yale camp, and commoners who denounced the wrongdoings of the Yankee explorers to the press."[50] Hacienda owners likewise did not necessarily look upon the expedition favorably, especially with regard to its hiring of workers. Although some hacendados such as Señora Carmen, the owner of Huadquiña Hacienda where Nelson measured Rodríguez, and the Duque family of Hacienda Santa Ana, coordinated with Bingham and formed friendships with him, in other cases the expedition disrupted long-standing patron–client relationships between hacendados and peons. According to Salvatore, when the expedition returned in 1914–1915, "landowners charged that the excavations were luring away their workers."[51] This was partly because the expedition paid higher wages at $1 per day and paid workers in cash. However, it also provided workers free medical care, which undermined hacendados' prestige and "showed indigenous peasants a side of modernity that local landowners were not ready to embrace." As a result, the Yale Peruvian Expedition, through its forms of recruitment, constituted a "menace, for they raised wages, defied traditional social hierarchies, and engaged peasants in the search for Inca artifacts."[52]

[47] Ibid., 133.
[48] Ibid.
[49] Ibid., 133–134.
[50] Salvatore, *Disciplinary Conquest*, 93.
[51] Ibid., 88.
[52] Ibid., 90.

Local Intellectuals, the "Indian Problem," and Expeditionary Science

Racist views on Indigenous peoples also formed part of the moral thinking of urban Cusqueños who interacted with Bingham and his team, and who initially welcomed and then debated and critiqued the expedition's work. As Mark Rice notes, "colonial era concepts of race and ethnicity remained stubbornly powerful"[53] in the early twentieth century. This was even true of intellectuals who campaigned to bring about Indigenous peoples' uplift and fought against their exploitation on haciendas. Intellectuals in Cusco fell roughly into two different camps, traditionalists and indigenistas. The former sought to maintain long-standing social and political structures, while the latter positioned themselves as vindicating the region's Indigenous peoples after centuries of exploitation. Such views among indigenistas, however, did not signify abandoning long-standing ideas about difference that characterized Peru's entrenched internal colonialism. According to Yazmín López Lenci, indigenismo was "a theory woven together on an essentialist opposition between what is ours (*lo propio*) and what is others' (*lo ajeno*), and thought about in essentialist racial categories."[54] Moreover, as they sought to defend and bring about Indigenous peoples' uplift, indigenistas engaged the logic of racial thinking and speculated about communities' potential for modernization. The language used in indigenista scholarship bears this out. Research shed light on the "Indian problem," the notion that Indigenous culture and living conditions hindered the region's progress.[55]

When Bingham arrived in Cusco in 1911, he deliberately sought to build connections and alliances with indigenista scholars and their students at the local university. Recent political events had given rise to a new generation of intellectuals and students there, who sought to transform thinking about Cusco's regional identity and its Indigenous peoples. Having formed the Asociación Universitaria (University Association) in 1909, students went on strike, closing the campus. This led President Leguía to appoint an American professor, Albert Giesecke, as the university's rector with instructions to modernize it. Giesecke reorganized the university and created a more democratic environment, in which research about Indigenous peoples was encouraged. López Lenci writes that his administration emphasized "promoting among students knowledge of the existing reality in and around Cusco, and the requirement that they write on the basis of what they have seen and verified."[56] Giesecke supported Quechua language classes and archeological

---

[53] Rice, *Making Machu Picchu*, 19.
[54] Lenci, *El Cusco, paqarina moderna*, 30.
[55] Rénique, *Sueños*, 58.
[56] Lenci, *El Cusco, paqarina moderna*, 93–94.

72 ADAM WARREN

and ethnological research expeditions.[57] The number of Cusqueños doing research in the countryside thus grew significantly,[58] leading to broader practices that Jorge Coronado, describing indigenismo in Peru more broadly, characterizes as representing and speaking for Indigenous people.[59]

By establishing ties at the local university and the Sociedad Geográfica de Lima (Geographical Society of Lima), Bingham hoped to gain information and connections to facilitate his research. Through meetings and public talks, he gave local intellectuals the sense that "they had been invited to participate in this project of knowledge."[60] Some specialists even traveled with the expedition and served as intermediaries. For example, a professor of Spanish and literature from the university, José Gabriel Cosío Medina, accompanied the 1912 expedition as the Peruvian government's and the Geographical Society of Lima's official delegate. Students also became involved. Among them were sons of hacendados and other families who held land in the Urubamba River Valley and along the Vilcabamba River. These students invited the expedition to visit their family estates, shared knowledge of ruins, and facilitated additional connections.[61] Such intermediaries proved crucial in enabling expedition members to conduct their work, despite doing so at a time when, according to López Lenci, "in confrontation with travelers' representations of Cusco, there existed a struggle over representation, which was a struggle over the construction of place."[62] Through collaboration, they influenced expedition members' assumptions about the acceptable treatment of workers and research subjects.

Heated debate soon arose in Cusco over the expedition's removal of artifacts from Peru. Nelson's 1912 anthropometric research, however, never generated controversy or concern among students and scholars. This is true despite the intrusiveness, violence, and questionable ethics that characterized Nelson's interactions. It may be partly attributed to the fact that, as others have shown, traveling research expeditions and local scientists in the Andes had already debated racial difference and employed anthropometric photography before the Yale Peruvian Expedition's arrival.[63] Moreover, Cusqueños had been

---

[57] Rénique, Sueños, 50–53.
[58] Ibid., 58.
[59] Jorge Coronado, The Andes Imagined: Indigenismo, Society, and Modernity (Pittsburgh: University of Pittsburgh Press, 2009). Scholars disagree as to whether indigenismo in Cusco was an elite project or one with significant popular dimensions. See de la Cadena, Indigenous Mestizos; Zoila Mendoza, Creating Our Own: Folklore, Performance, and Identity in Cuzco, Peru (Durham, NC: Duke University Press, 2007).
[60] Salvatore, Disciplinary Conquest, 84.
[61] Heaney, Cradle of Gold, 79–80.
[62] Lenci, El Cusco, paqarina moderna, 30.
[63] Poole, Vision, Race, and Modernity; Gabriela Zamorano, "Traitorous Physiognomy: Photography and the Racialization of Bolivian Indians by the Créqui-Montfort

carrying out racial scientific research for decades. One such scholar was Antonio Lorena, a physician who published extensively on racial differences and ideas about racial fitness beginning in the 1890s, and who invoked the logics of creole nationalism by questioning whether contemporary Indigenous peoples were related to the original inhabitants of the region responsible for its monumental ruins. As a member of the Sociedad Arqueológica Cusqueña (Cusco Archeological Society), his research continually emphasized the settler colonial idea that "pure" Indigenous peoples in and around Cusco were a race destined to vanish. In the years immediately preceding Bingham's first trip to the region, Lorena's work included comparing measurements taken from ancient skulls to the cranial dimensions of living Indigenous peoples thought to be free of racial admixture, who resided near ruins from which said skulls had been recovered. Lorena presented this study at the 1908 Pan-American Scientific Congress in Santiago, Chile, which Bingham attended.

Another scientist interested in questions of racial continuity or discontinuity, José Coello y Mesa, conducted research involving anthropometric and craniological practices not unlike those of the Yale Peruvian Expedition. As a delegate of the Asociación Pro-Indígena del Cusco (Pro-Indigenous Association of Cusco), Coello y Mesa analyzed skulls from other parts of Peru as well as human remains excavated near Cusco, comparing them to the region's living Indigenous people. He aimed to determine whether multiple migration patterns explained what he perceived to be differences among Indigenous populations from distinct regions. He published his findings in 1913, having traversed the countryside around Cusco with scholars in years prior to measure community members in Pantipata, Colquepata, and Chincheros. His work and Lorena's thus established the precedent of intrusive scientific research in these rural communities, at least one of whose residents Nelson later encountered and measured.[64]

The Yale Peruvian Expedition's anthropometric research goals overlapped with Lorena's and Coello y Mesa's work around questions of racial origins, racial continuity and discontinuity, and ideas about the vanishing Native. It is unclear, however, to what degree Cusqueño racial scientists influenced Bingham's and other expedition members' thinking about anthropometry. Yale researchers did not acknowledge local researchers' work in their publications. For example, H. B. Ferris, the physical anthropologist who analyzed

Expedition (1903)," *Journal of Latin American and Caribbean Anthropology* 16, no. 2 (2011): 425–455.

64 See José Coello y Mesa, "Crania peruana por José Coello y Mesa," *Revista Universitaria* 2, no. 5 (1913): 42–52; 2, no. 6 (1913): 2–30. Given that Nelson also measured at least one subject from Chincheros, and that Pantipata and Colquepata were not far from the expedition's worksites and connected to Cusco through trade, Coello y Mesa could have already introduced some of Nelson's research subjects to anthropometry.

Nelson's data back at Yale, instead described Nelson as conducting racial scientific work in response to the urging of the prominent Czech anthropologist Aleš Hrdlička.[65] Elsewhere, the expedition positioned itself as building upon other foreigners' photographic studies and anthropometric research, emphasizing works by Alcide d'Orbigny and Arthur Chervin on the Quechua and David Forbes on the Aymara.[66]

One way that the Yale Peruvian Expedition did differ significantly from local research efforts, however, was in terms of its structure and members' training. Bingham organized the 1912 expedition on a model developed in polar exploration, in which there was "a large, semipermanent, multidisciplinary team of experts organized like a naval expedition, whose captain won the lion's share of the credit."[67] This model relied on local Indigenous assistants, much like Naval Commander Robert E. Peary's expedition to the North Pole in 1909.[68] Looking back on the 1912 expedition's work, Bingham hailed it as a success and example of what multidisciplinary research based on friendship and camaraderie could achieve. The paleontologist's work, for example, had benefited from the civil engineer's discovery of a rare fossil. Similarly, the botanist had identified an important feature on an ancient monument that the expedition's archeologist had missed.[69] The contributions of local informants whom Bingham described through the language of friendship, on the other hand, were rarely acknowledged.

Despite receiving fame and credit for these exploits, most of the men from the United States working alongside Bingham possessed limited levels of specialized knowledge. None across the three expeditions had received extensive training in archeology, arguably the most important science for the expedition's work and reputation. They came from a world in which non-professional scientists were commonplace, and in which dabbling in science was not unusual among educated people. Lack of training, however, ultimately mattered little to Bingham, an explorer for whom ambition and outside sponsorship shaped and legitimated everything. While benefiting from local intellectuals' and others' research and insights, he portrayed his fellow

---

[65] Ferris, "The Indians of Cuzco and the Apurimac," 59.

[66] Relevant local studies nevertheless appeared in the *Revista Universitaria*. During the expedition's visit in 1912, for example, the journal published Humberto Delgado Zamalloa's "Apuntes etnográficos de los aborígenes del pueblo de Acomayo" [Ethnographic Notes on the Aboriginals of the Village of Acomayo], an anthropometric study of Indigenous peoples south of Cusco. Other research underway examined questions of racial classification, degeneration, and criminality and customs and traditions among urban and rural populations of Indigenous descent.

[67] Heaney, *Cradle of Gold*, 69.

[68] Ibid.

[69] YPEP, Series V, 1915, Box 28, Folder 19, "Methods Followed in the Field Work of the Peruvian Expeditions of Yale University and the National Geographic Society," 3.

members' work as building upon a longer history of scientific expeditions and travelers' accounts of the Andes, such as those of Alexander von Humboldt, Clements Markham, and others.[70] According to Salvatore, Bingham "was trying to outdo the work of William Prescott and Sir Clements Markham" as the Yale Peruvian Expedition's scope expanded.[71] Bingham critiqued these travelers and others, arguing that "too often in recent years expeditions had gone out with a very narrow viewpoint, equipped only to do astronomical or anthropological or paleontological or physiographic work."[72] By concentrating the Yale Peruvian Expedition's work on the Cusco region, on the other hand, he sought to benefit from taking "a relatively small, unexplored area and covering it as thoroughly as possible, in a way making it a type area to which other areas can be compared."[73]

Cosío Medina noted the limited expertise of several of Bingham's men while also acknowledging their ambition and role as part of a transnational system of information extraction. Having been sent to accompany and observe the expedition, he wrote of Nelson's work:

> In the Anthropology section, the same doctor has taken a great many measurements of native types, in different sections, of their size, physiognomical proportions, thoracic and pulmonary capacity, and visual faculty, as well as hundreds of photographic views of Indians, data from which he has not drawn a single mean proportion, because according to the contract he has with Yale University, he should take [the data] to that center so that it may be studied by a notable anthropologist.[74]

This was a system of imperial knowledge production, one Salvatore connects to the expansionist tendencies of US capital, technology, and culture.[75] However, as a participant in this system, Nelson, a knowledgeable surgeon, having been hired primarily to address the expedition's medical needs, mostly

---

[70] Cox Hall, *Framing a Lost City*, 7.

[71] Salvatore, *Disciplinary Conquest*, 60.

[72] YPEP, Series V, 1915, Box 28, Folder 19, "Methods Followed in the Field Work of the Peruvian Expeditions of Yale University and the National Geographic Society" (manuscript), 1.

[73] Ibid., 4.

[74] José Gabriel Cosío, "Informe elevado al Ministerio de Instrucción por el doctor don José Gabriel Cosío, Delegado del Supremo Gobierno y de la Sociedad Geográfica de Lima, ante la Comisión Científica de 1912 enviada por la Universidad de Yale, acerca de los trabajos realizados por ella en el Cuzco y Apurímac," *Revista Universitaria* 2, no. 5 (1913): 22.

[75] While the Yale Peruvian Expedition is known for receiving support from the National Geographic Society and Kodak, various businesses contributed to the expedition and shaped its work. Winchester Repeating Arms Company supplied weapons, while Waltham Watch Company provided astronomical watches and chronometers; Salvatore, *Disciplinary Conquest*, 78.

served as a poorly trained anthropometric data collector.[76] Ultimately, his research would generate tense encounters with Indigenous and Mestizo people compelled against their will to serve as research subjects.

Photography, Refusal, and Desire

How did Nelson undertake the expedition's racial scientific research? His diaries suggest he had no background in anthropometry prior to leaving for Peru, and that he had only studied Alphonse Bertillon's instructional works on standardized anthropometric measurements and photography during the ocean voyage to South America. Once in the Andes, the surgeon spent several months in 1912 photographing and measuring Indigenous and Mestizo people in urban and rural settings. He used camera equipment provided by Kodak along with a measuring set, which included long and short folding rods (rulers) for measuring height, a craniometer, two sliding compasses for large and small measurements, a steel tape, a grip dynamometer, and equipment for taking fingerprints. Numerous subjects measured and photographed were workers who accompanied the expedition and assisted in its excavations, while others had little or no relationship to it. In total, Nelson examined 145 people, all but one of whom were men, taking thirty-eight different measurements. His subjects ranged in age from seventeen to eighty-eight. Their photographs formed but a fraction of the 12,000 photographs taken between 1911 and 1915, of which 1,000 documented racial types and another 1,000 depicted Indigenous customs and social life.

As H. B. Ferris, the physical anthropologist who analyzed the data at Yale, described it, the research constituted an effort at "the acquisition of data for the study of the anthropomorphic and physiognomic characters of the Quichua."[77] A sense of urgency motivated this work, since "it is simply a question of a comparatively short time when there will be no race that has not suffered recent admixture."[78] Nelson's photographic records are notable, however, for deviating from the requirements and standards of anthropometric photography. He sought to mimic racial scientists' work in the United States and Europe, who in previous decades had made advances in its development. Indeed, building on the criminal identification method of Alphonse Bertillon in France, anthropometric photography had largely become standardized and was regarded as a reliable means to index precisely and scientifically the dimensions, features, and measurements of individuals. By following set

---

[76] Nelson's duties in 1912 also included studying coca's effects on Indigenous people, offering medical services to local communities, and studying health and disease prevalence.

[77] Ferris, "The Indians of Cuzco and the Apurimac," 59.

[78] Ibid., 64.

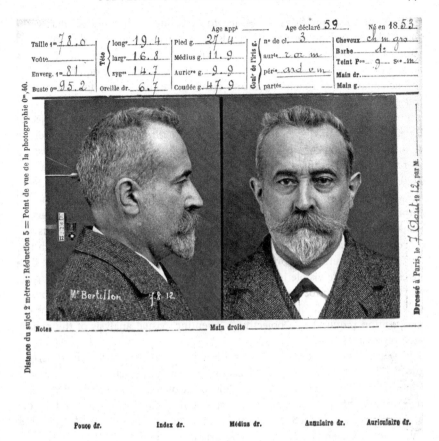

Figure 3.1   Self-portrait of Alphonse Bertillon, inventor of anthropometry, on anthropometric data sheet, dated August 7, 1912.
Source: Wikimedia Commons.

procedures for positioning subjects before the camera, by employing devices to hold them in place, by placing them in front of a neutral background, by controlling lighting, and by positioning rulers and measuring devices near them in the frame, racial scientists believed their photographs could reveal truths about the physicality of racial difference (Figure 3.1).[79]

[79] For analysis of Bertillon's criminal identification method and other approaches, see Josh Ellenbogen, *Reasoned and Unreasoned Images: The Photography of Bertillon, Galton, and Marey* (State College, PA: Penn State University Press, 2013); George Pavlich, "The Subjects of Criminal Identification," *Punishment and Society* 11, no. 2 (2009): 171–190.

Many of the conditions necessary for creating such photographs were unavailable to Nelson during his travels in the Cusco region. He did not have a formal photographic studio or a scientific laboratory for carrying out his work, but rather used a hotel room in Cusco and indoor and outdoor spaces on haciendas and at Machu Picchu to create makeshift studios for measuring and photographing subjects. His technology was also limited to tools and camera equipment brought from the United States. As a result, subjects appeared at various distances from the camera and often were not centered in the viewfinder properly, limiting the possibility of accurate comparison. Moreover, Nelson photographed them at angles in front of doorways, against stone walls or decorated adobe walls, or in exterior arcades with arches and cross-beams (Figures 3.2 and 3.3). His work thus failed to conform to one of anthropometric photography's key requirements: posing subjects against blank, neutral backgrounds without any depiction of depth as a means to prevent misperceptions of what constituted their physical form. To address this, Nelson sometimes improvised by hanging a sheet behind the subject, the edges of which are visible in some images (Figure 3.4). The photographs can thus at best be described as approximations of Bertillon's practices of "accurate" and "measurable" depiction. They failed to achieve the consistency of composition required for "objective," "scientific" visual analysis of the racialized physical form.

In journals and reports, Nelson and Bingham provided fairly curt descriptions of how the surgeon should acquire research subjects. These methods made use of local intermediaries while drawing on broader practices and assumptions about needing to engage and secure the cooperation of Indigenous people through coercion and force. For example, at the Huadquiña Hacienda where Justo Rodríguez refused to be measured and photographed, Nelson reported that "the administrator of the hacienda was the man who mustered Indians to be measured."[80] He did so on the orders of the hacienda owner, Señora Carmen, who befriended Bingham. In Cusco, on the other hand, Bingham's activities on July 5, 1912 included arranging for Nelson "to continue taking portraits and measurements of Indians. The Lieut. Sotomayor gets Indians and translates, Nelson measures and photographs and then gives the Indian [sic] a media. It takes from 40 minutes to 1.5 hr. to do the stunt."[81] According to Nelson, Lieutenant Sotomayor had been secured "through Mr. Bingham's influence with the prefect"[82] and spoke Quechua. He caught Indigenous men off the street and main square and brought them by force into the Hotel Central, where Nelson had a room for taking

[80] YPEP, Series III, 1912, Box 20, Folder 27, Journal of Luther T. Nelson, 11.
[81] YPEP, Series III, 1912, Box 19, Folder 15, Journal of Hiram Bingham, 6.
[82] YPEP, Series III, 1912, Box 20, Folder 27, Journal of Luther T. Nelson, 6.

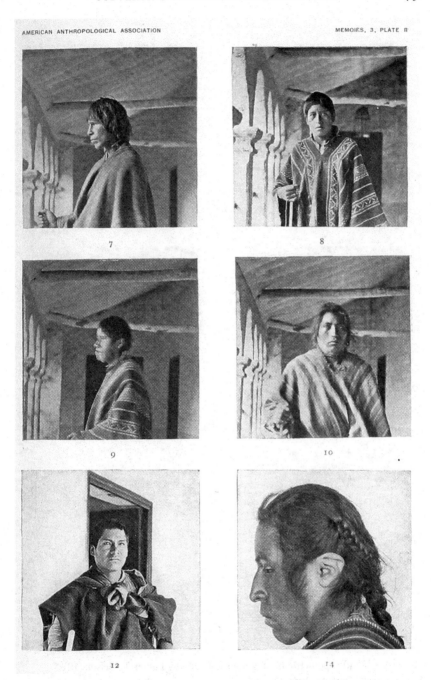

Figure 3.2 Anthropometric photographs 7–10, 12, and 14 by Luther T. Nelson, 1912
Source: National Library of Peru.

Figure 3.3 Anthropometric photographs 260–265 by Luther T. Nelson, 1912
Source: National Library of Peru.

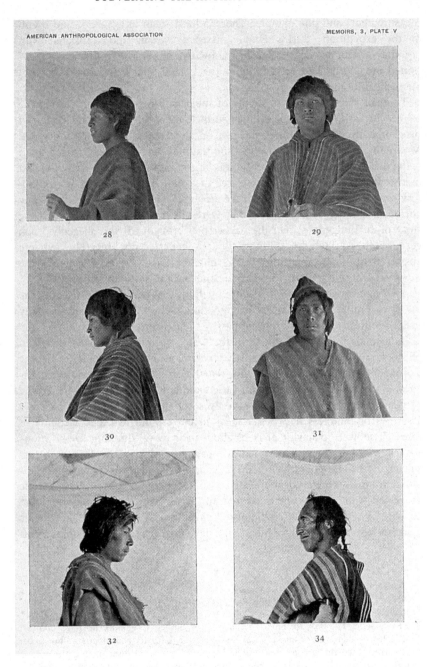

Figure 3.4   Anthropometric photographs 28–32 and 34 by Luther T. Nelson, 1912
Source: National Library of Peru.

measurements and photographs. There, a local "man of leisure"[83] and friend of the expedition, Carlos Duque, assisted Nelson and translated with Sotomayor. Duque was the son of the owners of Santa Ana Hacienda mentioned previously, had spent time in the United States, and spoke English, Spanish, and Quechua.

To guarantee an adequate supply of subjects, Bingham had instructed the soldier to "arrest any Indians that seemed to be of pure blood and who proclaimed by their costumes and general appearance that they were typical Mountain Indians."[84] Nelson noted the trauma some men experienced upon being detained unexpectedly, yet he was largely indifferent to it. He wrote that many captured in Cusco's square feared they were being recruited for military service "and not a few shed tears at the thought."[85] However, others "were only curious and much relieved when they were set free."[86] He even conveyed the story of an Indigenous man the soldier had brought in who allegedly "took very kindly to the idea. Military honors appealed to him. The teniente answered his many questions in the quichua [sic] tongue, and, when the measurements were taken, told him to come back in a month to be enlisted."[87] Suggesting a broader range of affective relations that may have shifted after the initial encounter and during the examination, Nelson claimed more generally that "The Indians are very fond of having their picture taken."[88]

Indigenous people and others in the Cusco region were not strangers to photography by the early twentieth century, as Deborah Poole and Jorge Coronado have documented.[89] However, the extent of their previous experience, if any, in front of the camera varied and few would have recognized the surgeon's measuring tools or his gaze. It is not surprising then, that Nelson's assertion of fondness does not appear to correspond to other accounts of research subjects' behavior or clues about their moral thinking. Descriptions suggest multiple ways of perceiving, engaging, and in many cases resisting, subverting, or co-opting the expedition's racial scientific work. The predominant sentiment conveyed, however, is reluctance. In Cusco, few people expressed willingness to be measured and photographed without the threat of force. A reflection of the ubiquity of violence as a form of encounter, this behavior became especially evident on a day when the soldier was absent. Left

[83] Ibid.
[84] Bingham, "Wonderland," 561.
[85] Nelson, quoted in Ferris, "The Indians of Cuzco and the Apurimac," 61.
[86] Ibid.
[87] YPEP, Series IV, Box 26, Folder 27, Report of Luther T. Nelson, no page number. If Nelson's identification of numbered negatives in the letter accompanying his report is correct, then this account corresponds to Lucio Tito from Pisac, in the province of Calca. Although Tito's age is difficult to discern, he was likely too old for military service.
[88] YPEP, Series IV, Box 26, Folder 27, Report of Luther T. Nelson, no page number.
[89] Poole, *Vision, Race, and Modernity*; Coronado, *Portraits in the Andes*.

on their own, Nelson and Duque "tried to round up some Indians but found they would not come. We succeeded in getting only one, and that one by aid of a policeman."[90] A similar phenomenon occurred at Huadquiña Hacienda, suggesting Rodríguez was not the only research subject there to refuse Nelson's efforts. Nelson noted that while the hacienda administrator had been assigned to bring men to be photographed and measured, "He was occupied with other business nearly all the time, the result being that it was rather difficult for me to get an Indian promptly when I was ready for him."[91] Conditions worsened, moreover, when the owner, Señora Carmen, departed for Cusco with her relatives, servants, and the administrator himself. Their absence "put a stop to measuring Indians," indicating the centrality of coercion and threats of violence to the expedition's research.[92]

In other cases, local officials came to Indigenous peoples' defense, objecting to the intrusive and coercive nature of the expedition's examination of reluctant subjects' bodies. In Arequipa, for example, city officials soundly rejected Nelson's efforts to set up a studio. Despite the expedition's lobbying, officials saw the work of measuring and photographing Indigenous research subjects as invasive and an affront to decency and privacy. According to Nelson, the subprefect informed the expedition that "the sentiment of Arequipa was different from that in Cuzco to the extent that the people would resent any action compelling an Indian to submit to measurements." Nelson also noted a further impediment to acquiring subjects, which was that "practically the only pure blooded Indians in Arequipa are those who come in for commercial purposes, driving their llama trains loaded with produce."[93]

Having failed to perform research in Arequipa and instead provided medical and surgical care to notable residents and a member of the Duque family, Nelson lamented the challenges he faced there and in the Cusco region as anthropometric data collector. Reflecting Bingham's and the expedition's racist thinking and broader disdain for South Americans, he expressed frustration with local officials and a dismal view of Indigenous Andean peoples. While he praised their apparent eagerness at other moments, in these cases he invoked racial fitness categories prevalent at the time, disparaging them as backward and asserting that "the Indians [sic] are dull mentally. How much coca has to do with this condition is hard to determine."[94] In other cases when traveling, he and others resorted to force to compel Indigenous people to obey

---

[90] YPEP, Series III, 1912, Box 20, Folder 27, Journal of Luther T. Nelson, 6.

[91] Ibid., 11.

[92] Ibid.

[93] Ibid., 8. References to Mestizo research subjects are largely absent from Nelson's journal and report, though they appear in Ferris's published study. In notebooks, Nelson lists them as likely possessing Spanish blood.

[94] YPEP, Series IV, Box 26, Folder 27, Report of Luther T. Nelson, no page number.

their instructions. For example, he lamented in the town of Arma that "We could not get any food from the Indians [sic] without using force, and then we would succeed in getting small amounts only."[95] Accounts thus reflect Nelson's and other members' failure to exert influence over Indigenous people in various settings, despite the assistance and support of powerful figures in Cusqueño society.

Photography itself proved especially provocative within these encounters. In at least one case, Indigenous people attacked expedition members and their camera equipment while they were traveling. Osgood Hardy wrote of encountering a group of Indigenous men who turned aggressive and specifically "laid violent hands on the tripod."[96] Resistance is also evident among research subjects in Nelson's anthropometric photographs. Rather than exhibit the emotionless, unfocused stares that typify the genre of standardized anthropometric portraits Bertillon had developed in France, and which Bingham expected Nelson to recreate, subjects frequently exhibited confusion, fear, and annoyance. They either focused directly on the camera or looked to the side when photographed head on (Figure 3.5). In some cases they appear to have refused to sit still. While the body remains in focus, the head appears blurred as if they moved to resist being photographed. Subjects' shoulders, moreover, were often positioned with one higher than the other, rather than at equal levels. In some cases, men kept bundles strapped to their backs and hats on their heads, thus reducing the viewers' ability to discern their physical form (Figure 3.6). In an ethnological picture of a woman, moreover, it appears that someone was positioned behind the subject, perhaps to hold her in place (Figure 3.7). In this way, research subjects' behavior undermined Nelson's work and its value as a comparative study of physiognomy and race; he could not exercise control fully in his improvised anthropometric studios.

These photographs' most revealing aspect, however, is the way research subjects articulated desire, as theorized by Eve Tuck and K. Wayne Yang. This becomes evident in subjects' use of a ruler to express their sense of self in the face of scientific practices that depicted them in a demeaning fashion. In anthropometric photography, the ruler should be positioned in exactly the same location and angle in every portrait. Likewise, the camera should be the same distance from every subject to make their portraits consistently measurable and comparable. In the Yale Peruvian Expedition photographs, however, research subjects held the upper portion or tip of a folding ruler (described in journals as a "folding rod") in their hand, rather than having it placed alongside their head. Although Nelson wrote in one of his anthropometric notebooks that "In the photographs each subject holds a metre rule so

[95] YPEP, Series III, 1912, Box 20, Folder 27, Journal of Luther T. Nelson, 16.
[96] YPEP, Series III, 1912, Box 19, Folder 22, Record of Assistant Osgood Hardy, 18.

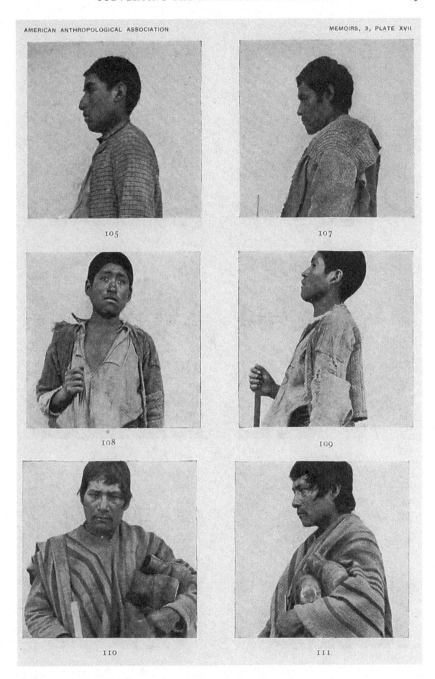

Figure 3.5 Anthropometric photographs 105 and 107–111 by Luther T. Nelson, 1912.
Source: National Library of Peru.

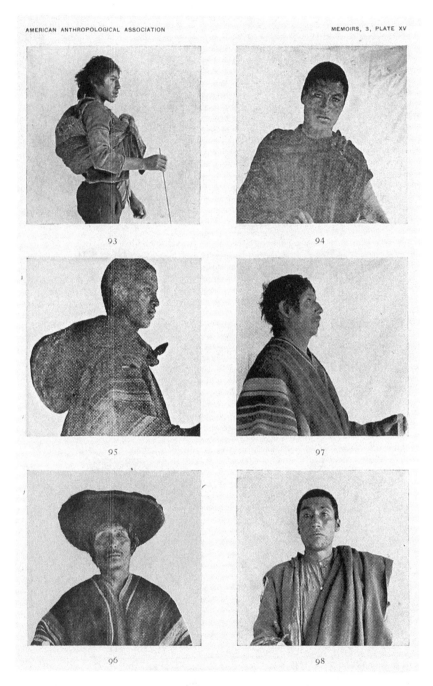

Figure 3.6   Anthropometric photographs 93–98 by Luther T. Nelson, 1912.
Source: National Library of Peru.

Figure 3.7 Anthropometric photographs 269–271 by Luther T. Nelson, 1912.
Source: National Library of Peru.

as to give an idea of height of the individual," the subjects he photographed positioned it at different angles and directions in their portraits, undermining its precision as an anthropometric tool.[97] Some held it directly outward in front of them at an angle, while others leaned it to the side and still others appeared to hold it upright or raise it off the ground (Figure 3.8). In some cases, moreover, they treated it like a cane, grasping it as if they would put weight on it or stabilize their balance with it when walking. In still other cases, they actually held it like a *vara*, a traditional ceremonial staff that officials in Andean communities used to connote authority[98] (Figure 3.9). Research subjects thus subverted the ruler's function, adopting poses with it that consciously or subconsciously challenged expedition members' authority and reflected their own sense of community, identity, and status. The ruler became little more than a gesture toward the expedition's goals of "scientifically" depicting racial difference. Research subjects used it to assert power and

[97] YPEP, Series III, 1912, Box 20, Folder 28, Notebooks of Luther T. Nelson, no page number.
[98] Thanks to Kris Lane for making this observation when I presented an earlier version of this chapter.

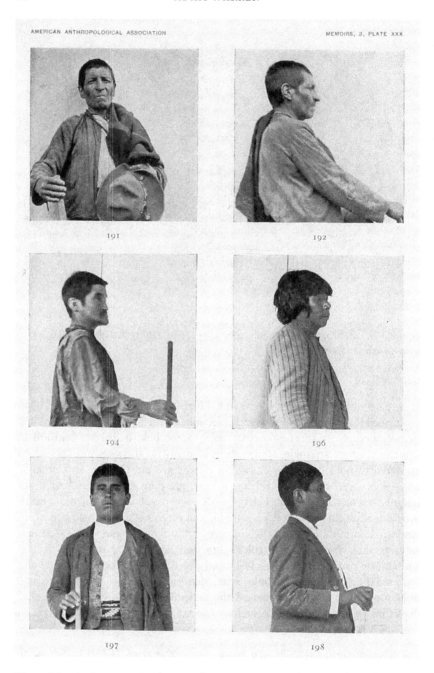

Figure 3.8 Anthropometric photographs 191–192, 194, and 196–198 by Luther T. Nelson, 1912.
Source: National Library of Peru.

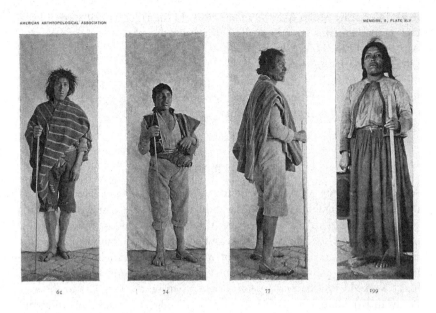

Figure 3.9   Anthropometric photographs 61, 74, 77, and 199, Luther T. Nelson, 1912.
Source: National Library of Peru.

fashion how they would be depicted in accordance with their desires and claims of local autonomy.

Finally, cases exist in which men appear to have gone to more extreme lengths to seize control of how they would be perceived in anthropometric photographs.[99] They, too, expressed what Tuck and Yang would describe as desire. For example, Bingham's "In the Wonderland of Peru" includes an account of an Indigenous man, who "when he found he could have his picture taken for free, dressed in his Sunday clothes. The next day he returned to the photograph. When he was shown the negative he refused to believe that it was his picture, because he could not see the colors and the spangles that decorated that Sunday coat he wore."[100] While Bingham no doubt included this story to entertain a US and global reading public and reinforce stereotypes of Indigenous unfamiliarity with photography, it can be

[99] The expedition's written records include no discussions of the one woman who was measured and appeared in an anthropometric photograph.
[100] Bingham, "Wonderland," 561. Nelson also mentions this account in the letter accompanying his medical report. If his identification of numbered negatives is correct, then this story corresponds to Mariano Larota, age twenty-seven, from the province of Canchis.

read differently. Much like the men who held the ruler as if it were a *vara*, this gentleman chose to co-opt the anthropometric photograph to make it a depiction of how he saw himself and wanted others to see him. He rejected the photograph's purpose as a demeaning, ostensibly "scientific" record of his physical form. The makeshift anthropometric studio thus became a site of negotiation and assertion of Indigenous identity and desire, in addition to being a site of attempted domination, coercion, violence, resistance, and refusal.

Conclusion

Although photographs and written records provide valuable clues and insights, there remains much that the archive cannot tell us about the research subjects Nelson sought to measure and photograph. For example, we know little about Justo Rodríguez and his reasons for refusing to comply beyond the fragmentary descriptions included in anthropometric notebooks and published materials. Given where he and Nelson encountered one another, he likely worked as an hacienda peon. If true, then he bore witness to the Yale Peruvian Expedition's use of Huadquiña Hacienda as a base camp and stopping point, from which they organized their efforts to excavate Machu Picchu. He may even have joined the expedition's workforce, one of many hacienda workers whose contributions went unacknowledged. That said, he may equally likely have had little, if any, connection to the expedition, his behavior an act of refusal by someone suddenly and unexpectedly caught within its scientific gaze.

Rodríguez, however, was one of many research subjects who pushed back against Nelson's efforts through forms of resistance, refusal, and subversion. Others embraced or co-opted aspects of the project for their own ends. In the process, such figures acted not within a single moral field that was widely shared across Cusqueño society. Rather, they acted on the basis of the various positions they occupied within a complex, highly unequal, rural agrarian society marked by internal colonialism and undergoing rapid expansion and economic growth. Furthermore, their ways of knowing, perceiving, and judging likely engaged and exceeded the categories expedition members and their Peruvian scientific and intellectual counterparts employed in written accounts. In this sense, the concept of the moral field need expanding and complicating in order to include the experiences of ordinary people and explain their decisions. This is especially true for colonial contexts where transnational science has typically operated. In writing about those settings, historians must avoid reconstructing moral fields in the singular and only in reference to the circulation of widely shared ideas and beliefs. To do so without considering different epistemologies and political, social, and economic realities is to suggest that those people historians and others have traditionally identified as intellectuals act within moral fields, while others do not.

Coda

How can this study of past scientific research encounters inform the present? Since 2000, two conflicts over scientific research have brought the Peruvian government, Indigenous communities, and other constituencies from the Cusco region into conflict with the institutions that sponsored the Yale Peruvian Expedition's research in the 1910s. First, under the governments of Alejandro Toledo and Alán García, tensions escalated between the Peruvian government and Yale University over the repatriation of artifacts and, to a lesser degree, human remains removed under Bingham's supervision. Efforts by Peru's First Lady, Eliane Karp, to draw attention to the dispute, lawsuits brought by the Peruvian government against Yale, and other factors resulted in Yale signing an agreement in November 2010 and transferring its Machu Picchu collection to Cusco in 2011. While doubts persist in Peru as to whether Yale returned all items, the collection now sits in a museum in Cusco along with tattered exhibition materials Yale provided.[101]

In 2011, a second conflict emerged between Indigenous peoples and a project organized by the National Geographic Society. As Kim TallBear describes, the community of Hatun Q'eros refused to cooperate with researchers from Genographic, a scientific project that sought samples of community members' DNA. Funded by the National Geographic Society, Genographic wanted this genetic material because the Q'eros claim to be descendants of the Incas and to lack Spanish admixture. Genographic's goal was to chart human populations' origins and development worldwide through modern genomics. In this sense, despite being led by scientists who framed their research from an antiracist stance, Genographic engaged questions of racial origins that had motivated the Yale Peruvian Expedition's work and Peruvian scientists like Lorena and Coello y Mesa a century earlier. Like Nelson, who meticulously recorded which research subjects appeared to be of racially "pure" Indigenous origins, Genographic pursued racial purity through science.[102]

The community of Hatun Q'eros refused Genographic scientists' requests for DNA samples and questioned how they engaged Indigenous communities. As reported in a communiqué, the Q'eros took this action based on concerns about the project's ethical dimensions, citing dubious methods employed to seek informed consent, questions about who owned or would have control over samples, questions about what the National Geographic Society might do with data, and doubts about the project's treatment of Indigenous knowledge. The communiqué outlined seven reasons for refusal

---

[101] Cox Hall summarizes these disputes in *Framing's* concluding chapter.
[102] TallBear, *Native American DNA*, 189–195.

that centered on ethics. Genographic ultimately suspended its project in the Cusco region.[103]

Comparing these recent conflicts and situating them in relation to the material in this chapter raises questions that complicate our understanding of moral thinking, the ethics of expedition science, and contemporary Indigenous rights and self-determination. Why has Peru's government struggled with Yale to repatriate artifacts and prioritized them over human remains while ignoring the possible return of reams of physiological data stored there? What, on the other hand, can be learned about the National Geographic Society by situating conflicts between the Q'eros and Genographic within a longer history of racial scientific research that it sponsored? By focusing on desire in this chapter and making the work available in translation, I hope this study of how past Indigenous and Mestizo research subjects resisted, refused, engaged, subverted, and co-opted racial scientific research can further inform Indigenous communities' already deep historical understandings of their relations with outside researchers.

---

[103] Ibid.; "Genographic Project Hunts the Last Incas: Resurrected 'Vampire' Project Brings Fears of Biopiracy to Cusco Region," *Andes Communiqué*, May 2011, www.slideshare.net/BUENOBUONOGOOD/andes-communique-genographicprojecthuntsthelastincas.

4

Modest Witnesses of Violence
Salvage Ethnography and the Capture of Aché Children

SEBASTIÁN GIL-RIAÑO

During the heyday of salvage anthropology – the ethnographic study of supposedly vanishing "races" – practitioners often praised the moral integrity of their own work. In a preface to the 1939 book *Une Civilisation du Miel: Les Indiens Guayakis du Paraguay* (an exemplar of the salvage genre), the French anthropologist Paul Rivet applauded the objectivity of the book's author: the French naturalist Jehan Albert Vellard. Rivet described his compatriot as a "biologist at heart" with "extraordinary knowledge" of "tropical nature." He also marveled at how quickly Vellard adapted to the demands of anthropological fieldwork and praised Vellard's monograph for its firm basis in "observations made in direct contact with reality and not summary impressions." For Rivet, Vellard's rigorous observations stood in sharp contrast to those of amateur ethnographers who produced a "superficial and hasty litera-ture[,] which the taste for exoticism and the age of communication have so annoyingly made fashionable." The clarity of Vellard's narrative was thus like a "documentary filmed on the spot, at the risk of his life" rather than a "fake film executed in a comfortable studio."[1] By championing Vellard's moral and scientific acumen, Rivet thus rendered him a quintessential "modest witness" – a type of observer that Donna Haraway classically described as an "authorized ventriloquist for the object world" with a "remarkable power to establish the [unadorned] facts."[2] By emphasizing Vellard's modesty and sacrifice, Rivet also implied that he was doing both the Aché and the anthropological community a favor by capturing the remnants of a disappearing people.

Rivet also revealed that Vellard's narrative went beyond mere representation. As another example of Vellard's virtue, Rivet highlighted one of *Une Civilisation du Miel's* most striking passages: the story of how, in 1932, the French biologist adopted a young Aché girl and named her Marie-Yvonne. "With what simpli-city," Rivet mused, "J. Vellard reports the most moving episode of the beautiful

---

[1] Jehan Albert Vellard, *Une Civilisation du Miel: Les Indiens Guayakis du Paraguay* (Paris: Gallimard, 1939), 6.
[2] Donna J. Haraway, *Modest_Witness@Second_Millennium. "FemaleMan_Meets_Onco Mouse": Feminism and Technoscience* (New York: Routledge, 1997), 24.

adventure he lived." Like many French anthropologists from the interwar period, Vellard's book represented a literary counterpart to the scientific monographs that emerged from his fieldwork.[3] With literary flair, Vellard's account of Marie-Yvonne's adoption framed the episode as a benevolent rescue mission. In his own telling, Vellard described how the "fugitive Indian guides" who assisted him during his fieldwork brought Marie-Yvonne to him when she was approximately two years old. The guides discovered the girl after they had fled from Vellard's group and stumbled upon an Aché camp with two women and a child. When the women fled, the guides seized the girl and brought her to Vellard who observed that she "had been badly abused and . . . was terrified." Once brought to Vellard, he claimed that Marie-Yvonne chose to stay with him and his family – "she came with us and has not left us since," wrote Vellard. By his own reckoning, Vellard assumed a benevolent paternal role that stood in stark contrast to his "fugitive" guides and the Aché women who supposedly abandoned Marie-Yvonne. Vellard thus presented himself as rescuing Marie-Yvonne from the clutches of his unruly guides and her neglectful family members.

Rivet, who was prominently involved in antifascist and antiracist struggles in interwar France, took Vellard's adoption story at face value and interpreted it as offering important correctives to scientific debates about race and heredity. Reflecting on these events several years later, Rivet noted that under the care of Vellard and his mother, Marie-Yvonne had grown into a charming, intelligent, and "pretty" ten- to eleven-year-old girl. Rivet also marveled at the fact that she spoke fluent Portuguese and French and had quickly adapted to an entirely new environment without "heredity" diverting her from the path of "civilized life." Had she not been adopted and cared for by Vellard, Rivet conjectured, Marie-Yvonne would likely have lived "the precarious and primitive life of which J. Vellard gives us a striking picture." In Rivet's estimation, her remarkable change in fortune offered a potent argument against racism – "I deliver [the facts of Marie-Yvonne's story] to the meditations of those who believe in the irreducible inequality of races and the imprescriptible laws of heredity." Marie-Yvonne's adoption story thus gave Vellard's "beautiful" book a "a human value, which should ensure the success it deserves in so many other ways."[4] From Rivet's perspective, Vellard's book and the adoption story it told did more than just preserve the fading remnants of a disappearing people, it provided ammunition for the international struggle against racism that he and other prominent anthropologists like Franz Boas had been involved with.[5]

[3] Vincent Debaene, *Far Afield: French Anthropology between Science and Literature*, trans. Justin Izzo (Chicago: University of Chicago Press, 2014).

[4] Vellard, *Une Civilisation du Miel*, 7.

[5] Christine Laurière, "Anthropology and Politics, the Beginnings: The Relations between Franz Boas and Paul Rivet (1919–42)," *Histories of Anthropology Annual* 6, no. 1 (2010): 225–252.

In the following decades, Marie-Yvonne's adoption story was retold by scholars and journalists who built on Rivet's humanitarian and antiracist framing. Marie-Yvonne's story was retold in public-facing articles published by the *UNESCO Courier* in 1950 and *Reader's Digest* in 1960. These articles continued to celebrate her adoption as a golden opportunity to escape primitivism and join civilization. Yet this humanitarian framing ignored the violent circumstances and colonial structures surrounding Marie-Yvonne's adoption. In Paraguay, Marie-Yvonne's story corresponds to a period when the reigning Liberal government introduced Indigenous assimilation policies inspired by the Jesuit missions – the *reducciones*, or reductions – of the seventeenth century. These policies promised Native land to religious organizations and others who could successfully "reduce" Indigenous people and thus created the conditions for an intensification of violence and abuse toward Indigenous peoples with the grim manhunts of the Aché serving as a prime example.[6] Through publications like the *UNESCO Courier* and *Reader's Digest*, Marie-Yvonne's story also circulated in settler colonial states such as the United States, Canada, and Australia. From this global perspective, the circulation of Marie-Yvonne's story coincides with a period when settler colonial governments, notably Australia and Canada, encouraged the forced separation and removal of thousands of Indigenous children from their families as part of state assimilation policies. In a global historical context that scholars have retrospectively described as genocidal, how is it that Marie-Yvonne's story became framed in the redemptive terms of humanitarianism and antiracism?[7]

By examining the various retellings of Marie-Yvonne's story and the many stories of captured children that populate ethnographic studies of the Aché, this essay tracks how colonial violence against Indigenous peoples was repackaged within a powerful conceptual framework that challenged the biological basis of race. I argue that the transnational celebration of Indigenous assimilation at play in Marie-Yvonne's story relied on a set of epistemic, affective, and moral dispositions that were shared by human scientists, and which ultimately authorized the removal of children from their families and territories in the service of science and international struggles against racism. While the redemptive accounts of her story did important work by challenging biological determinism, they also concealed how the practices of mid-century human scientists ignored and at times enabled the forced removal of children from their families and the dispossession of Indigenous territory. In fact, as this chapter demonstrates, until the 1960s ethnographic studies of the Aché

---

[6] René Harder Horst, *The Stroessner Regime and Indigenous Resistance in Paraguay* (Gainesville: University Press of Florida, 2007).

[7] David B. MacDonald, "Canada's History Wars: Indigenous Genocide and Public Memory in the United States, Australia and Canada," *Journal of Genocide Research* 17, no. 4 (October 2, 2015): 411–431, https://doi.org/10.1080/14623528.2015.1096583.

were based primarily on children captured under violent circumstances and attest to an established practice and *economy* of buying and trading Aché children as servants. Although researchers who studied captured Aché children positioned themselves as civilized men of science, they did not condemn the trafficking of Aché children that they benefited from and instead presented it as a fait accompli that they could only observe as modest witnesses. Thus, although Marie-Yvonne's story indexes important epistemic shifts in the trajectory of race science, it also reveals how human scientists' ethical horizons were beholden to colonial structures that persist from the Iberian conquest of the sixteenth century.

Civilization and the Science of Children

Rivet and Vellard's redemptive framing of Marie-Yvonne's life bears the imprint of major trends in the human sciences from the first half to the twentieth century, which amounted to a rejection of biological determinism and fixed racial hierarchies in favor of cultural and environmental approaches to human diversity. Beginning in the interwar period and intensifying after World War II (WWII), human scientists in North America and Europe moved away from conceptualizing human diversity through the prism of static typological races and instead adopted frameworks that emphasized how human differences are transmitted through cultural and social practices.[8] As part of this shift, which scholars have called the retreat of scientific racism, many experts turned to children and child-rearing in order to observe how values, attitudes, habits, and practices persist from one generation to another.[9] For instance, in the United States, adherents of the cultural and personality school such as the anthropologists Margaret Mead, Ruth Benedict, Otto Klineberg, Edward Sapir, and Ashley Montagu rejected biological explanations of human behavior in favor of cultural and linguistic studies that examined how specific groups transmit culture from one generation to the next through child-rearing practices and through language acquisition during infancy.[10] Through ethnographic studies of Indigenous groups in the South Pacific and the United States, culture and personality researchers turned child-rearing and children themselves into prized research objects that promised insights on

---

[8] Sebastián Gil-Riaño, *The Remnants of Race Science: UNESCO and Economic Development in the Global South* (New York: Columbia University Press, 2023).

[9] Elazar Barkan, *The Retreat of Scientific Racism: Changing Concepts of Race in Britain and the United States between the World Wars* (Cambridge: Cambridge University Press, 1996).

[10] Joanne Meyerowitz, "'How Common Culture Shapes the Separate Lives': Sexuality, Race, and Mid-Twentieth-Century Social Constructionist Thought," *The Journal of American History* 96, no. 4 (2010): 1057–1084.

how environmental factors such as cultural patterns and socioeconomic opportunities mold the personalities and intellectual abilities of individual subjects.

Examples of this research in North America abound, and they had implications both for domestic and international policies and for applied social scientific research conducted elsewhere. In his landmark experimental study of "racial" differences in intelligence, the Canadian psychologist Otto Klineberg administered an array of intelligence tests to "Native American" and "Negro" children and concluded that their comparatively low performance was due to having been raised in cultures that prioritized accuracy over speed.[11] With longer test-times, Klineberg's analysis suggested, the observed differences between Black and Indigenous children and their white counterparts would disappear. After WWII, Klineberg gave expert testimony in the Brown vs. Board of Education trial that ended segregated schooling in the United States and played an important role in the development of UNESCO's race campaigns, which challenged scientific racism. In the context of this ostensibly nonracial domain of knowledge in North America, conceptions of "culture" thus offered alternative ways to theorize human variation that prioritized the role of nurture in producing difference and lent themselves to liberal projects of reform.[12] In applied social science projects, this emphasis on culture and nurture also often aligned itself with projects of assimilation that identified Indigenous and other non-European cultures as backward and used anthropological insights for the purposes of attempting to reengineer these cultures in conformity with Western modernity.[13] By framing Marie-Yvonne's story as one that disproved racist hereditarian theories, Rivet's preface to Vellard's book thus echoed these interwar trends in the human sciences of North America.

To understand Rivet and Vellard's descriptions of Marie-Yvonne, however, we must also examine the influence of intellectual trends from Europe and South America. In Southern Europe (especially France and Italy) and South America, eugenicists mostly rejected rigid Mendelian approaches to heredity

---

[11] Otto Klineberg, *An Experimental Study of Speed and Other Factors in "Racial" Differences* (New York: Columbia University, 1928); see also Ellen Herman, *The Romance of American Psychology: Political Culture in the Age of Experts* (Berkeley: University of California Press, 1995).

[12] However, as Peter Mandler's work on Margaret Mead shows, the culture and personality approach experienced considerable challenges when scholars attempted to scale up its conclusions to the level of international relations. See Peter Mandler, "One World, Many Cultures: Margaret Mead and the Limits to Cold War Anthropology," *History Workshop Journal* 68, no. 1 (October 1, 2009): 149–172.

[13] Daniel Morrow and Barbara Brookes, "The Politics of Knowledge: Anthropology and Maori Modernity in Mid-Twentieth-Century New Zealand," *History and Anthropology* 24, no. 4 (2013): 453–471.

in favor of a neo-Lamarckian approach to heredity that emphasized racial improvement through environmental and sanitary reform. This environmentalist approach was often tied to pro-natalist politics concerned with the nurture and care of life.[14] Yet unlike the North American context where this environmentalist framing emerged out of the social sciences and through a rejection of eugenics, in "Latin" countries this environmentalist approach was part and parcel of the eugenics movement and one that was adopted by a wide array of experts including physicians, public health officials, and human scientists. As scholars of Latin eugenics have argued, it was also a style of eugenics that meshed well with Catholic values concerning reproduction and, in some cases, with fascist politics.[15] In France, eugenics grew out of the medical discourse of *puericulture*, which was broadly concerned with a scientific approach to child-rearing. The term was coined in 1865 by a French physician named Alfred Caron who studied the health of newborns out of a concern with "improving the species" and taught courses on the education of young children. Though the term did not initially gain much traction, it was revived and popularized in the 1890s by Adolphe Pinard, the Chair of Clinical Obstetrics at the Paris Medical School. Pinard adopted the term "puericulture" to describe a program concerned with prenatal care for pregnant women.[16] The term gained widespread support in the pro-natalist context of fin-de-siècle France and was also quickly adopted in other Southern European countries and Latin America where legislators and medical professionals adopted French practices for infant well-being and maternal protection as benchmarks for their own societies. By the early 1900s, for instance, physicians from Uruguay, Argentina, Colombia, and other countries established milk stations called "gotas de leche" that were based on French institutions and served as community-based clinics for infant and child health.[17]

In the settler colonial states of Australia, New Zealand, and Canada, government officials and experts also embraced environmentalist conceptions of culture and educability as part of an effort to assimilate Indigenous children

---

[14] Nancy Stepan, *"The Hour of Eugenics": Race, Gender, and Nation in Latin America* (Ithaca, NY: Cornell University Press, 1996); Marius Turda and Aaron Gillette, *Latin Eugenics in Comparative Perspective* (London: Bloomsbury, 2014); Sarah Walsh, *The Religion of Life: Eugenics, Race, and Catholicism in Chile* (Pittsburgh: University of Pittsburgh Press, 2021).

[15] Walsh, *The Religion of Life*; Turda and Gillette, *Latin Eugenics*.

[16] Pinard adopted and began popularizing the term after he noticed a higher average birthweight for babies born to mothers who had stayed at the "Maison maternelle" – a refuge for homeless pregnant women – that he had founded with philanthropic support. William H. Schneider, "Puericulture, and the Style of French Eugenics," *History and Philosophy of the Life Sciences* 8, no. 2 (1986): 265–277, 267.

[17] Anne-Emanuelle Birn, "Child Health in Latin America: Historiographic Perspectives and Challenges," *História, Ciências, Saúde-Manguinhos* 14, no. 3 (2007): 677–708, 688.

through forced reeducation. Policies of Indigenous assimilation in settler colonial societies were often informed by discourses of racial improvement that bore a similar logic to the arguments put forward by Rivet and Vellard. For instance, as Fiona Paisley has argued, at the beginning of the twentieth century, settler colonial states sought to replace Christian missionaries' concern with the spiritual salvation of Indigenous people with the "racial sciences of mind and body."[18] In Australia and Canada, educators, politicians, and church officials pushed this ideology to an extreme and created a system that forcibly removed Indigenous and mixed-descent children from their families and placed them in boarding schools and foster families, where they were often subject to neglect and abuse.[19] In Australia, this practice of taking Indigenous children from their parents – what are now referred to as the "stolen generations" – garnered strong support from late nineteenth and early twentieth-century anthropologists, physicians, and physiologists who viewed Australian Aboriginals as heading toward extinction and advocated for a policy of "racial absorption" that often targeted "half-caste" children.

Two of the most prominent ideologues of this system were the physician Cecil Cook, who served as chief medical officer and "chief protector of Aborigines" in Australia's Northern Territory, and the civil servant A. O. Neville, who also served as chief protector of Aborigines in Western Australia. Both men challenged policies of racial segregation and instead argued that the best hope for Australia's dark-skinned Aboriginals was full cultural and biological "absorption" into Australia's settler white community.[20] Their stances typified international approaches to Indigenous assimilation and modernization and bore similarities to questions that researchers were asking in Paraguay. Cook envisioned a path to absorption through a program of scientific breeding that targeted "half-caste" girls who he viewed as having inherited the best qualities of both white and Aboriginal "stock." Yet he also viewed aboriginal culture as exerting a negative influence on childhood development and advocated for the creation of a program offering domestic training for "half-caste" girls that would render them suitable housewives for white men in frontier regions. Neville similarly viewed Aboriginal families as incompetent and advocated for the creation of boarding institutions that

[18] Fiona Paisley, "Childhood and Race: Growing Up in the Empire," in *Gender and Empire*, ed. Philippa Levine (Oxford: Oxford University Press, 2007), 240–259, 241.

[19] *Canada's Residential Schools: The Final Report of the Truth and Reconciliation Commission of Canada*, McGill-Queen's Native and Northern Series (Montreal: Published for the Truth and Reconciliation Commission of Canada by McGill-Queen's University Press, 2015); Peter Read, "The Stolen Generations," Occasional Paper No. 1 Ministry of Aboriginal Affairs, Sydney, 1982.

[20] For more on Cook and Neville's approach, see Warick Anderson, *Cultivation of Whiteness: Science, Health, and Racial Destiny in Australia* (Durham, NC: Duke University Press, 2006) and Paisley, "Childhood and Race."

would remove Aboriginal children from the supposed negative influence of their families and offer technical and industrial training.[21]

In these settler colonial contexts, race experts thus framed the removal of Indigenous children from their families as a benevolent and even humanitarian civilizing mission that was necessary for the future prosperity of the nation. Such practices thus exemplify Robert Van Krieken's thesis of the barbarism and violence that inheres in civilization discourse.[22]

Captured Children as Research Objects

Anthropological studies of the Aché exemplify the scholarly interest in children and acceptance of removal that featured so prominently in the human sciences of the first half of the twentieth century. While many Indigenous groups in Paraguay and especially the majority Guarani established economic and political relationships with Europeans following the Iberian conquest, the Aché refused to establish relations with both Europeans and neighboring Indigenous groups. The distance that they chose to maintain, stoked speculation about their supposedly barbaric practices and exotic appearance. And it also thwarted ethnographic accounts based on direct observation. Up until the 1960s, instead of direct ethnographic observation scholars relied on captured children as evidentiary sources.

Before the late nineteenth century, the only written account of the Aché came from an eighteenth-century source – a seven-page summary of their culture written by Pedro Lozano, a Jesuit missionary. Lozano based his account of Aché culture on a group of about thirty Aché who had been captured by small groups of *Guarani* who had been sent out by Jesuit priests hoping to settle the Aché in one of the Jesuit reductions.[23] Jesuit reductions were one of the Iberian empire's key instruments of colonization. By establishing settlements in Indigenous territories and enticing or capturing Indigenous populations to live with them in the missions, Jesuits sought to transform the Native inhabitants of the Americas into a productive and Christianized workforce. The reduction where Lozano observed the Aché was one populated primarily by Guarani, who represented the majoritarian Indigenous group in the region and who had historically waged a war of extermination against the Aché. In Lozano's reduction, Jesuit missionaries sent out small parties of Guarani hunters to capture Aché prisoners and bring

[21] Anderson, *Cultivation of Whiteness*; Paisley, "Childhood and Race."

[22] Robert Krieken, "The Barbarism of Civilization: Cultural Genocide and the 'Stolen Generations'," *The British Journal of Sociology* 50, no. 2 (June 1999): 297–315.

[23] Alfred Métraux and Herbert Baldus, "The Guayaki," in Julian H. Steward, *Handbook of South American Indians: Vol. 1* (Washington, DC: United States Government Printing Office, 1946).

them back to the settlements where they could be brought up as neophytes.[24] Lozano's description of Aché culture was thus based on observations of children and teenagers who were captured and raised by Jesuit missionaries.

Given their hostile relations with Guarani groups and Iberian settlers and two major wars in Paraguay, the Aché remained forest-bound throughout the nineteenth and first half of the twentieth century. They remained purposefully at a distance from the encroaching agricultural settlements and they were known to Mestizos and settlers through the campfires and other remains they left in their forest or through occasional raids of livestock and tools. In fact, in the period between Lozano's study in the late eighteenth-century reduction up until Vellard's visit, direct observations of the Aché were based primarily on captured children who were raised in Paraguayan *estancias*, or cattle ranches.

Kidnapping Aché Children for Science

The research conducted on an Aché girl named Damiana Kryygi in the last decade of the nineteenth century serves as an iconic example of how early researchers relied on captured children as sources. Damiana's story, like that of many other Aché children, also attests to the existence of an informal market for Aché children that encompassed Paraguayan ranchers, neighboring Guarani groups, and European anthropologists. After being captured at the age of two by Paraguayan settlers who killed her parents to avenge the killing of a horse, Damiana became an object of fascination for European anthropologists. During an ethnographic mission in 1896 to study the Aché on behalf of the Museo de la Plata in Argentina, the French anthropologist Charles de la Hitte and his Dutch colleague Herman Ten Kate conducted anthropometric measurements of Damiana's head and took photographs of her while she was in the care of her parent's murderers and described her as sad and sickly.[25] In 1898, Damiana was sent to live in San Vincente, an Argentinian town close to La Plata, where she was raised as a "maidservant" by the mother of Alejandro Korn, the director of the psychiatric hospital Melchor Romero in Buenos Aires. Korn's mother was a German immigrant and in this period Damiana learned to speak Spanish and some German. By the time she was approximately fourteen or fifteen years old, Korn arranged for the director of the Museo de La Plata, the German anthropologist Robert Lehmann-Nitsche, to observe Damiana on two separate occasions. During these visits, Lehmann-Nitsche took photographs of her naked body and conducted a series of anthropometric measurements. Lehmann-Nitsche observed that she had

---

[24] Ibid.

[25] Katrin Koel-Abt and Andreas Winkelmann, "The Identification and Restitution of Human Remains from an Aché Girl Named 'Damiana': An Interdisciplinary Approach," *Annals of Anatomy – Anatomischer Anzeiger* 195, no. 5 (2013): 393–400.

followed a path of "normal development" until she hit puberty at which point she developed a sexual libido so "alarming" that "all education and punishment on behalf of the family proved ineffective."[26]

Not ones to tolerate such insubordination, the family resolved to send her to Melchor Romero – the psychiatric hospital directed by Korn – where she was looked after by the nurses. Damiana died from tuberculosis shortly after arriving at the hospital and her remains were quickly snapped up for further scientific studies. Lehmann-Nitsche sent Damiana's head and brain as a gift to the German anatomist Hans Virchow, son of Rudolf Virchow, at the Charité hospital in Berlin. Virchow quickly incorporated Damiana's head into the anatomical collection of the Charité and performed a series of dissections on it as part of a comparative study on facial muscle attachments. After publishing a series of papers based on these dissections, Virchow then handed over Damiana's skull to the Charité collection in 1911, where it was kept for the next hundred years as part of its anthropological collection. Lehmann-Nitsche preserved the rest of Damiana's body as a skeleton at the Museo de La Plata, where it was stored away in a cabinet that was only recently rediscovered and identified in 2010. This prompted the museum to return the remains to the Aché who gave her a traditional burial in their ancestral homelands. In 2012, the Charité restituted Damiana's skull, which Aché leaders buried alongside Damiana's remains that had been buried in 2010.[27]

Vellard's Mission: French Ethnology and the Capture of Marie-Yvonne during the Chaco War

Damiana's mistreatment in the name of science was unfortunately not an isolated incident. In fact, abducted children like Damiana feature prominently in the early ethnographic literature on the Aché. In this early literature from the late nineteenth century to the interwar period, anthropologists viewed children like Damiana through the frame of race science and regarded their prospects for improvement as limited by an innate biological inferiority or "savagery."

By the early 1930s, however, when Marie-Yvonne's story begins to appear in the literature, anthropologists had begun to question the biological determinism at play in race science and instead began to adopt redemptive stories that cast Aché children as objects of improvement. This approach to Aché children exemplified a new style of anthropology that emerged in France during the

---

[26] Robert Lehmann-Nitsche, "Relevamiento antropológico de una india Guajaquí," *Revista del Museo de La Plata* 15 (1908): 92–110.

[27] In 2014, the filmmaker Alejandro Fernández Mouján released a documentary titled *Damiana Kryygi*, which tells Damiana's story and documents the process by which her remains were restituted to her Aché kin.

interwar period and whose adherents proclaimed to have broken with the discipline's racist past. One of the most important figureheads and institution builders within this new anthropology was the socialist Paul Rivet. A military physician by training, Rivet decided to devote himself to the study of anthropology after participating in a geodesic mission to Ecuador (1901–1906), which was a joint venture between French and Ecuadorean militaries and gave Rivet the opportunity to conduct ethnographic and anthropometric work on Indigenous groups in the Andes and Amazon. Dissatisfied by the centrality of physical anthropology and race science in French anthropology from this period, Rivet devoted himself to transforming and modernizing the discipline upon his return to France. Instead of the narrow approach of anthropometry that predominated in turn-of-the-century French anthropology, Rivet worked toward professionalizing the discipline by bringing it in line with the four-field approach that had been institutionalized in the United States and other North Atlantic nations and by raising the profile of ethnographic fieldwork. In collaboration with Marcel Mauss and Lucien Lévy-Bruhl and with the financial and administrative support of the French colonial administration, Rivet created the Institut d'Ethnologie at the University of Paris in 1925. In the following years, Rivet, Mauss, and Lévy-Bruhl trained a new generation of anthropologists, including Alfred Métraux who was one of the Institut's first and most distinguished students, and taught them to combine the study of the "social facts" of non-Western cultures with museum work and physical anthropology.

During the interwar period, Rivet further established himself as a powerful figurehead of French anthropology by transforming the imperial nation's major ethnographic museum. In 1928, he was appointed Chair of the Trocadéro ethnographic museum and completely transformed it over the next decade and eventually converted it into the Musée de l'Homme, which opened its doors in 1938. As Alice Conklin has argued, Rivet and Mauss's ambition with the Musée de l'Homme was to create an institution with state-of-the-art research facilities that would gather all the branches of French anthropology under a single roof.[28] It was also meant to serve as an important civic institution that would teach the French public about the equality of all peoples and cultures and thus reflect Rivet and Mauss's socialist and antiracist commitments. Yet, like their previous endeavors, the Musée relied heavily on the financial and institutional backing of the French imperial nation-state and many of its collections came from colonial ethnographic missions like the Dakar to Djibouti mission, where anthropologists collected artifacts under dubious ethical circumstances.[29]

---

[28] Alice L. Conklin, *In the Museum of Man: Race, Anthropology, and Empire in France, 1850–1950* (Ithaca: Cornell University Press, 2013).

[29] See, for example, Phyllis Clarck-Taoua, "In Search of New Skin: Michel Leiris's L'Afrique Fantôme." *Cahiers d'Etudes Africaines* 167, no. 3 (2002): 479–498.

It was in the context of this ambitious project to reform French anthropology and transform its institutions that Rivet developed the idea for an ethnographic mission to Paraguay. An "Americanist" – the term used then to refer to those who studied the Americas – Rivet was a prominent member of the Société des Américanistes de Paris and a regular participant of the yearly meeting of the International Congress of Americanists.[30] Thanks to the extensive South American contacts and correspondents he cultivated through these Americanist networks, Rivet obtained financial and political support from Argentinian, Brazilian, and Paraguayan authorities to conduct the ethnographic mission in Paraguay. Rivet's main goals for the mission were to conduct ethnographic and natural historical studies of Paraguay, especially the Gran Chaco region, and to cultivate closer ties between the scientific communities of Paraguay and France.[31]

At this point, thanks in part to the efforts of Alfred Métraux, the lowland plains of the Gran Chaco (which spanned parts of Paraguay, Bolivia, and Argentina) and its Indigenous peoples were emerging as promising research objects for European Americanists. With Rivet's help, Alfred Métraux became the founding director of the Institute of Ethnology at the University of Tucumán in northern Argentina in 1928. As director of this institute, Métraux sought to emulate Rivet's institution-building efforts in France by turning the Tucumán Institute into the leading center for ethnographic study in South America. To this end, Métraux led numerous expeditions to the Gran Chaco region, where he collected artifacts for an ethnographic museum in Tucumán. Describing the newly created institute at Tucumán and Métraux's efforts, Rivet wrote that "colonial questions" were becoming more and more pressing every day and would only be solved if approached with "a scientific spirit." According to Rivet, it was due to these colonial circumstances that ethnology experienced a dramatic development in France. With the creation of the Institute of Ethnology at Tucumán, he also prophesied that Argentina was now poised to find solutions to its "indigenous problems."[32]

In contrast to Argentina, Paraguay did not have well-developed ethnological institutions at this time, which likely spurred Rivet's interest in sending a mission there. To lead the Paraguay expedition, Rivet chose Jehan Albert Vellard – a physician by training who had spent most of the 1920s in Brazil and had become known to anthropologists in France through his studies of

---

[30] On Rivet's involvement with the Société des Américanistes, see Christine Laurière, "La Société des Américanistes de Paris: une société savante au service de l'américanisme," *Journal de la Société des Américanistes* 95, no. 2 (2009): 93–115.

[31] Jehan Albert Vellard, "Une mission scientifique au Paraguay (15 juillet 1931–16 janvier 1933)," *Journal de la Société des Américanistes* 25, no. 2 (1933): 293–334, 293.

[32] Paul Rivet, "L'institut d'ethnologie de l'Université de Tucumán," *Journal de la Société des Américanistes* 25, no. 1 (1933): 188–189, 189.

spider poison, curare, and other Indigenous medicines in South America.[33]
Vellard traveled to Paraguay in 1931 and spent two years there. His trip to
Paraguay coincided with the Chaco War (1932–1935) – a conflict over the
northern part of the region thought to be rich in oil that was fought between
the landlocked countries of Bolivia and Paraguay. Many observers consider
this war to be the bloodiest armed conflict in Latin America during the
twentieth century. During his two years in Paraguay, Vellard made several
ethnographic trips to the Gran Chaco, which followed Rivet's goals of
gathering ethnographic and natural historical data and establishing links
between French and Paraguayan scholars.[34] More narrowly, Vellard's missions had the purpose of conducting ethnographic studies on the "least
known tribes" of Paraguay and gathering objects for the Trocadero
Ethnographic Museum and for the Musée de l'Homme.[35] Vellard's missions
thus advanced Rivet's agenda of updating France's ethnological institutions
and were similar to missions that Rivet helped to organize in other colonies
and regions where France exerted political influence. These missions include
the Dakar to Djibouti mission in Africa and the Easter Island (Rapa Nui)
mission in the South Pacific, which also enjoyed the patronage and support
of Paul Rivet.[36]

During his fieldwork in Paraguay, due to the Chaco War, Vellard required
guidance and help from Paraguay's president, the Minister of War, and
various military officials and armed guides, who advised him on his itineraries
and shepherded him through regions where armed conflict was erupting.[37] For
his first trip to the Gran Chaco region, Vellard collaborated with the Russian
general Juan Belaieff. Belaieff had been recruited by the Paraguayan minister
of war and navy to conduct a reconnaissance mission and ethnographic census
of the Chaco, which gave Paraguay a tactical advantage in its war with Bolivia.
Belaieff had previously trained in military science and ethnography and had

---

[33] Readers of Lévi-Strauss's *Tristes Tropiques* will recognize Vellard as the medical doctor
who accompanied him in his ethnographic mission to the Nambikwara. See Claude Lévi-
Strauss, *Tristes Tropiques*, trans. John Russel (New York: Criterion Books, 1961), https://
archive.org/details/tristestropiques000177mbp/page/n363.

[34] Diego Villar, "Les Expéditions du Doctor Vellard," in *Les Années folles de l'ethnographie:
Trocadéro 28–37*, eds. André Delpuech, Christine Laurière, and Carine Peltier-Caroff
(París: Muséum national d'histoire naturelle, 2017), 536–579.

[35] Alice L. Conklin, *In the Museum of Man: Race, Anthropology, and Empire in France,
1850–1950* (Ithaca, NY: Cornell University Press, 2013); Christine Laurière, *Paul Rivet, le
savant & le politique* (Paris: Publications Scientifiques du Muséum national d'Histoire
naturelle, 2008).

[36] Christine Laurière, "Un lieu de synthèse de la science anthropologique: histoire du musée
de l'Homme," in *Bérose – Encyclopédie internationale des histoires de l'anthropologie*
(Paris, 2019) www.berose.fr/article1680.html?lang=fr.

[37] Vellard, "Une mission scientifique au Paraguay," 293–334.

conducted several ethnographic studies in the Caucasus of Russia.[38] When Vellard joined him in the Chaco region in 1931, Belaieff had already established relations with Indigenous groups in the region and when war broke out he advised Paraguayan soldiers to rely on Indigenous people from the regions as guides. With Belaieff's input, Vellard drew up a plan to study the least known Indigenous groups from the region. With the support of the military barracks built in the region and a Mestizo sergeant who served as his guide, he spent three and half months collecting as much anthropometric and linguistic data, and material artifacts, as he could.[39]

Vellard did not last long in the Gran Chaco. His report of the mission described the region as desolate and neglected and during the month or so that he stayed he encountered regions where the fighting between the Paraguayan and Brazilian militaries impeded fieldwork. His only notable ethnographic achievement during this time was a brief encounter with the Maká tribe whom he described as very hospitable. According to Vellard, the Maká were also eager to exchange goods with his crew and would demand "big gifts" in exchange for the smallest object "without having the slightest notion of the value of things."[40] Despite these differences, Vellard eagerly reported that he quickly gained the Maká's trust and was thus able to gather "a nice collection for the Trocadero."[41] Yet after traveling to the town of Nanawa at an ill-fated time when fighting between the Paraguayan and Bolivian armies had killed hundreds of civilians in the region, Vellard grew increasingly disillusioned with his Gran Chaco fieldwork. He was eventually arrested by the Paraguayan military for reasons unknown to him and, after having his firearms taken away, escorted back to Asunción where he plotted a second expedition to a less turbulent region.[42]

After his unrewarding fieldwork in the Gran Chaco, Vellard decided to travel to the region inhabited by the "Guayaki Indians" (the name formerly given to the Aché by anthropologists) whom Rivet had flagged as a group that was "little known" and "highly interesting" since they were difficult to access.[43] For this trip, Vellard secured the help of the wealthy Balanza family who owned a large estancia (ranch) situated at the entrance of the forest where the Aché roamed. The Balanza family consisted of the sons of a French botanist named Benjamin Balanza, who had made several collecting trips to Paraguay

[38] Bridget María Chesterton and Anatoly V. Isaenko, "A White Russian in the Green Hell: Military Science, Ethnography, and Nation Building," *Hispanic American Historical Review* 94, no. 4 (2014): 615–648.

[39] Vellard, "Une mission scientifique au Paraguay."

[40] Ibid., 303.

[41] Ibid.

[42] Ibid., 307.

[43] Ibid.

for the Muséum National d'Histoire Naturelle in the late nineteenth century and bought property. His sons turned the terrain into an industrial farm.[44] To Vellard's good fortune, the Balanza brothers proved to be very generous hosts who supported his fieldwork in numerous ways. According to Vellard, they lent him horses to travel into the forest, helped him to recruit men who could serve as his guides, and shared all manner of useful information on the region. Thanks to their help, Vellard was able to settle in a small ranch even closer to Aché territory, which was owned by one of the Balanza brother's employees, and the brothers sent weekly provisions for Vellard and his guides.[45] The success of Vellard's fieldwork in Paraguay was thus dependent on the relations and exchanges he established with people who possessed knowledge of the region as well as material resources and social connections.

Though it proved more rewarding than his Gran Chaco fieldwork, Vellard's descriptions of his Aché research reveal a context similarly marked by violence. Through Rivet described Vellard's account of his Aché research as "beautiful," Vellard's narrative was often framed as a difficult "hunt" through the forest peppered with fleeting moments of violent contact.[46] Given his lack of experience with the territory, Vellard relied on a team of Paraguayan laborers from nearby ranches and Indigenous guides from a nearby mBwiha village that he described as "semi-civilized." At the time of Vellard's fieldwork, relations between the Aché, Paraguayan ranchers, and the mBwiha had grown tense due to a recent bout of Aché raids on neighboring ranches where they killed several cattle, horses, and sheep. Both the Paraguayan laborers and the mBwiha guides viewed the Aché with a combination of fear and hostility. As they pursued the Aché through the forest, Vellard and his crew thus resolved to always keep firearms and machetes on hand. Yet Vellard's crew found it incredibly difficult to observe the Aché in any meaningful way. The Aché were constantly foraging and hunting for food and incredibly skilled at concealing their tracks. Vellard and his crew thus found it nearly impossible to observe them directly or establish direct contact. Much of their time was spent searching aimlessly for signs of their presence while attempting to maintain their energy and morale.

After several failed attempts to establish "friendly relations," Vellard and his crew gave up on the idea of direct contact and instead resolved to observe the Aché camps from a safe distance and to raid them at opportune times. Yet the raids of the Aché camps did not always go as planned and even descended into

---

[44] Villar, "Les Expéditions du Doctor Vellard," 543.

[45] Vellard, "Une mission scientifique au Paraguay," 313.

[46] I have derived the following descriptions from Vellard's published accounts of this mission. See Vellard, *Civilization du Miel*, 41–67; Jehan Albert Vellard, "Exploration du Dr Vellard au Paraguay," *Journal de la Société des Américanistes* 24, no. 1 (1932): 215–218; Vellard, "Une mission scientifique au Paraguay," 293–334.

violent skirmishes on two occasions. Although they failed to establish amicable relations with the Aché, Vellard and his crew used these violent encounters to collect material artifacts that were sent back to the Trocadero Museum in Paris and offered a small window into the Aché's way of life. And it was during these two violent encounters that Vellard and his crew adopted two Aché children – Marie-Yvonne as well as a young Aché boy who they named Luis.

Vellard described the team's first violent encounter, which led to his adoption of Marie-Yvonne, as a significant breakthrough in their journey. On this occasion, Vellard's team found "fresh evidence" pointing to the presence of an Aché group nearby. By following this trail of evidence, Vellard's team approached the Aché without being detected and found a hiding spot that they surmised was about 100 meters away from the Aché camp based on what they could hear in the distance. In the light of the next day, Vellard's team drew closer to the Aché camp and hid in the forest, which prompted their mBwiha guides to flee fearing a violent response from the Aché. Vellard's team was able to observe the Aché from a distance over the course of a day. When they attempted to get even closer the next day, they were spotted by Aché hunters who quickly fired a sea of arrows in their direction. Before "orders could be given," Vellard's men began to fire their guns in response to the Aché, thereby injuring one of their men and prompting the entire group to flee. Once the group had fled, Vellard and his crew quickly descended upon their camp and collected as many objects as they could. At this point, the Paraguayan laborers accompanying Vellard became increasingly difficult to control and spoke of "massacring the lot of them."[47] Eager to avoid "unnecessary violence," Vellard resolved to bring his team back to their main camp. Over the course of their return journey during the next two days, they were closely followed by Aché hunters who occasionally fired arrows at them, which Vellard's men returned with gunfire. When they finally arrived at the small house that they were using as their base, Vellard's men were joined once again by the three mBwiha guides who had previously fled. Vellard's three "fugitive guides" brought back their own spoils from a raid of another Aché camp, namely, a pot of honey, a coati, and a baby girl whose mouth they had stuffed with dead leaves after "tying her feet and hands together."[48] After some prodding from Vellard, the mBwiha confessed that the inhabitants of the camp they raided consisted of two women and the baby, and that they intended to rape the two women and instead captured the girl when the women fled, with the intention of selling her.[49]

Vellard and his crew captured Luis, an Aché boy, during similarly violent circumstances on what would be their last encounter with the nomadic group.

[47] Vellard, *Civilization du Miel.*
[48] Vellard, "Une mission scientifique au Paraguay."
[49] Ibid.

Vellard's crew encountered Luis after a botched raid on an Aché camp that led the camp's inhabitants to flee after Vellard's men resorted to gunfire and injured one of the Aché men. Luis was not able to flee with the rest of the camp and instead decided to follow Vellard and his men on their return journey to their ranch. According to Vellard, Luis did not demonstrate any desire to find his relatives, and he left him at the Balanza ranch under the care of the family. However, Vellard concluded that Luis was not as intellectually gifted as Marie-Yvonne. Although Luis helped him to produce a vocabulary of Aché words, Vellard noted that the boy's attention would tire very quickly. Vellard similarly noted that although the boy had a highly developed "visual memory," he did not seem to have great skill in retaining "strange words and sounds" and was slow to learn Spanish and French.

Having decided to keep Marie-Yvonne and leave her in the care of his mother, Vellard also had to decide what to do with Luis. While the decision to keep Marie-Yvonne seemed easy, Vellard struggled to figure out what to do with Luis. Though he left him in the care of the Balanza brothers, he also offered Luis to Rivet and asked for his advice as to what to do with him. Yet when he wrote to Rivet, it was not the boy's well-being that seemed to be Vellard's main concern but rather how he might best be incorporated into an ethnological research program. Indeed, Vellard reported to Rivet that he planned to study the boy by taking photographs and x-rays of him and, if possible, by measuring and weighing him. Yet he also asked Rivet if he had any recommendations for specific studies to conduct on the boy and wrote "or do you want him? Is there any interest in keeping him with me for studying his development?"[50]

Vellard normalized the capture of Marie-Yvonne and her adoption by explaining that many ranchers from the region had purchased Aché children in similar fashion and that they treated them "very well." Indeed, Vellard explained that Aché children could be bought for 200–300 Paraguayan pesos. In his book on the Aché and in one of the scientific articles he published, Vellard devoted an entire section to descriptions of Aché children who were raised on ranches.[51] Here Vellard noted that in addition to the well-documented case of Damiana, at least four other captured children had been observed by anthropologists; he had also learned of at least five other children in nearby regions who lived with Paraguayan settlers. Although most of these children adjusted well to their new surroundings and often learned multiple languages, Vellard noted that in some cases they developed "unstable person-alities" later in life and were prone to disappearing for days to "vagabond in the forest."[52] On the basis of these observations, Vellard felt inclined to offer

[50] Quoted in Villar, "Les Expéditions du Doctor Vellard," 550.
[51] Vellard, *Civilization du Miel*, 131–139.
[52] Ibid., 134.

some general reflections on the "character" of Aché children raised in "more or less civilized environments."[53] In general, Vellard concluded, Aché children raised in these circumstances tend to be "soft, docile, affectionate, very fearful, generally intelligent (several speak two languages and frequent school)." Yet after puberty, in some instances, they become "volatile, restless, and inclined to take flight."[54]

From his earliest publications on the Aché, Vellard expressed a keen interest in studying Marie-Yvonne and Luis's physical and intellectual development and how they adjusted to their new environments. And like many human scientists from the interwar period, Vellard framed the observations and measurements he made on captured Aché as ones that offered insights on the nomadic group's distinctive racial type and immunological profile. In his first article on the Aché, Vellard explained that he was closely following Marie-Yvonne's development now that she was under his mother's care and that he was confident that this would yield some "valuable observations." He also mentioned that the Balanza brothers would keep him up to date on Luis's development and on his adaptation to "civilized life."[55] In another article, where he offered detailed notes on the physical type, character, and state of health of the captured Aché children, Vellard also summarized his observations and measurements of another Aché girl living on a ranch whom he called Fortunata. Based on the direct observations and measurements he made of Marie-Yvonne, Luis, and Fortunata as well as the facts he could glean from the accounts of previous researchers who had studied Aché children like Damiana, Vellard affirmed the prevailing orthodox view that the Aché represented a highly homogenous racial type. "Coloration of the skin, hair, and irises," were the only traits that Vellard discerned to be "highly variable."[56] After observing that Marie-Yvonne and Luis, like many other captured Aché children, quickly succumbed to various respiratory illnesses after encountering non-Aché, Vellard also confirmed the views of previous writers that the Aché were highly sensitive to "contact with civilized people." Yet in contrast to previous writers who speculated that such sudden onset in illness stemmed from dietary changes, Vellard insisted that it stemmed from a much "deeper cause," namely, ". . . the absolute lack of immunity against many germs, in particular the pneumococcus, against which the civilized have a great resistance."[57]

Decades later, Vellard returned to the observations he made on captured Aché children in an article on the "Biological Causes of the Disappearance of

---

[53] Ibid., 138.
[54] Ibid.
[55] Vellard, "Une mission scientifique au Paraguay," 139.
[56] Jehan Albert Vellard, "Les Indiens Guayaki," *Journal de la Société des Américanistes* 26, no. 2 (1934): 223–292.
[57] Ibid., 276.

American Indians." In this article, Vellard argued that the demographic collapse of Indigenous populations in the Americas occurred primarily due to biological forces and that the use of systematic violence and cruelty by Iberian colonizers played a minimal role. According to Vellard, the disappearance of the Indigenous peoples of the Americas occurred due to a lack of immunity to European diseases that was itself a by-product of geographic and biological isolation. One of the key pieces of evidence that Vellard used in these accounts were the stories of the captured Aché children he had encountered and written about. As he looked back to his experience with Marie-Yvonne and Luis, Vellard framed the diverging trajectory of these two children as a sort of natural experiment. According to Vellard, Marie-Yvonne's fortune stood in stark contrast to those of other captured Aché children. Although Marie-Yvonne contracted pneumonia shortly after her capture, she was also vaccinated against tuberculosis and ever since enjoyed a normal and illness-free development even after traveling to major city centers. Marie-Yvonne's story thus demonstrated that if Indigenous children are "artificially" placed in civilized conditions and vaccinated for diseases then their development will be equivalent to those of civilized children. By contrast, Vellard described Luis's fate as a tragic one. Like Marie-Yvonne, Luis also contracted pneumonia a few days into his "life with civilized people" yet he did not receive a vaccine against tuberculosis. As a result, Luis suffered from poor health for the rest of his life and eventually ended up succumbing to tuberculosis.[58] For Vellard, the stories of Marie-Yvonne and other Aché children thus served as object lessons for how vaccination could reduce the biological distance between Indigenous and European populations who had acquired immunity through hereditary mechanisms, in other words the "biological shock of European conquest." Through these historical narratives, Vellard ultimately identified biological forces as the main causes of demographic collapse and thus minimized and denied the historical responsibility of European colonizers, an early version of what David Jones has aptly called "immunological determinism."[59]

Marie-Yvonne's Life as a "Lesson for Humanity"

For Vellard, Marie-Yvonne's story thus served as an object lesson for the immunological adaptation of Indigenous groups to modern civilization. By contrast, for Vellard's colleagues like Rivet and Métraux, who were plugged into internationalist antiracist networks to a much greater degree, and for

---

[58] Jehan Albert Vellard, "Causas biológicas de la desaparición de los indios americanos," *Boletín del Instituto Riva-Agüero* (Pontifica Universidad Católica del Perú) no. 2, 1956.

[59] David S. Jones, "Virgin Soils Revisited," *William and Mary Quarterly* 60, no. 4 (October 1, 2003): 703.

American journalists, Marie-Yvonne's story was framed as a striking example of the benefits of cultural assimilation. While Vellard's narrative displaced European agency through immunological determinism, these contrasting accounts of Marie-Yvonne's story amplified European agency by casting European scientific experts and educators in the role of civilizers.

In the hands of Rivet and later in a UNESCO article, Marie-Yvonne's story stood in stark contrast to Damiana's. Unlike Damiana who was described as reverting to a state of "savagery" once puberty hit, Marie-Yvonne's story was one of permanent transformation. In the case of the UNESCO article, Marie-Yvonne's story was further couched in a language that suggested the influence of the individualized and psychologically oriented research program of the behavioral sciences. Within this ostensibly nonracial regime of truth Marie-Yvonne's story became a morality tale about how to escape the trappings of a culture doomed to poverty and stagnation. This redemptive framing of Marie-Yvonne's story can be most clearly observed in an article published in the *UNESCO Courier* in 1950, which was titled "An Indian Girl with a Lesson for Humanity." The article was penned by Alfred Métraux – one of Paul Rivet's most accomplished students – who directed UNESCO's campaign against scientific racism during the 1950s. In his article, Métraux argued that had she not been adopted at the age of two, Marie-Yvonne would have been condemned to a "primitive and rudimentary culture" that wanders "at large in the forest" hunting animals and gathering fruits and whose way of life is "very little different from that of the first bands of men who colonized the empty spaces of South America thousands of years ago."[60] Yet by virtue of being "brought up exactly as a white girl," Marie-Yvonne became "an attractive, intelligent girl of twenty and a typical product of the cultural environment in which she has lived for 18 years." For Métraux, Marie-Yvonne's story was precisely the kind of evidence that could be used to convince the "layperson" of the arguments put forward in the 1950 Statement on Race. Above all, Métraux argued, Marie-Yvonne's story proved one of its central arguments, namely, that ". . . given similar degrees of cultural opportunity to realize their potentialities, the average achievement of the members of each ethnic group is about the same."[61] Despite the violent circumstances described in Vellard's account, journalists, educators, and Métraux thus adapted the story of Marie-Yvonne's capture into a feel-good tale about bridging the temporal chasm between modern and primitive life.

This redemptive framing also persisted in other popular accounts of Marie-Yvonne's life. For instance, in 1960, *Reader's Digest* published a four-page story penned by the California journalist Reese Wolfe, which described Marie-

---

[60] Alfred Métraux, "An Indian Girl with a Lesson for Humanity," *UNESCO Courier* 3, no. 8 (1950): 8.

[61] Ibid.

Yvonne Vellard's life as "the remarkable story of a child who, born of Stone Age people, bridged a gap of 5,000 years to become a twentieth-century scholar."[62] Like Métraux in the *UNESCO Courier*, Wolfe's story relied on a framing that equated social, spatial, and temporal distance – an imperial trope that postcolonial theorist Anne Mclintock refers to as the production of "anachronistic space." Yet in a postwar context in which a booming US economy gave rise to the liberal ideal that everyone could prosper regardless of their race, Wolfe framed children as potent symbols of upward social mobility who had the ability to escape a damaging culture. For instance, Wolfe's story noted that "ethnologists" like Vellard knew that an "infant from a primitive environment, brought up as a modern child, readily adapts to civilization."[63] As such, Vellard's encounter with Marie-Yvonne offered an unprecedented chance to "witness this remarkable transformation" and Wolfe mused that when Vellard first saw her she was a rare "baby born into a tribe still living in the Stone Age!" Like Métraux, Wolfe also obfuscated the violent circumstances surrounding this encounter. Indeed, Wolfe's story did not mention Vellard's guides' hostility toward the Aché and instead implied that Marie-Yvonne had been rescued from tragic circumstances. "Her emaciated body, the tell-tale bloat of her belly, the long red weals on her coffee-brown skin, bore eloquent testimony to hunger and abuse," lamented Wolfe. By appealing to temporal tropes such as being born into the Stone Age and highlighting signs of ill-health, Wolfe, like Métraux, implied that Marie-Yvonne was destined to suffer if she were to remain with her birth family.

Wolfe's article also offered a much fuller portrait of Marie-Yvonne's intelligence and astonishing transformation into a "twentieth-century scholar." During her first few days in Asunción where she was looked after by Vellard's mother Amélie Vellard, Marie-Yvonne was silent and fearful and "clung fiercely" to all of her possessions even while asleep. She also resisted Vellard and his mother's attempts to teach her French. Yet after weeks of effort, her adoptive grandmother noticed her repeating the same word in a low voice and then later burst into her room to "triumphantly" declare "Grandmère!" multiple times. From this point on, Marie-Yvonne quickly acquired more and more words and soon became fluent. She became a fixture in the Vellard family and stayed with them as they relocated first to Brazil, where she became fluent in Portuguese, and then Peru, where she learned Spanish. With her grandmother's support she learned to read and write at an early age and "adapted herself to numerous schools, never failing to receive high marks."

---

[62] Reese Wolfe, "The Girl from the Stone Age," in *Reader's Digest*, vol. 76 (August 1960), 43–48.
[63] Ibid., 45.

According to Wolfe, it was at the age of fourteen, when Marie-Yvonne accompanied Jehan Vellard on a field trip to study an Aymara community on the shores of Lake Titicaca that her transformation into a scientist began in earnest. During the first day of Vellard's mission, the Aymara community shunned the team and did little to make them feel welcome. "They stood in the doorways of their hovels in hard-eyed silence and watched their intruders," wrote Wolfe. Deflated by the poor reception, Jehan decided that they were best off camping for the night outside the village. Marie-Yvonne, on the other hand, "sensed the Indian's hidden pride and fears" and became determined not to be refused by them. She thus approached a group of "sullen-faced Aymaras" and told them in a mixture of Spanish and the few words of their language she had learned that "she came from an Indian tribe in the far-off jungles to the south" and that she had been adopted and raised by the "white man" who had come to study them.[64] According to Wolfe, after Marie-Yvonne's impassioned plea the "sullen" Aymaras "stirred uncertainly" until one of the women finally beckoned them to come in. Having established a rapport, Jehan and Marie-Yvonne then spent the summer living among "their new friends" and learning about "their daily lives, their fiestas, [and] their sacred ceremonies."[65] After this first taste of fieldwork, explained Wolfe, Marie-Yvonne became certain that ethnology was her calling. "She began to learn the painstaking art of scientific observation and note taking," wrote Wolfe. And she also quickly became adept at learning to communicate with "the tribes she studied." By the age of twenty-one, after four years of "distinguished work" at the Instituto Riva Agüero, a Peruvian research institution, Marie-Yvonne obtained a degree in ethnology. With her training complete, she continued traveling with her adoptive father on ethnographic trips that purportedly took her as "far afield" as Tierra del Fuego and "to Eskimo Villages near the Arctic circle."[66] And in the spring of 1959 Marie-Yvonne began her own series of independent studies, which took her to a remote village in the Peruvian Amazon where she told the Indigenous inhabitants that "their special ways of weaving, cooking and pottery-making would soon be lost if no one made a record of them."[67]

Territorial Confinement and the Golden Age of Aché Research

Almost a decade after Métraux published his story in the *UNESCO Courier*, a Paraguayan rancher named Manuel de Jesus Pereira succeeded in "pacifying"

---

[64] Ibid., 47.
[65] Ibid., 48.
[66] Ibid.
[67] Ibid.

two Aché groups, who came to live on his ranch. Pereira captured the first group in 1959 and the second group in 1962, doing so with the help of neighboring settlers known for "hunting" the Aché. Pereira also persuaded the Aché who were living with him to track down and capture members from other forest-dwelling Aché groups and bring them back to his ranch where they were "pampered and then released to bring in the others of their group."[68] By 1963, Pereira had convinced the Aché from two major groups to live under his protection and was given a government post and salary to administer this newly created "reservation."

After these groups of Aché settled on Pereira's ranch, more than half of their population died from disease under Pereira's watch. Yet despite these violent circumstances, anthropological research on the Aché blossomed thanks to urgent appeals made by Alfred Métraux. From this moment forward, anthropologists and human biologists began to visit the Aché more regularly and for much longer periods. They no longer had to rely on captured children as sources of evidence. Despite this change in circumstances, subsequent experts continued to center children and the study of childhood in their research regimes. This is particularly evident in Carleton Gajdusek's genetic studies, which he conducted while serving as director of the "Study for Child Growth and Development and Disease Patterns in Primitive Cultures" at the National Institutes of Health (NIH) in Maryland.[69]

Perhaps not surprisingly, scientific research on the Aché spiked in the subsequent years. Upon the recommendation of Alfred Métraux, anthropologists Pierre Clastres, Helene Clastres, and Lucien Sebag – three of Claude Lévi-Strauss's most promising students – traveled to Pereira's reservation to conduct detailed ethnographies of the Aché. For Pierre Clastres, although he lamented that they were destined to disappear, the Aché became exemplars of societies "against the state" and the kernel for a new research program in "political anthropology" that would help transform French political thought after 1968.[70] In the period that the Clastres's and Sebag did their fieldwork on the Aché reservation, Métraux also met the human biologist and Kuru researcher Carleton Gajdusek and encouraged him to visit the Aché.[71] With the help of

[68] Kim Hill and H. Magdelena Hurtado, *Aché Life History: The Ecology and Demography of a Foraging People* (New York: Aldine de Gruyter, 1996), 49.

[69] Daniel Carleton Gajdusek, *Paraguayan Indian Expeditions to the Guayaki and Chako Indians, August 25, 1963 to September 28, 1963* (Bethesda, MD: National Institute of Neurological Diseases and Blindness, National Institutes of Health, 1963).

[70] Pierre Clastres, *Chronicle of the Guayaki Indians* (New York: Zone Books, 1998); Pierre Clastres, *Society against the State: Essays in Political Anthropology* (New York: Zone Books, 1987); Samuel Moyn, "Of Savagery and Civil Society: Pierre Clastres and the Transformation of French Political Thought," *Modern Intellectual History* 1, no. 1 (2004): 55–80, doi:10.1017/S1479244303000076.

[71] Gajdusek, *Paraguayan Indian Expeditions to the Guayaki and Chako Indians.*

the Clastres's and Sebag, Gajdusek collected numerous blood samples of Aché children and adults, which he later published as part of a larger study comparing the genetic markers of Indigenous groups throughout Paraguay.[72]

Yet as more researchers began to visit the Aché and spend longer stretches of time with them they also grew less passive in the face of violence. During the 1970s, the plight of the Aché began to attract international scrutiny when several anthropologists and legal scholars accused the Paraguayan government of promoting an intentional government policy of genocide against the Aché.[73] Although activists warned of their imminent demise, the Aché endured and began attracting a new set of US-based biological anthropologists in the 1980s, who were trained in human ecology and interested in understanding the fertility and mortality patterns of "hunter-gatherer" and other "unacculturated, technologically primitive" populations in an effort to understand the degree to which their demographic curves differ from those of "modern populations."[74] Since the early 2000s leading Aché specialists and human ecologists, Ana Magdalena Hurtado and Kim Hill, have played a prominent role in combating what they view as "antiscientific" attacks on anthropology (such as the controversies sparked by the publication of journalist Patrick Tierney's *Darkness in El Dorado*) by articulating more robust ethical guidelines for anthropological fieldwork on Indigenous communities. Yet their interventions have often been guided by the colonial narrative of Indigenous peoples as "vanishing" populations in need of salvage. A key component of their ethical strategy has been to question the self-determination framework adopted by many nations toward Indigenous peoples and to instead emphasize the need for greater "epidemiological surveillance" of Indigenous groups and "controlled contact" with "isolated" or "uncontacted groups." In advancing these proposals, Hill and Hurtado often cite the case of the Northern Aché as a model of "well-designed" and controlled contact led by missionaries, anthropologists, and physicians who were able to provide medicine and care to Aché members who became ill.

Conclusion

In August of 2008, at the height of the pink tide (leftist electoral victories) that swept through Latin America, a forty-seven-year-old Aché woman named

---

[72] Stephen M. Brown et al., "Genetic Studies in Paraguay: Blood Group, Red Cell, and Serum Genetic Patterns of the Guayaki and Ayore Indians, Mennonite Settlers, and Seven Other Indian Tribes of the Paraguayan Chaco," *American Journal of Physical Anthropology* 41, no. 2 (1974): 317–343.

[73] Richard Reed and John Renshaw, "The Aché and Guaraní: Thirty Years after Maybury-Lewis and Howe's Report on Genocide in Paraguay," *Tipití: Journal of the Society for the Anthropology of Lowland South America* 10, no. 1 (2012): 1–18.

[74] Hill and Hurtado, *Aché Life History*.

Margarita Mbywangi was appointed Minister of Indian Affairs by President Fernando Lugo. After her ministerial appointment, Mbywangi attracted, much like Marie-Yvonne, considerable international attention and her story was retold in a wide range of news outlets including the *Guardian*, the *Financial Times*, *El Pais Uruguay*, and *Indian Country Today*. She had also been abducted in childhood; unlike Marie-Yvonne, however, Mbywangi's circumstances gave her much greater control over her own narrative. Though there were certainly outlets that describe Margarita's life as a straightforward redemption story, she also gave extended interviews where she told her story with greater complexity. For instance, Margarita gave an interview published in the *Financial Times* magazine where her narrative struck a delicate balance between describing the damage she suffered at the hands of Paraguayan settlers and the desires that prompted her to eventually become a leader within her community and a national politician.

In her biographical interview with the *Financial Times*, parts of Mbywangi's story share some troubling similarities to the stories of captured children that populate the Aché ethnographic archive. At the age of five, she was captured by Paraguayan ranchers who sold her for 5,000 *guaraníes* to a family of ranchers with ten children, who adopted Margarita as their servant. With help from one of her adopted "sisters" who worked as a teacher, Margarita enrolled in school and attended until the fifth grade yet was unable to continue because she had no birth certificate. At the age of sixteen, Margarita escaped from her adopted family and began working as a domestic servant. In contrast to the stories of other Aché children, Margarita's account offers a revealing portrait of her attitude toward her adopted family, which is something we can only infer in Marie-Yvonne's story. In her narrative, Margarita strikingly describes her relationship to her adopted family in strictly transactional terms. "I called the lady who bought me 'Mum'," she explains in her interview "but I cleaned the house and looked after the grandchildren." Throughout her interview she refers to her adopted siblings in quotes – "I called them my 'brothers' and 'sisters'" – and she also points out that, unlike her, "none of them worked, they studied." Her narrative also describes a stark contrast in the affection she received compared to her adopted siblings. "I wore their hand-me-downs but I never had any presents and no one ever showed me love," she explained. "I was a servant."[75] From Margarita's perspective, a story that non-Indigenous actors like Vellard and Métraux might have framed through the benevolent terms of adoption and kin-making is revealed instead to be a harsh economic project of forced removal and servitude.

---

[75] "First Person: Margarita Mbywangi," *Financial Times*, July 3, 2009, www.ft.com/content/1f4e11f4-6475-11de-a13f-00144feabdc0.

Yet Mbywangi's account also departs from the stories of captured Aché children like Damiana and Marie-Yvonne insofar as it describes her escape from Paraguayan society. After fleeing her adopted family at the age of sixteen she found a job at a bar where she was eventually recognized by one of her "brothers" who then attempted to have her arrested. After this incident, Margarita returned to her adopted family but then escaped for good and spent the next two years tracking down her birth village with the help of a priest. She was then able to return to her village at the age of twenty. When she returned, one of her "real brothers" recognized her, but she was no longer able to speak Aché. She struggled to adapt to sleeping around a fire without blankets and questioned why she had come back. She "became an alcoholic." Yet she eventually re-learned her language and completed a nursing course. By the time she gave her interview she had proudly become the cacique (chief) of her community, Kuetuvy. Leading up to the 2008 elections, leaders of the Tekojoja movement that backed the leftist President Fernando Lugo asked her to run for the senate and although she did not win, Lugo asked her to become Minister of Indigenous Affairs. Yet she did not thrive in this position. "I consider myself a leader but I think politics is dirty," she explained "and it was hard being in an office all day." After a few months in office, she left her post yet continued working with the Tekojoja movement as well as with an association of Aché communities. Although she found it difficult to balance her political career with family life – she is the mother of three kids – she also believed that her service would pay future dividends for her community. "But this work gives me strength to give to others what I never had: love and a family."[76]

Mbywangi's trajectory thus marks an important departure from the stories of captured children that populate the Aché ethnographic archive. Whereas her political and personal desires feature prominently in her story, the prevailing assumption of most early ethnographic experts was that the Aché would be irrevocably damaged, or even destroyed, through contact with "Western" civilization. As this chapter has shown, the fact that Aché children were routinely captured and sold as domestic servants was one that ethnographers from the first half of the twentieth century described as matter of fact but did little to oppose. Such passivity in the face of violence, this chapter argued, factored prominently into the ways that ethnographic experts constructed their moral and professional duties. As modest witnesses, they framed their moral duty as one to observe, preserve, and collect the traces of what they presumed to be a vanishing people and thereby implied that there was little they could do to stop the Aché from disappearing let alone thriving. Such fatalistic narratives persisted well into the 1960s and can be observed in the

[76] Ibid., 2.

work of Pierre Clastres, who described how the Aché population had collapsed after his stay and was "eaten away by illness and tuberculosis, killed by a lack of proper care, by lack of everything." For Clastres, the dwindling of the Aché's population demonstrated that the "whole enterprise that began in the fifteenth century is now coming to an end" and that "an entire continent will soon be rid of its first inhabitants."[77] The example of Clastres, who was an anarchist, as well as his French predecessors such as Rivet and Métraux who were prominent antifascists and antiracists, demonstrates that such fatalistic damage narratives coexisted with progressive and internationalist politics. Mbywangi's personal narrative thus accentuates and renders visible the colonial ideologies that persisted within well-intentioned scientific discourses well through the latter half of the twentieth century.

---

[77] Pierre Clastres, *Chronicle of the Guayaki Indians. Translated by Paul Auster* (New York, Cambridge, MA: Zone Books; Distributed by MIT Press, 1998), p.96.

PART II

Institutional Encounters, Discipline,
and Settler Colonial Logics

5

Replacing Native Hawaiian Kinship with Social Scientific Care

Settler Colonial Transinstitutionalization of Children in the Territory of Hawai'i

MAILE ARVIN

In 1929, Dorothy Wu, a multiracial Native Hawaiian and Chinese teenager, suddenly found herself a ward of the Salvation Army Girls' Home in Honolulu, Hawai'i, then a US territory.[1] Until her institutionalization, Dorothy had worked as a maid in the home of Princess Abigail Campbell Kawānanakoa, a member of the ali'i or ruling, royal class of the formerly independent Hawaiian Kingdom. To Dorothy's bewilderment, the Princess abruptly fired Dorothy for "associating with boys," and recommended her commitment to the Girls' Home, where she could be "better cared for." Dorothy found the Salvation Army matrons mean. She complained that they discriminated against "darker girls" like her. After six months, Dorothy ran away. Her escape soon landed her in juvenile court, which sentenced her to the Kawailoa Training School for Girls, a recently opened juvenile detention facility, where matrons worked to shape "wayward girls," the majority of whom were Native Hawaiian and/or Asian, into "proper" and modern American women.[2]

Dorothy's narrative is a particularly striking example of what Susan Burch terms "transinstitutionalization," meaning "the process of moving individuals from one variety of institution to another – as part of sustained containment, surveillance, and slow erasure," which she argues is a potent type of "settler colonial removal."[3] Dorothy Wu's story gives one example of the path this

---

[1] Dorothy Wu is a pseudonym. Throughout this essay, I do not use the real names of children who are named in archival records given. I have not (at least not yet) been able to contact descendants or other kin who may have particular desires about sharing these stories.

[2] This brief version of Dorothy's story is based on the transcript of an interview she did with sociology graduate students, Doris Lorden and Margaret Lam, in 1931. Accessed at the University Archives, University of Hawai'i at Mānoa, Romanzo Adams Social Research Laboratory Records, 2013.

[3] Susan Burch, *Committed: Remembering Native Kinship in and Beyond Institutions* (Chapel Hill: University of North Carolina Press Books, 2021), 16.

transinstitutionalization could take. Forced to work at a young age due to the financial circumstances of her family, her perceived impropriety of "associating with boys" got her fired and referred to the Salvation Army Girl's Home; and when she ran away, her court-ordered punishment was to be sent to the government training school.[4] Though Dorothy's story ends at the Kawailoa Training School in the archival record I have accessed so far, many other children found themselves on even longer paths of transinstitutionalization. Based on intelligence quotient (IQ) testing conducted regularly at the training schools, children deemed "feebleminded" would often be transferred to the Waimano Home for the Feeble-Minded. And especially for young men incarcerated at the government Waiale'e Industrial School for Boys, their "delinquency" would often get them sent on to O'ahu Prison. The traffic between these institutions (the training schools, home for the "feebleminded" and the prison) was eased by the fact that, beginning in 1939, they were all administered under the same Territorial Department of Institutions, along with the Territorial Hospital for the Mentally Ill and divisions of parole.[5] As Burch argues, the frequency of transfers between such institutions alerts us to the "dynamic, interlocking and far-reaching ... processes, practices, and experiences" that characterize transinstitutionalization as a method of settler colonialism, factors that we might miss if we focus only on histories of particular institutions in isolation.[6]

This essay is about how social scientific knowledge production structured the logic and practices of transinstitutionalization in Territorial Hawai'i. Specifically, I look at the history of three institutions operated by the Territory of Hawai'i: the Waiale'e Industrial School for Boys (opened in 1902), the Kawailoa Training School for Girls (opened in 1929), and the Waimano Home for the Feeble-Minded (opened in 1921). The combined rhetorics of correction and care for Hawai'i's children at play in these carceral institutions echoed the broader, paternalistic justifications for annexing Hawai'i as saving the islands both from other colonial empires and a Hawaiian Kingdom that white settlers characterized as uncivilized and childish. In this view, US settler colonialism could be portrayed as a progressive uplift of Native Hawaiians, rather than the structural cause of their increasing losses of land, language, and culture. Social science provided the Territorial government language and "evidence" with which to argue that their practices

---

[4] A striking detail of Wu's story is the complicity of Princess Kawānanakoa in Wu's institutionalization. This point is too complex to fully explore here, but is perhaps one example of how some settler colonial ideologies about gender and sexuality had, by the 1920s and 1930s, been thoroughly internalized by some Native Hawaiians.

[5] Hawaii Department of Institutions, *Department of Institutions, Territory of Hawaii, 1942* (Honolulu, 1942).

[6] Burch, *Committed*, 16–17.

of institutionalization were modern, progressive, and humane, even when the official reports of those institutions suggested otherwise. The Territorial government repeatedly used the scientific imprimatur of work by those like psychologist Stanley Porteus as well as models of training schools and homes for the "feebleminded" in the continental United States as justifications for institutionalizing Native Hawaiians and immigrants of color.[7]

Rather than providing a comprehensive account of the history of any one of the institutions I reference here, or attempting to fully reconstruct what life might have been like for the children incarcerated in these spaces, this essay focuses on pulling two main threads through its analysis of the histories of Waialeʻe, Kawailoa, and Waimano. The first thread tracks the settler colonial process of pathologizing Native Hawaiian and other non-white forms of kinship and care, and attempts to replace it with institutionalized care, as determined by white American social science and public policy.[8] Woven into this thread are the broader questions raised by this edited collection in considering how and why we study the history of science in relation to settler colonialism. The second thread examines how a critical history of these institutions offers a different picture of the Territorial period in Hawaiʻi. As I detail below, conventional histories of this time in Hawaiʻi have largely overlooked Native Hawaiian stories, focusing more on labor histories of immigrant populations, and presented the political shifts from Hawaiian Kingdom to Republic of Hawaiʻi to the Territory of Hawaiʻi and then to Hawaiʻi as the fiftieth state of the union as largely smooth and inevitable. Taking seriously the colonial violence and damage done through institutionalization offers a more complex and troubling picture of the Territorial period. The next two sections provide more context on these two threads, before moving into further analysis of Waialeʻe, Kawailoa, and Waimano.

A Structural Approach to Histories of Race, Settler Colonialism, and Science

Social sciences have long played crucial roles in disseminating ideas and practices that shore up white supremacy and settler colonialism. This is not especially surprising. How and why, then, should we seek to understand the ways that social scientists have contributed to structuring settler colonialism?

---

[7] S. D. Porteus, *The Institutions of the Territory of Hawaii and Their Policies, Plans and Needs for Sound Institutional Practices* (Honolulu: University of Hawaiʻi Press, 1949).

[8] I largely reference Native Hawaiian customs of kinship in this essay, in part because even though Asians were also incarcerated at the institutions I discuss here, many of the official reports focus on pathologizing Native Hawaiian culture in particular. However, I make note of "other non-white forms of kinship" to be analytically inclusive of Asian and other non-white cultures in Hawaiʻi, many of whom also made kin with Native Hawaiians.

I argue that such histories are most meaningful when they grapple with the ways social scientific research and constructions of so-called scientific truths contributed to building enduring systems of injustice, rather than seeking to ascertain and judge the personal orientations of individual social scientists. Accordingly, the most meaningful historical question is not: "were they racist?" It is not that such judgments are entirely irrelevant but simply that the effects of racism are easily dismissed if recognized only as an individual trait. In the case of Stanley Porteus, one social scientist I consider in this essay as discussed further below, there is little question that he was racist: he published numerous studies claiming that Native Hawaiians and Filipinos had much lower intelligence than white people. My point in studying his work is not to prove his racism but to demonstrate how his research and reasoning created abiding racist ideas about Native Hawaiians, Filipinos, and others in Hawaiʻi. These ideas continue to help uphold the structure of settler colonialism, and thus continue to deserve rich historicization and critique far beyond critiquing Porteus as an individual.

In this approach, I depart from historians who worry about being "presentist," "objective," or judging historical actors "who were just from a different time" by our contemporary standards (this excuse almost always made in reference to white people, despite the contemporaneous existence of people of color who acted differently).[9] Some historians of science emphasize, for example, that social scientists who produced racist social scientific theories were not necessarily explicitly "rabble-rousing, hate-mongering" racists themselves.[10] Indeed, many racists were, and are, elite, liberal, and progressive for their times. My point is that it is not particularly useful or interesting to write histories of science in which we see racism as merely an individual trait. Rather, racism is systemic; as Ruth Wilson Gilmore has written, "Racism, specifically, is the state-sanctioned or extralegal production and exploitation of group-differentiated vulnerability to premature death."[11] This definition shifts ideas of racism away from the bigoted ideas of any one person to the devastating generational and structural effects of racial discourses on those who are racialized. One way the history of science might better reckon with racism as a matter of "group-differentiated vulnerability to premature death" is to more frequently put the work of scientists in context with the effects of

---

[9] James Sweet, "Is History History? Identity Politics and Teleologies of the Present," August 17, 2022, www.historians.org/publications-and-directories/perspectives-on-history/september-2022/is-history-history-identity-politics-and-teleologies-of-the-present.

[10] Jan E. Goldstein, "Toward an Empirical History of Moral Thinking: The Case of Racial Theory in Mid-Nineteenth-Century France," *The American Historical Review* 120, no. 1 (February 1, 2015): 1–27, https://doi.org/10.1093/ahr/120.1.1.

[11] Ruth Wilson Gilmore, *Golden Gulag: Prisons, Surplus, Crisis, and Opposition in Globalizing California* (Berkeley: University of California Press, 2007), 28.

their "science" on the people it impacted, rather than remain with telling the life story, however complex, of a scientist. Then the weight of any moral judgment of past scientists must fall not only on their individual shoulders but also, and perhaps more importantly, on the structures that their work upheld. So too, the complexity often granted to white scientists in historical treatments should be extended just as robustly to those communities impacted by their scientific ideas and practices.

My analysis of Porteus's work is accordingly motivated by excavating the impact of his work within the settler colonial logics and policies he helped institute in the Territory of Hawai'i, and telling a complex story about the lives of the youth who were impacted by Porteus's work and the broader system of carceral institutions he supported. Hawai'i Territorial officials sought out Porteus's opinions, and created policies based on his views, precisely because they took him to be an objective, and presumably apolitical, scientist. As I explore further in this essay, Territorial officials valued Porteus's conclusions about the mental inferiority of Native Hawaiians and Filipinos, among other people of color. Territorial officials institutionalized Native Hawaiians and others based on such "scientific" truths, couching the missions of the training schools as the uplift and civilization of inferior races. Such discourse fit with Progressive-era sentiments that lauded modernization and assimilation of the less fortunate into mainstream American norms. This is another reason why it is not particularly illuminating to assess the personal opinions or "intent" of past social scientists. So-called progressive ideas have been just as violent toward Indigenous peoples as more explicitly racist ones, and as Porteus's case shows, they often went hand in hand.

To put it another way, academic knowledge production about Indigenous peoples and people of color is far from innocent, no matter the tenor of individual researchers' intentions. Therefore, in the practice of the history of science, my goal is to historicize how settler colonialism operates, always alongside how Indigeneity has also endured. As Kēhaulani Kauanui argues, through the exegesis of Patrick Wolfe's work, settler colonialism is an enduring structure, but so is Indigeneity: "indigenous peoples exist, resist, and persist."[12] The goal of doing history of science scholarship from a critical Indigenous Studies perspective is to not shy away from the damage wrought by science and settler colonialism, but also to not spectacularize that damage such that we lose sight of the complex humanity of Indigenous peoples. One way I do so in this essay is by attending to the desires of the students at the Territorial training schools, who often frustrated the government's intended "rehabilitation" by running away, staging strikes, or maintaining Hawaiian

---

[12] J. Kēhaulani Kauanui, "'A Structure, Not an Event': Settler Colonialism and Enduring Indigeneity," June 1, 2016, http://csalateral.org/issue/5-1/forum-alt-humanities-settler-colonialism-enduring-indigeneity-kauanui/.

culture. Such acts remind us that despite the enormous power of the settler colonial Territorial government to restructure Hawaiian life, Native Hawaiians also lived in ways that were never wholly determined by that structure. Indigenous historian Damon Salesa writes that "an ethical and full engagement with an indigenous past is through an indigenous present."[13] The politics of seeking desire not just damage, as Indigenous feminist scholar Eve Tuck puts it, is one way to do ethical historical research because it opens up monolithic views of Indigenous peoples today as only tragic.[14] The idea is not to simply switch this idea for one in which Indigenous peoples are mono-lithically heroic. Rather, the goal is to get at the complexities: at describing the structure of settler colonialism and its violences while also showing how Indigenous peoples' lives operate both within and beyond that structure.

From "Inevitable" to Contested: Reframing the History of Territorial Hawai'i

This history of transinstitutionalization must be understood within the broader context of Hawai'i being annexed as a US territory in 1898. White settlers had made homes in Hawai'i since American Protestant missionaries arrived in 1820, but for over seventy years, they lived under the Indigenous-led government of the Hawaiian Kingdom. A powerful cadre of white settlers overthrew the Hawaiian Kingdom in 1893, seizing power away from Queen Lili'uokalani who had pushed to shore up the rights of Native Hawaiians in the face of the increasing might of American and British sugar plantation owners. The US federal government did not immediately approve of the Hawaiian Kingdom's overthrow. Under President Grover Cleveland, the US sent a representative to investigate the overthrow, who ultimately declared it an illegal "act of war." Cleveland nevertheless left any remedial actions up to Congress, which did nothing to restore the Hawaiian Kingdom's sovereignty, despite the substantial lobbying efforts of Queen Lili'uokalani and others. The subsequent US president, William McKinley, promoted US expansionism, and allowed Hawai'i to be annexed under the Newlands Resolution in 1898, which justified the need for annexation in order to secure Hawai'i as a refueling station for US naval ships on their way to fight in the Philippines.

What did it mean for Hawai'i to become a US territory? Popular under-standings of Hawai'i often assume the inclusion of Hawai'i within the United States as a predestined fate, written along a teleology that ends in US state-hood. Conventional historiography has similarly often perpetuated simplistic

---

[13] Damon Salesa, "The Pacific in Indigenous Time," in *Pacific Histories: Ocean, Land, People*, eds. David Armitage and Alison Bashford (London: Bloomsbury, 2013), 40.
[14] Eve Tuck, "Suspending Damage: A Letter to Communities," *Harvard Educational Review* 79, no. 3 (Fall 2009): 409–428.

ideas about Territorial Hawaiian history, such as the notion that Native Hawaiians welcomed becoming a US territory and that there was no organized resistance to the annexation of Hawai'i as a territory by the United States. Noenoe Silva and other Hawaiian scholars have roundly disproved such assumptions, documenting widespread, organized efforts that were in fact successful in delaying annexation.[15]

Ronald Williams Jr. has also importantly complicated our understandings of the transition between the Republic of Hawai'i (the government formed by those who overthrew the Hawaiian Kingdom in 1893) and the Territory of Hawai'i, which officially formed a government in 1900.[16] Many of the same white settlers who held power in the Republic of Hawai'i maintained their power in the Territory of Hawai'i (Sanford B. Dole, for example, the president of the Republic of Hawai'i became the first Territorial governor). However, Williams Jr. argues that the formation of a Territorial legislature posed a significant threat to the white oligarchy. In the Republic of Hawai'i, only white men and certain Native Hawaiian men of wealth were allowed to vote. Though Dole and others fought against it, the incorporating documents of the Territorial government re-enfranchised many Native Hawaiians and other non-white residents. When the first Hawai'i Territorial legislature convened in 1901, the independent Home Rule party won majorities in both the Territorial Senate and House of Representatives.

As Williams Jr. points out, these Native Hawaiian, Home Rule representatives proceeded to introduce policies that would restore rights to the people. These efforts were blocked by the Territorial governor's veto and other obstructionist measures, while the press propagated explicitly racist depictions of the Home Rule party, whom they called the "Heathen Party," as monkeys.[17] Conventional historiography of the Territorial period, even when critical of the racism and inequality of the Territorial leaders, downplays the efforts of the Home Rule party as "frivolous."[18] Similarly, while well-known histories like Gavan Daws' *Shoal of Time* (1968), often remark on the injustices of Hawaiian history, Daws and others tend to paint Native Hawaiians' position as a sad, lost cause and the rise to power of the so-called Big Five conglomeration

---

[15] Noenoe K. Silva, *Aloha Betrayed: Native Hawaiian Resistance to American Colonialism* (Durham, NC: Duke University Press, 2004); Jon Kamakawiwo'ole Osorio, *Dismembering Lāhui: A History of the Hawaiian Nation to 1887* (Honolulu: University of Hawai'i Press, 2002).

[16] Ronald Jr. Williams, "Race, Power, and the Dilemma of Democracy: Hawai'i's First Territorial Legislature, 1901," *Hawaiian Journal of History* 49, no. 1 (2015): 1–45, https://doi.org/10.1353/hjh.2015.0017.

[17] Ibid., 22, 28–29.

[18] Gavan Daws, *Shoal of Time: A History of the Hawaiian Islands* (Honolulu: University of Hawai'i Press, 1974), 295.

of sugar companies as inevitable.[19] Most Territorial histories are labor histories that focus on the successive waves of immigrant labor from China, Japan,
Korea, the Philippines, and more.[20] Native Hawaiians are rarely mentioned in
such labor histories, even though they did work on plantations during the
Territorial period. Here again, the history of the Territory's training schools
offers a different picture of this time. The Territorial training school for boys
(largely populated by Native Hawaiians) lent out male youth as plantation
labor. The management of the Boy's School saw a plantation job as a fitting
and proper job for its graduates.

In this context, it is important to move away from understanding the
Territorial period as a totally smooth transition from the Republic of
Hawai'i or as a relatively benign period where Native Hawaiians faded into
the background. The Territorial government's attempts at assimilating
Native Hawaiians and other non-white residents of Hawai'i into white
American culture during the Territorial period were far from guaranteed
to succeed. As Williams Jr. also points out, the white settler elite of the
Territory were an incredibly small minority: only about 6 percent of the
population.[21] Further, Dean Saranillio has pointed out how much various
and diversely constituted labor movements of plantation and dockworkers
in Hawai'i during the Territorial period threatened "white settler hegemony" to such an extent that "white supremacy was no longer capable of
governing a heterogenous nonwhite population, and a liberal multicultural
state began to emerge."[22] The government training schools thus can be
understood as one concerted effort on the part of a threatened white settler
elite to eliminate Hawaiian culture and train Native Hawaiian and other
non-white children of the Territory to understand themselves as part of a
multiracial and multicultural Hawai'i nonetheless structured by white,
heteropatriarchal norms.

These institutions can also be understood more broadly as a kind of
violence that worked to keep Native Hawaiians from more explicit resistance
to American colonization. Gender Studies scholar Laura Briggs, for example,
has argued that "child taking" is a time-honored "counterinsurgency tactic"
with a deep history in the Americas. "Child taking . . ." she writes, "has been
used to respond to demands for rights, refuge, and respect by communities
of color and impoverished communities, an effort to induce hopelessness,

[19] Ibid., 312.
[20] Ronald T. Takaki, *Pau Hana: Plantation Life and Labor in Hawaii, 1835–1920* (Honolulu: University of Hawai'i Press, 1983).
[21] Williams, "Race, Power, and the Dilemma of Democracy," 8.
[22] Dean Itsuji Saranillio, *Unsustainable Empire: Alternative Histories of Hawai'i Statehood* (Durham, NC: Duke University Press, 2018), 14–15.

despair, grief, and shame."[23] I agree that, as with the contexts of Native American and First Nations children removed by the US and Canadian settler states to boarding or residential schools in the nineteenth and twentieth centuries, the Hawai'i Territorial institutions that incarcerated Native Hawaiian and Asian children operated as a way to discipline and terrorize not only those children but their immediate and extended families.[24] So too, these practices were not limited to the United States and Canada. Similar institutions existed in other settler colonial contexts such as Australia.[25]

Such schools also existed across Latin America, as other scholars in this book show. Often Latin America is seen as outside the scope of settler colonial studies, but many scholars have shown that it can be productive to make connections between the colonialisms of Latin America and the rest of the world. For example, boarding schools for Indigenous peoples operated in Mexico from 1926, when the *Casa del Estudiante Indígena* opened in Mexico City. Soon a larger system of such schools, called *internados indígenas*, was established, with institutions across the country. Alexander Dawson argues that like the boarding schools to the north, the ones in Mexico sought to "produce individual subjects who spoke the national lingua franca, dressed and acted in ways that were deemed modern, and contributed to a dynamic national economy as workers or farmers."[26] Nonetheless, the schools in Mexico were also significantly different, Dawson argues, in the sense that they attempted to prove that Indigenous peoples in rural areas could contribute to national development. In doing so, they tended to "cultivate rather than break down ethnic affiliations."[27] Tanalís Padilla further argues that teacher training schools for the rural population of Mexico actually encouraged a sense of social justice in rural schools, based on a "socialist education principle that teachers be advocates for worker and campesino interests."[28] So too, "many found the rural normal [school] to be the path by which they escaped poverty."[29] While at times, then, education of Indigenous

---

[23] Laura Briggs, *Taking Children: A History of American Terror* (Oakland: University of California Press, 2020), 12–13.

[24] Brenda J. Child, *Boarding School Seasons: American Indian Families, 1900–1940* (Lincoln: University of Nebraska Press, 1998).

[25] Margaret D. Jacobs, *White Mother to a Dark Race: Settler Colonialism, Maternalism, and the Removal of Indigenous Children in the American West and Australia, 1880–1940* (Lincoln: University of Nebraska Press, 2009).

[26] Alexander S. Dawson, "Histories and Memories of the Indian Boarding Schools in Mexico, Canada, and the United States," *Latin American Perspectives* 39, no. 5 (2012): 83.

[27] Ibid., 84.

[28] Tanalís Padilla, "Memories of Justice: Rural Normales and the Cardenista Legacy," *Mexican Studies/Estudios Mexicanos* 32, no. 1 (February 1, 2016): 113, https://doi.org/10.1525/mex.2016.32.1.111.

[29] Ibid., 114.

peoples in Mexico opened up much broader possibilities than was the case with Native American boarding schools, Dawson argues that after 1940, the efforts toward a pluralistic inclusion of Indigenous cultures and communities within the Mexican nation were largely transformed. As he notes: "Indigenismo was ... divested of its mandate to undertake revolutionary and egalitarian social reforms, and instead became a tool for promoting capitalist accumulation and social control."[30]

The schools in Hawai'i never embraced a revolutionary class consciousness, even temporarily. But the largely white leaders and policymakers who established the schools did, undeniably, see themselves as progressive. The Territory framed the process of forced assimilation and constant threat of child removal as a form of care and rehabilitation. Social science played a key role in undergirding the Territory's assertion that institutionalization was a modern, progressive, and even caring process for Hawai'i's children. Psychologist Stanley Porteus was at the center of producing such social scientific knowledge. A founding member of the Psychological Clinic at the University of Hawai'i in 1922, Porteus is perhaps most well-known outside of Hawai'i for the creation of the Porteus Maze Test, a nonverbal, spatial intelligence test designed to measure mental capacity for planning, as I will discuss further below. In Hawai'i, he was involved not only with the development of psychology as an academic discipline at the University of Hawai'i, but was also frequently consulted by the Territorial government, particularly in regards to the training schools, Waimano Home, and other public institutions run by the Territorial government.

In justifying why civilized society required such expensive institutions, in a policy recommendation report to the Territorial government, Porteus painted Indigenous societies as necessarily less compassionate. "There are no homes for dependents in primitive life," wrote Porteus in a report published in 1949. "Also, the native who goes really crazy does not last long. There are no hospitals for the insane, nor for anyone else. Things are different in civilization."[31] To Porteus and many white Territorial leaders, this was the necessary burden of civilizing the relatively new US territory of Hawai'i: the high cost of the responsibility of the state to care for and discipline a seemingly ever-increasing number of "dependents," "delinquents," criminals, and disabled people. The cost of the Territorial institutions was a long-standing and ongoing concern, often remarked upon in the annual reports of the schools, along with suggestions about how to lower costs. Porteus's report was in part a defense of the need to continue funding these institutions. To Porteus, this

---

[30] Alexander S. Dawson, *Indian and Nation in Revolutionary Mexico* (Tucson: University of Arizona Press, 2004), 158.
[31] Porteus, *The Institutions of the Territory of Hawaii and Their Policies, Plans and Needs for Sound Institutional Practices*, 9.

expensive responsibility was a sign of "civilization" both having higher standards of its citizens and being more compassionate toward them. Porteus portrayed Indigenous peoples as brutal out of necessity, noting, for example, that "among the Australian aborigines, roaming over huge stretches of almost foodless country, there is no room either for the weakling or the rebel."[32] Thus, Porteus reasoned, "the higher the standards of living – and by this, I do not mean more food and more conveniences but also higher standards of work and effort – then the greater number of people who cannot attain to them."[33] Notably, Porteus believed that "civilization" also required greater intelligence, and that many Indigenous peoples would never be able to attain that intelligence.

The reference to Aboriginal Australians stemmed from Porteus being born in Australia, of "Irish-Scottish stock."[34] This is notable for the purposes of this essay in the sense that Porteus was familiar with institutionalization and settler colonialism in the different but resonant settler colonial contexts of Australia, the continental United States, and Hawai'i. He started his career as a teacher in 1913 at the Bell Street Special School "for the feebleminded" in Fitzroy, outside of Melbourne, Victoria.[35] It was at Bell Street School that Porteus began developing his own intelligence test, which would be known as the Porteus Maze Test. In 1918, he moved to the United States, where he took up a position as director of Research at the Vineland Training School, an institution also designed for the "feebleminded," in New Jersey. Vineland was already a well-known institution for its "research program on mental retardation, clinical psychological testing, and child development" under psychologist Henry Goddard who was director of Research at the school from 1906 to 1918, and whom Porteus replaced.[36] At Vineland, Porteus continued to

---

[32] Ibid., 2.

[33] Ibid., 3.

[34] Elizabeth Dole Porteus, *Let Us Go Exploring: The Life of Stanley D. Porteus, Hawaii's Pioneer Psychologist* (Honolulu: Ku Pa'a, 1991).

[35] Bell Street School was also sometimes referred to as Fitzroy School for the Feeble-Minded. It is likely that the students of this school were white and working-class, not Aboriginal. While he appears to have completed some academic coursework at the University of Melbourne, he never received a diploma. As noted in the vanity biography of Porteus written by his daughter-in-law Elizabeth Dole Porteus, without any training beyond his own interest and study, Porteus gradually moved from the role of a teacher to one that focused more on administering intelligence tests to children. His daughter-in-law describes the role he took up at Bell Street School in grand terms: he became "a practicing clinical psychologist" (though again, he had no training or credentials), she argues, "the first and only one in Australia."

[36] Sandra Moss, "Vineland Training School," in *Encyclopedia of New Jersey*, eds. Maxine N. Lurie and Marc Mappen (New Brunswick, NJ: Rutgers University Press, 2004), https://login.ezproxy.lib.utah.edu/login?url=https://search.credoreference.com/content/entry/rutgersnj/vineland_training_school/0?institutionId=6487.

develop and tout his Maze Test and did further research on "cranial capacity and intelligence."[37]

Thus, the Hawai'i Territorial government took Porteus as an expert because he had a long history of being understood as an expert at institutions for the "feebleminded," despite no formal education or training in psychology. Dean Saranillio puts it bluntly: "There is no shortage of evidence that he was making things up as he went."[38] Certainly he came to Hawai'i steeped in settler forms of knowing and "care" from other settler colonial contexts. By framing the Territorial government and its institutions as compassionate to "dependents," Porteus ignored how "civilization" in fact created "dependents" in Hawai'i in large part through the settler colonial dispossession of Native Hawaiians from land and ways of life that had sustained them for centuries before white settlement. The common Native Hawaiian adoptive practice of hānai, in which children were raised by family or friends who were not the biological parents, was especially targeted as a cause of needy and delinquent children. White reformers argued that hānai caused confusion and anxiety in children, though there was little evidence of this.[39] In fact, many of the children who wound up in Territorial institutions had not participated in hānai practices, but arguably could have benefited from them, given that they were often orphaned or had run away from untenable home situations. Yet, like many other practices of Native Hawaiian culture, hānai was discouraged in the Territory and white leaders sought to replace extended families with nuclear families, and communities' own forms of care with modern institutions run by charities or the Territorial government, where "dependents" would be subject to additional training and "rehabilitation" to conform to white US culture. The "cause" of delinquency and dependency was often attributed to children's families, whose cultures were pathologized as atavistic and incompatible with modern life.[40]

As Susan Burch writes in the context of the history of Native Americans incarcerated at the Canton Asylum for "Insane Indians" in South Dakota between 1902 and 1934, "the violence of Canton Asylum was *collective* as well as *individual*" and did (and continues to do) much to break a "fundamental tenet" of many Indigenous communities – that of "being a good relative."[41] The Hawai'i State Archives records on the training schools, which form the foundation of my research on this topic, are filled with official reports and

[37] Porteus, *Let Us Go Exploring*.
[38] Saranillio, *Unsustainable Empire*, 79.
[39] Janine Richardson, "Keiki o Ka 'Aina: Institutional Care for Hawaii's Dependent Children," PhD dissertation, University of Hawai'i, 2005, 61, 76.
[40] Patricia McMahon Wallace, "A Study of the Role of the Family in the Lives of Some Honolulu Girl Delinquents," Thesis for the Degree of Master of Arts, no. 179, University of Hawai'i, Honolulu, 1940.
[41] Burch, *Committed*, 111, 60.

correspondence, but also with letters from parents and other kin asking for their children back. When examined from this angle, the "child taking" of the Territorial-era institutions, and their afterlives, perhaps helps to explain why so many Native Hawaiians in the Territorial period and beyond felt compelled to make sure their children spoke only good English (rather than Hawaiian or pidgin), and to think of themselves as American rather than Hawaiian.[42] Though these sentiments have often been presented as robust consent to Hawai'i's annexation by the United States and eventual statehood, I suggest that they are also evidence of families desperate to keep their children, and terrified of losing them to indeterminate sentences in a government institution. In the following sections on Waiale'e, Kawailoa, and Waimano, I attempt to read parts of each institution's archive of largely official reports, social scientific studies, and other ephemera, against the grain to reveal the intertwined threads of the settler colonial process of replacing Hawaiian forms of care and kinship with institutional, "scientifically proven" ones, and the ways that critical histories of these institutions offer a more complex and contested history of Territorial Hawai'i than many recognize.

The dominant discourses at each institution that each of the following sections examines – "delinquency" at Waiale'e, "immorality" at Kawailoa, and "feeblemindedness" at Waimano – were deeply intertwined, and appeared at each institution. Yet social scientific knowledge claimed to be able to precisely identify, isolate, and/or (in some cases) rehabilitate these different categories as distinct pathologies that these different, specialized institutions would help care for or cure. Such knowledge, which in turn structured and justified the Territory of Hawai'i's policies of transinstitutionalization, was fundamentally grounded in settler ideas of race, Indigeneity, gender, sexuality, labor, and ableism. The discovery of "feeblemindedness" at Waiale'e or Kawailoa could be the cause for transfer to Waimano, but as I discuss more below, the discourse of "feeblemindedness" also undergirded the particular fears white settlers had about "delinquent" boys and "immoral girls." In the archives of these institutions, however, the self-assured ability of settler social science and Territorial public policy to effectively care for and "cure" children of delinquency, immorality, and feeblemindedness is shown to be, in practice, deeply contested and constantly challenged by the children incarcerated at these institutions, and their families, as well as the institution's own adminis-trations and governing boards.

---

[42] This references my personal experience of hearing stories of how my mother and her generation of Native Hawaiian people grew up before and shortly after Hawai'i became a state in 1959. For more on the tensions around statehood, see Saranillio, *Unsustainable Empire*.

"Delinquency" at Waiale'e Industrial School for Boys

Indeed, though the training schools were designed to "cure" delinquency and other bad behavior, many were concerned that the Waiale'e Industrial School for Boys only further trained boys in crime. "We call this place a 'training school.' It is a school and it does provide training but one subject high on the curriculum is vice," wrote Porteus of the Waiale'e Industrial School for Boys in his 1949 report.[43] Other reports frequently remark on the number of boys incarcerated at Waiale'e who went on to serve sentences at O'ahu Prison.[44] Located on 733 acres on the remote North Shore of O'ahu, distant from Honolulu, Waiale'e sought to definitively remove youth from what officials understood to be the ill-effects of urban living. Boys at the school trained and worked largely in agriculture, both producing food to sustain the community of the school and on sugar and pineapple plantations located nearby. Yet, from its earliest days since its opening in 1902, Waiale'e's isolation, along with the school being routinely short-staffed (in part because it was difficult to recruit teachers to live so far from town at Waiale'e), was also a weak point. Boys frequently ran away, and could be difficult to recapture.

The boys incarcerated at Waiale'e were commonly committed because of charges of "delinquency." A "delinquent" was defined by Territorial law as "Any minor who violates any law of the Territory or any county or city and county ordinance or who is incorrigible, vicious or immoral, or who is growing up in idleness or crime, or who is an habitual truant from school, or who habitually wanders about the streets in public places during school hours without lawful occupation or employment."[45] Note that such a definition was particularly concerned with those who did not stay in the roles allotted to them – whether it was being truant from school or being in a public space without a job. Such a broad definition of a delinquent left enormous subjective judgment to law enforcement and courts. Notably, the majority of the young men sent to Waiale'e were Native Hawaiian.[46]

The school largely blamed the "delinquency" of the boys on the failures of their families. "In the main, these charges have come from homes which have

---

[43] Porteus, *The Institutions of the Territory of Hawaii and Their Policies, Plans and Needs for Sound Institutional Practices,* 39.

[44] "[T]oday of the five-hundred and eight inmates of Oahu Prison one-hundred and thirty-eight have been at Waialee School. A sad record!" Quote from "Biennial Report of the Board for the Waialee Training School for Boys and Kawailoa Training School for Girls 1938–39," Hawai'i State Archives.

[45] Walton McWilliams Gordon, "Some Educational Implications of Juvenile Delinquency," MA thesis, University of Hawai'i, Honolulu, 1934, 3–4.

[46] Official reports on Waiale'e submitted to the Board of Industrial Schools and later the Department of Institutions, all accessed at the Hawai'i State Archives, attest to this fact year after year.

failed in the proper training of their children," noted a 1921 report.[47] This was perhaps the key difference in how Territorial officials understood children, still understood as innocent to some extent, as different from adults – that their bad behavior could be attributed to their families or a "poor home," instead of being seen as only an individual character flaw. In the view of administrators of the training schools, the sentencing of a "wayward" boy at Waialeʻe from a "bad" family was not punitive to the boy, for whom it was a kind of rescue, though in effect, it punished the family (by taking away their child). "If his home is bad or his environment debasing, he should not be punished, he should be replanted," reads a memo to the Board of Industrial Schools circa 1930. The memo further defines the job of an industrial school as "to reduce the population of the schools by constant (and of course careful) transplanting of the bulk of its entrants back into normal social environments." The "transplanting" would not return boys to their "poor homes" but to parole jobs and "home placements," after "the desirable disciplining of the mind and body."[48] Such comments were at least partially concerned with the cost of running the training schools, and submitted along with arguments for how the schools could be maintained at lower cost. However, in practice, many children were still kept at Waialeʻe and Kawailoa for indeterminate sentences.

The other reason often remarked upon to explain the cause of the "delinquency" of the boys incarcerated at Waialeʻe was their "low mentality." In a 1934 report, the superintendent of Waialeʻe notes the need for more "psychological" testing and the transfer of "morons and feeble-minded" boys to Waimano Home.[49] A major section of Porteus's comments in his 1949 report on the boys at Waialeʻe similarly focused on their intelligence. Undoubtedly, Porteus saw what he called the "Hawaiian and Oriental" boys (the vast majority of the Waialeʻe population) as mentally inferior to "Caucasians," a finding consistent with his observations in his 1926 book, coauthored with Marjorie Babcock, *Temperament and Race*, where he argued that Hawaiʻi's non-white population (including the Portuguese) averaged 73.3 percent of the intelligence of white people, with Filipinos and Hawaiians having the most "racial defects."[50] Thus, he noted that spending much money on attempting to seriously educate such populations would be akin to "helping lame dogs over

---

[47] Ernest Bryant Hoag, *Survey of the Boys' Industrial School, Hawaii* (Honolulu: Printed by Honolulu Star-Bulletin, 1921), 3.

[48] "Memorandum for the Members of the Board of Industrial Schools." Undated [circa 1930s]. Accessed Hawaiʻi State Archives.

[49] "Biennial Report of the Board of Industrial Schools of the Territory of Hawaii, For the eighteen month period beginning July 1st 1933 and ending December 31st 1934." Accessed Hawaiʻi State Archives.

[50] Stanley David Porteus and Marjorie Elizabeth Babcock, *Temperament and Race* (Boston: R. G. Badger, 1926), 110.

stiles, and when they are over they are still lame."[51] Notably, this book was based on research produced for the Hawaiian Sugar Planters' Association after the 1924 Sugar Strike on Kaua'i that resulted in the deaths of sixteen Filipino workers, known as the Hanapēpē Massacre.[52] As Dean Saranillio notes, such research was "supposed to help white settler leaders better understand and control Hawai'i's nonwhite population."[53] Because of their inferior intelligence, Porteus argued more specifically in his 1949 report on the Territory's institutions, that the boys should receive basic instruction in the three R's (reading, writing, and arithmetic). Porteus also put little stock in industrial training, however, noting that "practically speaking, they are as much outside the boys' interests and abilities as ordinary school subjects."[54]

The one thing Porteus thought the boys were fit for was "outdoor work such as gardening," which justified his recommendation that the older students at Waiale'e should be transferred to a forestry camp. Yet, he also makes a long digression in the report disagreeing with a previous report's characterization of "nearly forty percent" of the Waiale'e population as being of "very low mental classifications." Instead, he cites his own Porteus Maze Test findings on the Waiale'e population to note "an unsatisfactory average" in "practical intelligence" but distinguishing much of the population as merely "delinquent" rather than "feebleminded."

Such classifications all had particular histories and nuances specific to the practices of early to mid twentieth-century intelligence testing, psychology, and psychiatry in the United States. Porteus's Maze Test was one intelligence test among many; the most widely used at the time being the Stanford-Binet. The Porteus Maze Test purported to uniquely measure what Porteus called "planfulness," or planning capacity and foresight. It consisted of a graduated series of printed mazes (a total of eleven designs for the series for children aged three to twelve and a similar advanced version for those aged thirteen and up), which subjects would take by tracing a route through a maze printed on the exam sheet. The test was considered an "unsuccessful trial" as soon as the subject drew their line into a blind alley, and the test was considered failed after two trials for each test. Like other intelligence tests, the test scores resulted in a "mental age," which was compared to the subject's actual age to classify the subject's intelligence. Porteus argued that his test provided an important supplement to the Stanford-Binet, allowing for a more accurate measure of a subject's "social fitness" as compared to educational ability. Tests that gauged educational ability, Porteus wrote in touting his own test, "do not

---

[51] Ibid., 112.
[52] Saranillio, *Unsustainable Empire*, 77.
[53] Ibid.
[54] Porteus, *The Institutions of the Territory of Hawaii and Their Policies, Plans and Needs for Sound Institutional Practices.*

detect the mentally unstable, who form a large proportion of the socially unfit."[55]

In short, then, the difference to Porteus between the "delinquent" and the "feebleminded" was that the delinquent was capable of some industry, though they lacked the ability or desire to fit in with society, as opposed to the feebleminded who was intellectually unable to participate in society. In this distinction, Porteus seems to both promote his own intelligence test as more accurate and reserve the Waialeʻe population as proper subjects for agricultural labor, rather than subjects fit for the Waimano Home for the Feeble-Minded. As I will discuss further in the following sections, these distinctions Porteus made about the male youth of Waialeʻe fit with broader trends across the United States that understood female "feebleminded" people as greater threats to the national "gene pool" and thus more deserving of indefinite institutionalization and/or sterilization.[56] Thus, alongside his explicitly racist view of Native Hawaiian and Asian immigrants in Hawaiʻi as mentally inferior to white people, Porteus held out a progressive hope for rehabilitating these "delinquent" but not really "feebleminded" boys into a productive labor force for the Territory. I would draw attention to how this racist sentiment supplements Porteus's injunction to the Territorial government that "we" are responsible for rehabilitating such populations into proper adjustment to civilized society. In other words, the racism of seeing Native Hawaiians and Asian immigrants as inherently less than white people informed Porteus's sense of social duty to these groups. In this way, racist views were not incompatible with reformers' sense that they were doing charitable and civilized work; rather, this sense was dependent on scientifically argued racism that upheld the anti-Indigenous and anti-immigrant sentiments that the ideological structure of settler colonialism in Hawaiʻi operated through.

At times, the families of children sent to the training schools were also subjected to intelligence testing. The results of their tests would be included in the files kept by social service agencies and used as rationale for removing children from their families. For example, as noted in a Master's thesis on "Truancy in the Schools of Honolulu," by Esther Roberts Holmer in 1935, multiple case studies note the "mental age" of "truant" boys' mothers. Mrs. Salazar, for instance, the mother of Will, a "Portuguese-Hawaiian boy of 12 who has been truant . . ." is described as having a "Binet mental age of 10 years, I.Q. 72; Porteus mental age of 9 years – good rote memory, though

[55] Stanley David Porteus, Porteus Tests – the Vineland Revision (Vineland, NJ: Vineland Training School, Department of Research, 1919), 17.

[56] Molly Ladd-Taylor, Fixing the Poor: Eugenic Sterilization and Child Welfare in the Twentieth Century (Baltimore, MD: Johns Hopkins University Press, 2017); Michael A. Rembis, Defining Deviance: Sex, Science, and Delinquent Girls, 1890–1960 (Urbana: University of Illinois Press, 2011).

retarded ability – showed very little planning, is easily confused and disturbed . . ."[57] Holmer also notes that Will has a "low mental age, (10 years, at 12)" and that "Since his mother is definitely 'retarded' and at least two other children are 'slow,' it is reasonable to suppose that none of the family is very brilliant."[58] Overall, intelligence testing of Native Hawaiian parents provided the Territorial government and its agencies seemingly hard, scientific evidence that poor and working-class Native Hawaiian families were not deserving of having children and that the Territory should intervene.

Such rationales that pathologized Native Hawaiian families and perpetuated an inherently racist notion that Native Hawaiians had lower "mental ages" and intelligence than white people could always be used as an implicit or explicit explanation as to why Waiale'e did not often succeed at reforming "delinquent" boys. Though Waiale'e incarcerated youth who were Native Hawaiian as well as Filipino, Japanese, Chinese, Puerto Rican, and other members of immigrant communities in Hawai'i, Native Hawaiians were often singled out as especially irredeemable. A 1938 Master's thesis in Education from the University of Hawai'i did a "follow-up study" of fifty men who had been enrolled at Waiale'e. The author, Elizabeth Miller, determined that thirty men were "well-adjusted" to society after their time at Waiale'e, and twenty of the men were "poorly-adjusted."[59] Miller pointedly comments that "fourteen of the poorly adjusted men were Hawaiians or Part-Hawaiians."[60] Additionally, nine of these twenty men were in prison, three on parole and one in the Territorial Hospital. Miller made her judgments about social adjustment based on a survey that tallied a range of categories including profession, wages, quality of their neighborhood and home's furniture, neatness of their children, and if they drank alcohol or not. Overall, she concluded that her study demonstrated the difficulties of readjustment for Waiale'e graduates and argued for "the sympathetic understanding and help of the more fortunate members of the community."[61]

Thus, Miller, like Porteus, saw herself as merely relaying the facts about Native Hawaiian men's tendency to be more poorly adjusted than other men previously enrolled at Waiale'e, though of course her survey judgments were subjective. Without investigating more into why Native Hawaiian men might be more poorly adjusted than others, she leaves readers with the sense that they are just inherently more prone to criminality. Yet, she argues that her

---

[57] Esther Roberts Holmer, "Truancy in the Schools of Honolulu," MA thesis, University of Hawai'i, Honolulu, 1935, 176.

[58] Ibid., 181.

[59] Elizabeth Ruley Miller, "A Follow-up Study of Fifty Former Waialee Training School Boys," MA thesis, University of Hawai'i, 1938.

[60] Ibid., 62.

[61] Ibid., 65.

finding is simply evidence of the need for more charity toward Native Hawaiian and other men of color who are paroled from Waialeʻe, thereby distancing herself from the possibility that her findings are racist. This rhetorical move, conscious or not, on the part of Porteus and Miller, is a powerful one in the context of science and the structure of settler colonialism. It allows the damage of settler colonialism (Native Hawaiians's dispossession of land, culture, and nation) to be erased in the discourse of white settler leaders and scientists "helping" Native Hawaiians, who, in the evidence of settler science, are inherently backward, inferior, and poorly adjusted. The operation of this move does not depend on whether Miller or Porteus are individually racist or personally supportive of all the aims of the Territorial government in regard to Native Hawaiians. Rather, such rhetoric naturalized the structure of settler colonialism in Hawaiʻi, by both painting the Territorial government's institutions as solely charitable and portraying Native Hawaiians, Filipinos, and others as mentally inferior to white settlers.

"Immorality" at Kawailoa Training School for Girls

While Waialeʻe was often considered a failure even by Territorial officials and consultants like Porteus, the Kawailoa Training School for Girls (also called the Maunawili School for Girls or the Girls' Industrial School) was commonly seen as more successful and modern. Before opening the campus in a rural area on the windward side of Oʻahu near the towns of Kailua and Waimānalo in 1929, a previous girls' school was located in Mōʻiliʻili in urban Honolulu. The school was moved due to space constraints and the idea that the girls would benefit from a rural environment, instead of an urban campus where, as noted a 1925 report by the school's superintendent, "men of questionable character are continually getting in touch with our girls."[62]

Instead of reforming the petty crimes of truancy or delinquency focused on at the Boys' school, the crimes and methods of rehabilitation at the Girls' school were focused on perceived sexual immorality. Girls were committed to Kawailoa in large part for perceived "sex offenses" or "sex delinquencies." As with Waialeʻe, the blame for this "immorality" was the child's family and, by extension, Native Hawaiian culture more broadly. "The majority of the girls are committed because of sex delinquencies, are part Hawaiian and are from families known to many social agencies," notes a 1938 report.[63] As with Dorothy Wu whose story this essay began with, a "sex delinquency" did not need to involve actual proven sexual relations but merely the appearance of

---

[62] "Reports of the Board of Industrial Schools of the Territory of Hawaii," *Hawaii Gazette*, 1931.

[63] "Report of Kawailoa Training School for Girls," January 1938. Accessed Hawaiʻi State Archives.

improper "association" with boys or men. The primary end goal of rehabilitation at Kawailoa was to train girls to be more moral and equipped to perform domestic labor so they could work as maids in wealthy homes until they married and kept homes for themselves. As a 1938 report puts it, "The purpose of a correctional institution is twofold: <u>Vocational training</u> and <u>social readjustment</u>. In a girls' school, vocational training should emphasize <u>home making</u>, which is the probable and hoped-for ultimate profession of any normal woman."[64]

One acceptable reason for parole from Kawailoa, and to some Territorial officials the most desired outcome, was if a girl was to be married. Such marriages required the prior permission of the juvenile court and/or the Board of Industrial Schools.[65] We can think of this transfer, from Kawailoa to marriage, as a kind of transinstitutionalization that is different from but related to the transfers from the training schools to Oʻahu Prison or Waimano House. Marriage was a Territorial institution too, which sought to "rehabilitate" common Native Hawaiian forms of intimacy and kinship into a heteropatriarchal structure that aligned with white, settler colonial norms. Legal marriage was not uniformly valued by Native Hawaiians and "common-law" relationships, which could be easily dissolved as desired by either party, were (and still are) widely accepted. So too, same-sex relationships and nonbinary gender identities and roles were traditionally accepted and valued. Concerns over sexual relationships between the young women incarcerated at Kawailoa are often commented on in the archival records. A 1938 report makes note of a recent policy change allowing "the presence of friends other than relatives on visiting days ... [so that] Boy friends may call on girls and correspond with older girls. This situation has practically eliminated homosexual practices which used to be so rampant at the school."[66] Still, this concern was very active in 1945, as noted in the handbook given to girls on their arrival at Kawailoa, which emphasizes as a rule that "Holding hands, kissing other girls, and walking with arms around each other are forbidden. 'Boy haircuts' are not permitted. Girls are expected to be friendly with all, but not too intimate with any of the other students."[67] Such a rule makes clear how potential lesbian or queer relationships among those incarcerated at Kawailoa were threatening because of how they went against heteronormative expectations but also

---

[64] Emphasis in original. "Kawailoa Biennial Report to Board of Industrial Schools, 1937–1938." Accessed Hawaiʻi State Archives.

[65] "Girls who marry with the full permission of the Board, shall be granted a RELEASE from the girls' school." Letter, March 2, 1934. Folder Kawailoa Training School, General Correspondence, 1934–35. Accessed Hawaiʻi State Archives.

[66] "Superintendent's Report: Kawailoa Training School for Girls, December 1938." Accessed Hawaiʻi State Archives.

[67] Nell Elder, ed., *Girl's Handbook for the Kawailoa Training School for Girls* (Department of Institutions, Territory of Hawaiʻi, 1945), 18. Accessed Hawaiʻi State Archives.

because pleasure and intimacy (sexual or not) could threaten to break down the discipline of the school.

Though a heterosexual marriage was presented as the model of success for the future of a girl incarcerated at Kawailoa, presumably one of the reasons why the sexual "immorality" of Native Hawaiian and Asian girls was of such concern at Kawailoa was a prevalent assumption by social scientists and the general public at this time that they would reproduce similarly immoral children who would also not be beholden to the white, settler colonial norms of Territorial Hawai'i. This concern attached to so-called "feebleminded" women in particular ways. As Michael Rembis puts it in his study of a girls' training school in Illinois, they "reproduced at a much higher rate than their 'normal-minded' counterparts, swelling the ranks of society's 'unfit'."[68] Such concerns nationally spurred many involuntary or coerced sterilization campaigns in many states in the early twentieth century.[69] However, this eugenic concern also justified indefinite segregation of feebleminded women, especially of child-bearing age, from society. In my research so far, these concerns are not explicitly stated in the records of Kawailoa or Waimano Home (similarly I have not yet seen records of sterilization), but this discourse about the danger of feebleminded women reproducing at high rates certainly helps explain why the "immorality" of Native Hawaiian young women would have been so important to check and rehabilitate.

The official staff manual of Kawailoa in 1945 describes "the building up of a desirable attitude toward work" as key to the rehabilitation of delinquent girls. "Most of our wards have never been taught that work is an important part of every satisfactory life, and so regard it as punishment," the manual notes.[70] The school placed a heavy emphasis on getting the girls accustomed to domestic and agricultural labor along with building skills such as sewing, and weaving lauhala mats, fans, and baskets. Many could be paroled into jobs as cooks or in laundries. In the manual given to those newly incarcerated at Kawailoa, the superintendent Pearl McCallum warned: "You will find that everyone works, and you will be expected to work. Do not try to shirk when you are given a job to do, but remember that the more you learn while you are here, the more successful you will be when you leave the school to make your own living."[71] Such training and systems emphasized the necessity of the girls fitting into the capitalist, heteropatriarchal society of the Territory.

---

[68] Rembis, *Defining Deviance*, 24.

[69] Alexandra Minna Stern, *Eugenic Nation: Faults and Frontiers of Better Breeding in Modern America*, 1st ed. (Berkeley: University of California Press, 2005); Ladd-Taylor, *Fixing the Poor*.

[70] Nell Elder, ed., *Staff Manual for the Kawailoa Training School for Girls* (Department of Institutions, Territory of Hawai'i, 1945). Accessed Hawai'i State Archives.

[71] Elder, *Girl's Handbook for the Kawailoa Training School for Girls*.

Though Kawailoa generally enjoyed a better reputation than Waialeʻe, Kawailoa nonetheless faced many of the same challenges. The archival records of Kawailoa attest to the many challenges the institution faced in reforming the children incarcerated in its cottages into "useful women." Archival materials note many incidents of runaways, strikes, and even occasionally violence against school teachers or staff. In February 1939, a report notes eighty-four instances of girls running away from Kawailoa within one month (this included many multiple attempts by the same girls; the total population of Kawailoa at this time was 133).[72] A 1938 report notes that in one cottage of the school, "The girls were noisy, unruly and quarrelsome. There were two runaways and many more had the notion in their heads. They broke window panes and yanked nails off from other windows. They used lipstick and rouge and smoked in the house. This last offense was by a group of six girls who continued their misbehavior by a so-called 'strike' in class."[73] Apparently the "strike" meant that the girls refused to go to class. Their punishment was "two mornings of isolation at home in holokus." A holoku is a missionary-styled Hawaiian dress, which despite its colonial foundation, likely would have signaled particular connections and affinities to Native Hawaiian culture, in contrast to contemporary American-styled clothing. That wearing a holoku was considered a punishment underscores the extent to which Native Hawaiian culture was pathologized and stigmatized at Kawailoa. In the reports, being made to wear a holoku and stay in isolation was presented as a kinder punishment than previous protocols enacted at Kawailoa of corporal punishment, locking girls in a "punish" cell, or shaving a girl's hair.

A later report from 1938 includes an attempt to explain the frequency of runaways from Kawailoa, amid apparent bad press. Superintendent Edith Field notes that she has:

> interviewed each runaway returned after I took office. The Board might be interested to know that they all run away for one of given reasons:
>
> 1. The excitement of the chase. (These are in the majority and most of these expressed themselves as glad they were caught.)
> 2. Quarrels among the girls.
> 3. Bad handling on the part of the staff.
> 4. Boredom from dull routine, and there is too much of that left.
> 5. The boy-friend they must see.[74]

---

[72] "Superintendent's Report: Kawailoa Training School for Girls, February 1939." Accessed Hawaiʻi State Archives.
[73] "Kawailoa Training School for Girls: Report for the Month of July, 1938." Accessed Hawaiʻi State Archives.
[74] Edith Field, "Supplementary Report, October 8, 1938." Accessed Hawaiʻi State Archives.

Field's interpretation of the reasons for those incarcerated at Kawailoa running away acknowledges minor forms of responsibility on the part of the institution: that there has been some "bad handling on the part of the staff" and that the school has a "dull routine." But largely, she sees running away as merely bad behavior or immorality – the wildness of the "excitement of the chase" or the loose nature of girls wanting to see a "boy-friend," underscoring the kind of perceived "sex offenses" that often got girls committed to Kawailoa in the first place.

However, Superintendent Field's account of the causes of running away contrasts with first-hand accounts by the "runaways" themselves. In a 1938 "Red Roof Weekly" newsletter produced by girls at Kawailoa, a recent runaway named Violet who was returned to the school is interviewed. Asked why she ran away, she answers first, "One reason because of being my baby's birthday. Another reason was that Mrs. Hoke said I had to behave for four or five weeks. So I was burnt up and took off." The interviewer, another girl at the school, asks if it was really so hard to behave for four or five weeks. Violet admits that it is not so hard, and when asked again why she ran away, she says, "Oh, for pleasure." This small glimpse into the life of a young woman at Kawailoa notes that her reasons for running away had primarily to do with returning to and reconnecting with her family and simply wanting to experience pleasure and live a free life. Violet is referred to in the newsletter as a girl, but actually she is also a mother. She has a baby that she is apparently prevented from seeing regularly; the note of it being her baby's birthday suggests that it may have been her baby's first birthday, traditionally an especially important milestone that was and still is marked with a large lū'au, or feast and party. From the archival record, there were many pregnant girls committed to the school. In some cases, it appears Kawailoa became the place the juvenile court would send any unmarried pregnant girl, despite objections from Kawailoa superintendents.[75] It is unclear from the records I have accessed so far exactly how babies of those incarcerated at Kawailoa were handled. It appears that sometimes babies were allowed to remain with girls at Kawailoa for at least a short time after birth, but were likely separated from their mothers and sent to extended families or other "home placements" within their first year of life. This is another example of how institutionalization broke families and non-settler forms of care; in this case, we see the intergenerational impact as Violet is separated not only from her own parents but also her own child.

We know that like at Waiale'e, intelligence tests and other social scientific studies were conducted at Kawailoa, and that transfers were regularly made to

---

[75] Letter to Superintendent of Kawailoa from DH Case, Judge, Circuit Court October 21 1934. Kawailoa Training School General Correspondence, 1934–35. Accessed Hawai'i State Archives.

Waimano Home.[76] Yet, from the record I have accessed so far, the emphasis in the institution's official reports is less on the low intelligence of those incarcerated at Kawailoa and more with using psychology to "treat problem cases." In fact, in at least one instance, high IQs are cited as a reason for running away: "It is unfortunate that we have so little to offer our brighter girls," a 1939 report remarks. "Among the 10 chronic runaways from Aloha are 6 girls with I.Q.'s of better than 100. One of these has an I.Q. of 123, had completed 3 years of high school and should be in college."[77] Instead of low intelligence, at least through the reports filed from the late 1930s, "over-emotionalism" or "emotional problems" are cited as the driving factor that explains the girls' immoral behavior. Accordingly, "mental hygiene" and other "psychological" care are posited as an appropriate treatment to help rehabilitate those at Kawailoa. An extended anecdote in a report by Superintendent Frances Brugger in 1938 describes the desired outcome of promoting "mental hygiene":

> What is happening to the girls is the most important. One girl who has run away 7 times and burned her mattress after her last return, asked to have a special talk with the superintendent. Quite tongue-tied she finally "got out" that she thought she was trying to run away from herself. She asked if she could have a note-book and pencil so that she could write her thoughts . . . Following is a passage copied from her note-book: "For after all I am not helping Angela Roberts but the one that is in Angela Roberts. The real me, the me that I lost and only now found it. I had to go into despair to find it. But it was worth it. I am trying to straighten the 'me' up."[78]

This account encourages viewing Kawailoa staff as kind and caring therapists, who just want to facilitate self-discovery and self-reflection in those they keep locked up. This account, however genuine, emphasizes the image of this girl as troubled and troublesome (a self-image she has internalized and desires to "straighten up") instead of the fact that this girl is being kept indefinitely, against her will in an institution. She was not running away from herself but from Kawailoa and the settler colonial, heteropatriarchal future that institution imagined for her.

---

[76] Porteus was well-acquainted with the institution; in January 1938, a report notes that Porteus "showed his movies of Africa and Australia one Saturday night." A routine description of the institution from this same report notes "the intelligence level [of those incarcerated at Kawailoa] is from I.Q. 58 to 130, though the majority are between 70 and 80 I.Q." "Report to the Board of Industrial Schools, January 1938." Accessed Hawai'i State Archives.

[77] "Superintendent's Report: Kawailoa Training School, March 1939." Accessed Hawai'i State Archives.

[78] "Report to the Board of Industrial Schools, January 1938." Accessed Hawai'i State Archives.

"Feeblemindedness" at Waimano Home

When children incarcerated at Waiale'e or Kawailoa were determined, often through the results of intelligence testing, to be "feebleminded," they could be transferred on to the Waimano Home for the Feeble-Minded, which opened in 1921. Waimano was situated on 612 acres of Territorial land overlooking Pearl City and Pearl Harbor, on the leeward side of O'ahu.[79] Waimano held both children and adults. Commitment required the ruling of a circuit court judge, but "any adult relative, guardian, or custodian of the individual sought to be committed, or any authorized social agency, or an agent of any governmental department or bureau, may file in court, an application for the commitment of the individual into Waimano Home."[80] A "certificate stating that individual is in need of institutionalization," was required by the court, "verified by a committee consisting of a psychiatrist, a clinical psychologist and a psychiatric social worker."[81] Before the opening of Waimano, so-called feebleminded children, or "juvenile defectives," as one study put it, would have been committed to the Territory's training schools. But as those schools became overcrowded and their superintendents argued that "feebleminded" wards made their gendered "rehabilitation" of "delinquent" boys and "immoral" girls more difficult, they advocated for a specialized institution, which would become Waimano. As with Waiale'e and Kawailoa, Native Hawaiians were overrepresented at Waimano.[82]

Waimano Home, throughout its history, was largely focused on "custodial care," though some administrations attempted more concerted efforts toward education and training than others. Tamotsu Ishida, writing in 1955 about Waimano, notes that the proximity of the institution to the military base at Pearl Harbor and the beginning of World War II guided some of these changes: "It was seen that the Home could play a valuable part in the war by the production of vegetables on the farm. Thus, the agricultural program was accelerated and at the same time, it was seen that much help could be given the defective in other fields of vocational training."[83] It is unclear how much actually changed in practice. Reports from the mid late 1940s continued to

---

[79] Francis Tamotsu Ishida, "A Comparison of the Activities of the Social Service Department at Waimano Home, Territory of Hawaii, with Present Day Activities in Social Service Departments in Institutions for the Mentally Defective," Thesis for the Degree of Master of Social Work, no. 25, University of Hawai'i at Manoa, Honolulu, 1955.

[80] Ibid., 36.

[81] Ibid.

[82] Ibid., 39; C. K. Szego, "The Sound of Rocks Aquiver?: Composing Racial Ambivalence in Territorial Hawai'i," *Journal of American Folklore* 123, no. 487 (2010): 58.

[83] Ishida, *A Comparison of the Activities of the Social Service Department at Waimano Home*, 31.

describe Waimano's primary function as "custodial care," a characteristic which distinguished Waimano from the Territorial Hospital for the Mentally Ill, which was more "equipped for making diagnoses and therapeutic treatment" of the "insane."[84]

Many of those incarcerated at Waimano were kept there for their entire lives. A 1942 report estimates 68 percent of the total population "will remain there for their entire lifetime" because they do not fall into the "higher class" of the "feebleminded" who are "taught many things and, when possible, are placed on parole."[85] This report further broke down how the "feebleminded" were categorized, largely according to intelligence test scores: as "moron, imbecile, idiot, and incorrigible. The last named are few in number, fortunately, but require constant vigilance since they are dangerous both to themselves and the attendants. A moron's I.Q. is 50–69 inclusive – the imbecile's, 20–49 inclusive and the idiot's, below 20."[86] Porteus and the staff of the Psychological Clinic appear to have been very involved with such testing and categorization at Waimano, as with the training schools.[87] Porteus, in the same 1949 report where he castigates Waialeʻe for training boys in "vice," notes that Waimano is "in a healthy state, with, at present, adequate accommodations except for obsolete buildings in the girls department." He does recommend better treatment plans and the creation of a new director of research (similar to the position he held at Vineland).[88]

Similar to, and inextricably intertwined with, official explanations of the causes of "delinquency" and "immorality" described by staff and administrators at Waialeʻe and Kawailoa, the cause of "feeblemindedness" was understood to rest with the child's family. Though "feeblemindedness" was generally understood as a genetic handicap which was hereditary, and thus not always the intentional fault of a family, there was also a sense that the family was irresponsible for having children to whom such traits would be passed on. This was shaped by the fears, mentioned above in relation to pregnancies at Kawailoa, about "feebleminded" women reproducing at greater numbers than average and dragging down the "racial stock" of the nation. In a 1942

---

[84] Robert G. Dodge, "Mentally Ill and Defectives," Report 1948, no. 1 (Legislative Reference Bureau, Honolulu, 1949), 6.

[85] Hawaii Department of Institutions, Zaida Nelson, *Unto the Least of These* ... (Honolulu, 1942), 10.

[86] Ibid.

[87] "As a result of the thorough work of the Psychological Clinic of the University of Hawaiʻi under Dr. Stanley D. Porteus, one of the outstanding men of the United States in his field, it is probable that no Institution in any state has more exact information on the mental capacity of its patients." Hawaii Department of Institutions, 10.

[88] Porteus, *The Institutions of the Territory of Hawaii and Their Policies, Plans and Needs for Sound Institutional Practices*, 102.

report on Waimano, the pathologization and presumed genetic nature of "feeblemindedness" as well as "delinquency," "immorality," and criminality more broadly is clear in this quote:

> One case was cited in which there were thirteen members in the family originally. There have been four deaths in the past few years. Four are at Waimano Home. Two are in Oahu Prison and both Kawailoa Training School for Girls and Waialee Training School for Boys have representatives in the same family. Does not this one example – and there are others comparable to it! – prove the necessity of indexing the feebleminded in the Territory and of institutionalizing the most urgent cases?[89]

This quote presents the institutionalization of many of members of the same family, throughout the carceral institutions of the Territory of Hawai'i as evidence of the supposed dangers of "feeblemindedness." Though a race is not mentioned in reference to this (and "others comparable"), for many reading this report, given the overrepresentation of Native Hawaiians at all of these institutions, it would likely have been a Native Hawaiian family they pictured.

This use of the discourse of the "feebleminded" as a danger to modern, settler society is a potent form of settler ableism, a term used by American Studies scholar Jessica Cowling to draw attention to the ways that settler colonialism is structured by ableist logics.[90] Ableism, as Susan Burch defines it, is "a system of power and privilege that hierarchically organizes people and societies based on particular cultural values of productivity, competitive achievement, efficiency, capacity and progress."[91] Settler ableism is evident in the ways that settler ideas of "normality, fitness, and competency" have been used to judge Indigenous peoples, and so often designate them in need of "rehabilitation," assimilation, or other forms of institutionalization and correction.[92] The assumption that Waimano Home would segregate and properly care for the "feebleminded" was a settler ableist one that took for granted that everyone should aspire to white, settler forms of intelligence (especially including white settler ideas of productivity, given how Waimano emphasized training and preparing those incarcerated for jobs or at least work in the gardens at Waimano). Settler ableism also structured the assumptions that dominated Waimano, Waiale'e and Kawailoa, that non-white forms of care (including Native Hawaiian families caring for their children) were not sufficient, modern, or appropriate, as with the references to the low intelligence of Hawaiian mothers whose sons were "delinquent" and therefore sent to Waiale'e.

---

[89] Hawaii Department of Institutions, *Unto the Least of These* . . ., 12.
[90] Burch, *Committed*, 9.
[91] Ibid.
[92] Ibid.

While I have not been able to find much in my research so far on how Native Hawaiians or Asians in Hawai'i may have understood different forms of intelligence, certainly categorizing and institutionalizing people based on their scores on Western-designed intelligence tests was not a traditional practice. Permanent institutionalization in other contexts was heartily resisted by Native Hawaiians; namely, in the case of leprosy or Hansen's disease during this same period, we know that many Native Hawaiian families fought the Territorial government taking their family members who had been diagnosed with Hansen's disease to the so-called leper colony of Kalaupapa on the island of Moloka'i.[93] Resistance or reluctance to commit children or adult family members to Waimano appears to have been a common enough reaction that a study on Waimano from 1955 notes that one of the main tasks of a caseworker at Waimano would involve explaining to parents that "institutionalization is usually long-term," and further that "... they must leave the patient in the institution until such time that the staff feels that the child is ready to return home, to go on a work placement, or to go to family care. It is highly probable too that in many instances the family must be helped to sever ties for the good of the patient."[94] This description of what it meant to institutionalize a child at Waimano Home showcases how cruel and carceral this practice could be. Not only would children not be released until "the staff feels that the child is ready" but also, "in many instances," Waimano would recommend completely cutting the child off from their family. Again, in many explicit and implicit ways, Waimano, like Waiale'e and Kawailoa, emphasized through settler ableism that Native Hawaiian and other non-white families were incompetent, and that their children were often better off without them.

Conclusion

The histories of Waiale'e, Kawailoa and Waimano have rarely been featured in official histories of Hawai'i, but many people in Hawai'i are familiar with these institutions because of family ties and stories lived and passed down to them. These institutions also live on in literal and haunting ways. Kawailoa is now the site of the Hawai'i Youth Correctional Facility, a state juvenile correctional facility. The ruins of some of the buildings at Waiale'e remain

---

[93] Adria L. Imada, "Lonely Together: Subaltern Family Albums and Kinship During Medical Incarceration," *Photography and Culture* 11, no. 3 (September 2, 2018): 297–321, https://doi.org/10.1080/17514517.2018.1465651; Kerri A. Inglis, *Ma'i Lepera: Disease and Displacement in Nineteenth-Century Hawai'i* (Honolulu: University of Hawai'i Press, 2013).

[94] Ishida, *A Comparison of the Activities of the Social Service Department at Waimano Home*, 10.

visible from the highway and are well-storied as a haunted site; officially, after Waialeʻe closed, the campus was used by the University of Hawaiʻi as a "livestock experiment station." Waimano is also storied as a haunted place, and has even been used as the setting for spooky TV shows; officially, the state continues to use some of the buildings as offices. By telling the histories of these institutions together, my hope is that we can move away from both uncritical histories of the Territorial period in Hawaiʻi and passing interest in these sites as spooky. This chapter has attempted to show how focusing on the ways these institutions deployed interlocking discourses of "delinquency," "immorality," and "feeblemindedness," offers a critical understanding of the ways that the Territory of Hawaiʻi instituted settler colonial hierarchies (of race, gender, sexuality, and ability, among others) by pathologizing Native Hawaiian and other non-white forms of care and family. That children and their families resisted institutionalization and attempted to continue to care for each other demonstrates not only their humanity, but a powerful if overlooked critique of settler colonialism.

   I hope this chapter also provides readers a different perspective on social science and its complicity with settler colonialism, with specific reference to Stanley Porteus and the Psychological Clinic and their work across these Territorial institutions, but also more broadly. In 1974, shortly after Porteus's death, the University of Hawaiʻi at Mānoa named one of its buildings "Porteus Hall." Ethnic Studies and allied students immediately protested, but these efforts were initially unsuccessful.[95] In 1998, the university's Board of Regents finally agreed to rename the building, though the change did not occur until 2001.[96] At the heart of the protests was Porteus's major published text, *Temperament and Race* (1926), which as discussed above, portrayed white people as the most intelligent race, while Filipinos and Native Hawaiians were the least intelligent. This history of the ethics of commemorating Porteus demonstrates that though the psychologist was a prominent leader and well-respected in Hawaiʻi by the white elite during his lifetime, his legacy has been significantly contested by Native Hawaiians and other students of color. However, because this fight over the name of Porteus Hall focused so much on the individual man, I would argue that less attention has been paid to the broader legacies of his and his colleagues' work in applying intelligence testing and encouraging the Territory to use the results of those tests to institutionalize or transinstitutionalize people who did not readily accede to white, settler norms of gender, sexuality, and ability. This chapter is one step toward beginning that broader work,

---

[95] "ISAR – Stanley Porteus Biography," www.ferris-pages.org/ISAR/bios/Porteus/stannard .htm.
[96] "Porteus Hall | Building Names | University of Hawaii at Manoa," https://libweb.hawaii .edu/names/porteus.html.

which continues to be urgent precisely because of the often painful inter-generational legacies of Waiale'e, Kawailoa, and Waimano that continue to reverberate across many Hawaiian families today, and the still prevalent assumption that the Territorial period was benign, as well as the legitimacy many social sciences still lend the settler colonial state as they take and institutionalize children.

6

Port of Epistemic Riches

Social Science Research and Incarceration in Mid Twentieth-Century Puerto Rico

ALBERTO ORTIZ DÍAZ

On April 24, 1945, a twenty-four-year-old Black former bartender and mechanic from Caguas named Marcial Hernández García (alias "Yombe") entered the Insular Penitentiary at Río Piedras (popularly known as Oso Blanco).[1] A week prior, the Caguas district court sentenced him to eight years for a bundle of offenses ranging from breaking and entering and multiple counts of theft to a crime against nature, all of which he claimed to have committed under the influence of drugs and alcohol. Behind bars, he was interviewed and studied by penitentiary health professionals, who were interested in his life before prison and his health while incarcerated.[2] Psychiatrist José R. Maymí Nevares, for instance, concluded that Yombe suffered from mental maladjustment, or more specifically, episodes of bewilderment during which he observed "abnormal" and "undisciplined" conduct.[3] Several individuals familiar with him in Caguas before he was jailed confirmed his erratic state of mind. These potential advisors had since distanced themselves from him, telling parole officials he was "half-crazy."[4]

To treat Yombe's mental instability, he was isolated in a cell and regularly observed. An intelligence exam (Wechsler-Bellevue test) administered by psychologist Juan B. Picart, on which the prisoner scored a 63 (indicative of

---

[1] "Yombe" is likely a reference to African ancestry. The Yombe people reside primarily in Zambia, the Congo, and Angola. See "Yombe [multiple entries]," Art and Life in Africa, Stanley Museum of Art, University of Iowa, Spring 2014, https://africa.uima.uiowa.edu.

[2] By "health professionals" I mean medical and social scientists, as well as social workers, parole officers, and other technical personnel. Many of these individuals participated in Oso Blanco's Classification and Treatment Board, an interdisciplinary medico-legal entity founded in the mid-1940s that imagined and implemented rehabilitation programs for prisoners. Alberto Ortiz Díaz, *Raising the Living Dead: Rehabilitative Corrections in Puerto Rico and the Caribbean* (Chicago: University of Chicago Press, 2023), chapters 2–3.

[3] Expediente del confinado Marcial Hernández García, Caja 138, Serie Junta de Libertad Baja Palabra, Fondo Departamento de Justicia, Archivo General de Puerto Rico (hereafter SJLBP, FDJ, AGPR).

[4] Expediente del confinado Marcial Hernández García.

mental deficiency), corroborated "signs of organic cerebral disease."[5] Picart's clinical and vocational analysis confirmed Yombe's cognitive and applied shortcomings, leading prison technocrats to believe that mental infirmity was why he neglected to study or learn a new craft while in Oso Blanco. This could become a problem, as reform logic prevailed in the penitentiary alongside carceral capitalism.

When Yombe was finally eligible for parole between late 1948 and early 1949, officials rejected the prospect given that his mental state and behavior had not improved much. Parole officer María Casasnovas cosigned their reservations, and in the end, Yombe remained imprisoned. Still, penitentiary health professionals sensed that the man was treatable and therefore salvageable. They insisted that solitary confinement, psychiatric observation, psychological examination, social work interviews, an appropriate labor assignment, and sports would achieve his rehabilitation. In fact, Oso Blanco Classification and Treatment personnel forecasted Yombe's rehabilitation as "regular," so they went about preparing him for parole reconsideration. Health professionals viewed his impaired mind as surmountable, especially if addressed across rehabilitative techniques, and more importantly, as but a snapshot of a complicated, panoramic whole that included both the tangible and intangible.

This brief foray into Yombe's experience, who was interviewed, studied, and deemed worthy of rehabilitation by an interdisciplinary group of penitentiary health professionals, reveals myriad, overlapping encounters between different health practitioners and convicts and the frequent exchanges and loops between them in a mixed-race society under US rule. At the microlevel, the interaction between Yombe and Picart was mediated by science and destabilized homogenous notions of Blackness. Prison authorities categorized Yombe as Black. Picart was also of color.[6] Sharing skin tone did not mean the two men were entrenched kin folk, though, for they had diverging class backgrounds. Whereas Yombe had worked in bars and gas stations and expressed wanting to be his own boss, ship logbooks indicate Picart frequently traveled to New York City, where he earned an MA degree at Columbia University Teachers College in 1946.[7] Both men

[5] Ibid. Also see Corwin Boake, "From the Binet-Simon to the Wechsler-Bellevue: Tracing the History of Intelligence Testing," *Journal of Clinical and Experimental Neuropsychology* 24, no. 3 (May 2002): 383–405; Patti L. Harrison and Alan S. Kaufman, "History of Intelligence Testing," in *Encyclopedia of Special Education*, 3rd ed., eds. Cecil R. Reynolds and Elaine Fletcher-Janzen (New York: John Wiley & Sons, 2007), 1128.

[6] Federal census records between 1910 and 1940 interchangeably refer to Picart as a Mulatto and a person of color. His World War I draft registration card identifies him as Black. "Juan B. Picart [born 1897]," Ancestry Library Edition, www.ancestrylibrary.com/search/?name=Juan+B._Picart&event=_puerto+rico-usa_5185&birth=1897.

[7] American Psychological Association, *Directory* (Washington, DC: American Psychological Association, 1957), 351.

negotiated settler colonial scenarios, albeit from very different positions and with different goals and outcomes in mind.

More broadly, Yombe's case illustrates that multiple settler colonial traditions shaped how members of Oso Blanco's Classification and Treatment Board (i.e., psychologists and those trained in social science disciplines like social workers) coevaluated incarcerated people in the 1940s and 1950s.[8] Spanish colonial ethnographic approaches to criminal science informed mid-century social scientists' compilation of prisoners' histories. These, in turn, allowed them to better "see" the entirety of the people under their scrutiny and in their care.[9] This happened in a laboratorial setting, however, which reflected eugenic American approaches to science, such as psychometrics, that arrived in Puerto Rico after 1898.[10] The result was a "creole" (racially mixed, race-neutral) nationalist science that blended Spanish and American colonial traditions, while contradictorily utilizing the resulting knowledge to reclaim Puerto Rico's future.[11]

---

[8] Violence, dichotomous race relations, and eugenic science are central to settler colonialism. But in this chapter, I view settler colonialism as unfixed and emphasize how Puerto Ricans recast colonial logics and practices to restructure relationships beyond and within the archipelago; when within, resulting in a racially elastic internal colonialism. Nancy Leys Stepan, *"The Hour of Eugenics": Race, Gender, and Nation in Latin America* (Ithaca, NY: Cornell University Press, 1991); Severo Martínez Peláez, *La patria del criollo: An Interpretation of Colonial Guatemala* (Durham, NC: Duke University Press, 2009 [1970]).

[9] Arthur Kleinman, "What Is Specific to Western Medicine?" in *Companion Encyclopedia of the History of Medicine*, vol. 1, eds. William F. Bynum and Roy Porter (New York: Routledge, 1993), 15–23.

[10] By "psychometrics" I mean the measurement of the human mind and its functions and how these determine social behavior. I use the terms "psychometric," "psychological," and "mental," especially regarding testing instruments, interchangeably. On the concepts at the core of psychometrics, see Bonnie A. Green and Harold Kiess, *Measuring Humans: Fundamentals of Psychometrics in Selecting and Interpreting Tests* (San Diego: Cognella, 2017). On the consequences of the War of 1898, see the essays in *Hispanic American Historical Review* 78, no. 4 (November 1998): 577–765; Noenoe K. Silva, *Aloha Betrayed: Native Hawaiian Resistance to American Colonialism* (Durham, NC: Duke University Press, 2004); Paul A. Kramer, *The Blood of Government: Race, Empire, the United States, and the Philippines* (Chapel Hill: University of North Carolina Press, 2006).

[11] I use the term "creole" as a marker for the racially diverse group of health professionals involved in imagining and materializing rehabilitation in modern Puerto Rico. Originally, creole referred to the descendants of enslaved and free people born in American colonies. Later, the term connotated miscegenation and distinct ethnic identities, including "Puerto Rican." The racial landscape and terminology of Puerto Rico is complex and subject to a class-color pyramid that has long prevailed across the Caribbean and Latin America. Binary ways of understanding race in Puerto Rico have been and continue to be entwined with viewing race through a prism of racial fusion, which has resulted in inconsistent notions of Blackness, whiteness, and mixture marked by silences, affirmations, denials, and stereotypes. José Luis González, *Puerto Rico: The Four-Storeyed Country and Other*

The blended creole science apparent in Yombe's case and others like it in modern Puerto Rico substantiates insights about the multidirectionality of science and medicine in Latin American and Caribbean history.[12] In the mid twentieth century, the Partido Popular Democrático (Popular Democratic Party, or PPD) and the statesman-turned-governor Luis Muñoz Marín presided over the rise and fall of Puerto Rico as a social laboratory.[13] During this era of multifaceted reform, health professionals promulgated social science research that was cognizant of prisoners' humanity, showcasing that racialized colonial subjects created their own scientific expertise and used it for the purpose of national rehabilitation in the midst of unequal relations with the United States. Critical ethnography and psychometric tests enabled Puerto Rican social scientists to forge their own "objective" assessments of prisoners. The creole-appropriated projects of mapping convict minds, lives, and behaviors, rehabilitating them, and (re)integrating them into the Puerto Rican political economy and body politic helped mute and overturn long-standing colonial assumptions about tropical unfitness. "Repairing" prisoners positioned privileged creole technocrats to contribute to a broader project of rationalizing domestic self-government, a vision that came to fruition in 1951 via commonwealth status. The irony was that they aspired to rehabilitate inmates while reinforcing common tropes of their dysfunction.[14]

Using a variety of sources, this chapter traces Puerto Ricans' pursuit of a decolonized science, one that responded to the archipelago's evolving health

Essays (Princeton: Markus Wiener Publishers, 1993); Jay Kinsbruner, *Not of Pure Blood: The Free People of Color and Racial Prejudice in Nineteenth-Century Puerto Rico* (Durham, NC: Duke University Press, 1996); Ileana Rodríguez-Silva, *Silencing Race: Disentangling Blackness, Colonialism, and National Identities in Puerto Rico* (Basingstoke: Palgrave Macmillan, 2012); Kathryn R. Dungy, *The Conceptualization of Race in Colonial Puerto Rico, 1800–1850* (New York: Peter Lang, 2015); Isar P. Godreau, *Scripts of Blackness: Race, Cultural Nationalism, and U.S. Colonialism in Puerto Rico* (Champaign: University of Illinois Press, 2015).

[12] Juanita de Barros, Steven Palmer, and David Wright, eds., *Health and Medicine in the circum-Caribbean, 1800–1968* (New York: Routledge, 2008); Mariola Espinosa, "Globalizing the History of Disease, Medicine, and Public Health in Latin America," *Isis* 104, no. 4 (December 2013): 798–806; Mariola Espinosa, "The Caribbean Origins of the National Public Health System in the USA: A Global Approach to the History of Medicine and Public Health in Latin America," *História, Ciências, Saúde-Manguinhos* 22, no. 1 (2015): 241–253; Gabriela Soto Laveaga, "Largo Dislocare: Connecting Microhistories to Remap and Recenter Histories of Science," *History and Technology* 34, no. 1 (2018): 21–30.

[13] Michael Lapp, "The Rise and Fall of Puerto Rico as a Social Laboratory, 1945–1965," *Social Science History* 19, no. 2 (Summer 1995): 169–199.

[14] Eve Tuck and K. Wayne Yang, "R-Words: Refusing Research," in *Humanizing Research: Decolonizing Qualitative Inquiry with Youth and Communities*, eds. Django Paris and Maisha T. Winn (Thousand Oaks: Sage, 2014), 230.

challenges despite biomedical progress and the reduction of mortality rates.[15] Penitentiary psychologists and social workers used psychometrics and ethnography to measure the intelligence, skill sets, and personalities of prisoners, and to signpost how to best uplift them. These diagnostic and descriptive tools revealed that convicts required discipline, tutelage, and treatment, but that they also had redemptive potential regardless of social difference. Socioscientific interaction of this kind troubled linear notions of human progress, tempered professional preferences for certain technologies of power, and formed and made visible the value commitments of prisoners *and* the experts charged with evaluating and shepherding them.[16] Instead of seamlessly accepting the results of mental tests, social scientists put them into dialogue with inmate ethnographies to forge what to them were forward-looking treatment programs, illustrating how racialized racelessness and intersubjective exchanges transformed Puerto Rican corrections, at least for a time.

Blended Science

The history of science in Puerto Rico is as long as the history of the archipelago.[17] Puerto Rican science acquired cross-imperial and polyglot features in the late eighteenth and nineteenth centuries through vertical health interventions and studies of botany, disease, forensic pathology, and medicine.[18] In many of these accounts, characters of flesh, bone, and blood were relegated

---

[15] Manuel Quevedo Báez, *Historia de la medicina y cirugía en Puerto Rico*, 2 Volumes (San Juan: Asociación Médica de Puerto Rico, 1946-1949); Oscar G. Costa Mandry, *Apuntes para la historia de la medicina en Puerto Rico: breve reseña histórica de las ciencias de la salud* (San Juan: Departamento de Salud, 1971); José G. Rigau Pérez, "Historia de la medicina: la salud en Puerto Rico en el siglo XX," *Puerto Rico Health Sciences Journal* 19, no. 4 (December 2000): 357-368; Raúl Mayo Santana, Annette B. Ramírez de Arellano, and José G. Rigau Pérez, eds., *A Sojourn in Tropical Medicine: Francis W. O'Connor's Diary of a Porto Rican Trip, 1927* (San Juan: Editorial de la Universidad de Puerto Rico, 2008); Nicole E. Trujillo-Pagán, *Modern Colonization by Medical Intervention: U.S. Medicine in Puerto Rico* (Chicago: Haymarket, 2014); Gil G. Mendoza Lizasuain, "Desarrollo del sistema de salud pública de Puerto Rico desde el 1900 al 1957: 'una visión salubrista hacia las comunidades aisladas'," PhD dissertation, Universidad Interamericana de Puerto Rico-Recinto Metropolitano, 2017.

[16] Warwick Anderson, "Objectivity and Its Discontents," *Social Studies of Science* 43, no. 4 (August 2013): 557-576.

[17] Lydia Pérez González, *Enfermería en Puerto Rico desde los precolombinos hasta el siglo XX* (Mayaguez: Universidad de Puerto Rico, 1997), chapter 1; Reniel Rodríguez Ramos, *Rethinking Puerto Rican Precolonial History* (Tuscaloosa: University of Alabama Press, 2010); Frances Boulon-Díaz and Irma Roca de Torres, "Formación en psicología en Puerto Rico: historia, logros y retos," *Revista Puertorriqueña de Psicología* 27, no. 2 (July-December 2016): 231.

[18] Fray Iñigo Abbad y Lasierra, *Historia geográfica, civil y natural de la isla de San Juan Bautista de Puerto Rico* (San Juan: Editorial Universitaria, 1966 [1778]); André Pierre

to the background. By the eve of the twentieth century, people-centered ethnographies increasingly characterized local scientific knowledge production. This shift was evident in the growth of social science studies, particularly a steady stream of anthropological and sociological literature about peasants, mental patients, and criminals.[19]

Among the studies published in the late nineteenth century that shed light on Spanish colonial ethnographic research in Puerto Rico are those of the physician-criminal anthropologist José Rodríguez Castro. In the early 1890s, he published several medico-legal reports about crime and madness on the southern coast. One of his investigations spotlighted a Black adolescent named Isidora Gual who had strangled her infant son. Rodríguez Castro applied European ideas and ethnographic methods to Gual's case to distinguish between her criminality and insanity. His narrative revolved around her youth, sensuality, lack of education, class and race background, regional origins in Guayama (a Black part of the island associated with sugar plantation slavery and witchcraft), and family history of mental instability. In addition to only possessing fifty-three centimeters of cranial circumference, the generally somber Gual endured numerous spells of hunger and sickness with her child.[20] The precise answer as to why she killed the boy lay somewhere between her individual decision-making and the social conditions shaping

Ledru, *Viaje a la isla de Puerto Rico en el año 1797, ejecutado por una comisión de sabios franceses, de orden de su gobierno bajo la dirección del capitán Nicolás Baudín,* translated by Julio L. de Vizcarrondo (Río Piedras: Instituto de Literatura Puertorriqueña, Universidad de Puerto Rico, 1957); René de Grosourdy, *El médico botánico criollo* (Paris: F. Brachet, 1864); Enrique Dumont, *Ensayo de una historia médico-quirúrgica de la isla de Puerto Rico,* 2 Volumes (La Habana: Imp. "La Antilla," 1875–1876); Agustín Stahl, *Estudios sobre la flora de Puerto Rico* (San Juan: Tip. "El Asimilista," 1883–1888); José G. Rigau Pérez, "El Dr. Francisco Oller y el inicio de la salud pública moderna en Puerto Rico, 1790–1831," XXVII Congreso Internacional de Historia de la Medicina (August 31–September 6, 1980): 199–202; José G. Rigau Pérez, "The Introduction of Smallpox Vaccine in 1803 and the Adoption of Immunization as a Government Function in Puerto Rico," *Hispanic American Historical Review* 69, no. 3 (August 1989): 393–423; Henri Alain Liogier, "Botany and Botanists in Puerto Rico," *Annals of the New York Academy of Sciences* 776, no. 1 (June 1996): 42–45.

[19] Francisco del Valle Atiles, *El campesino puertorriqueño: sus condiciones físicas, intelectuales y morales, causas que las determinan y medios para mejorarlas* (San Juan: Tipografía de José González Font, 1887); José Rodríguez Castro, *La embriaguez y la locura, ó, consecuencias del alcoholismo* (San Juan: Imp. del "Boletín Mercantil," 1889); José Calderón Aponte, *El crímen de Bairoa* (Puerto Rico: Imprenta del Heraldo Español, 1903); Salvador Brau, *Ensayos: disquisiciones sociológicas* (Río Piedras: Editorial Edil, 1972).

[20] "Noticias de la isla," *La Correspondencia de Puerto Rico,* September 2, 1891, 3; José Rodríguez Castro, *Infanticidio: causa contra Isidora Gual, informe médico-legal* (Ponce: Imprenta "El Telégrafo," 1892), 6–10.

her circumstances, a position that contrarian colleagues contested.[21] Truly understanding the case, Rodríguez Castro wrote, necessitated examining "all her life," her decisions over time, and moral and physical factors.[22]

Resolving health problems, such as the prevalence of alcoholism and anemia, in part instigated the turn toward multidimensionalizing people and using them to explain crime, disease, and other infirmities in Spanish colonial Puerto Rico. US military scientists subsequently recast Puerto Ricans as disease-ridden rural subjects.[23] In the early twentieth century, scientific knowledge production in Puerto Rico primarily revolved around biomedicine and the focus of the economy, agriculture. During the interwar years, nutrition became a public health concern and biochemists, home economists, agronomists, and social workers promoted rural hygiene programs.[24] By the mid twentieth century, the wombs of women increasingly garnered the attention of scientists.[25] An exception to this biomedical-corporeal rule surfaced in the mid-1910s, with the anthropological works linked to the initial phase of the New York Academy of Sciences' *Scientific Survey of Porto Rico and the Virgin Islands*. Although Franz Boas and John Alden Mason composed thick descriptions of Puerto Rican Indigeneity, rural oral folklore, and Blackness while nurturing liberal-reformist colonial interests to remake an "empty" colony in the image of a "civilized" metropole, it was not until the 1940s and 1950s that ethnography fused to modern Puerto Rican statecraft.[26]

Human science in Puerto Rico, then, has historically responded to colonial-imperial politics and prerogatives. Under Spanish rule, ethnographic science attentive to socioeconomic and spiritual conditions prevailed. Under

---

[21] Teófilo Espada Brignoni and Ashley Rosa Jiménez, "Entre lo individual y lo social: debates sobre lo psicológico en el caso de Isidora Gual, 1890–1892," *Umbral* 18 (December 2022): 3, 8, 18.

[22] Rodríguez Castro, *Infanticidio*, 9.

[23] Puerto Rico Anemia Commission (Bailey K. Ashford and Pedro Gutiérrez Igaravídez), *Uncinariasis en Puerto Rico: un problema médico y económico* (San Juan: Bureau of Supplies, 1916 [1911]); Francisco A. Scarano, "Jíbaros y médicos a comienzos del siglo XX: los cuerpos anémicos en la ecuación imperial," in *La mascarada jíbara y otros ensayos* (San Juan: Ediciones Laberinto, 2022), 161–183.

[24] Elisa M. González, "Food for Every Mouth: Nutrition, Agriculture, and Public Health in Puerto Rico, 1920s–1960s," PhD dissertation, Columbia University, 2016.

[25] Laura Briggs, *Reproducing Empire: Race, Sex, Science, and U.S. Imperialism in Puerto Rico* (Berkeley: University of California Press, 2002).

[26] Eugenio Fernández Méndez, *Franz Boas y los estudios antropológicos en Puerto Rico* (México: Editorial Cultura, 1963); George W. Stocking, Jr., *After Tylor: British Social Anthropology, 1888–1951* (Madison: University of Wisconsin Press, 1995), xiv, chapter 8; Simon Baatz, "Imperial Science and Metropolitan Ambition: The Scientific Survey of Puerto Rico, 1913–1934," *Annals of the New York Academy of Sciences* 776, no. 1 (June 1996): 1–16; Rafael Ocasio, *Race and Nation in Puerto Rican Folklore: Franz Boas and John Alden Mason in Porto Rico* (New Brunswick: Rutgers University Press, 2020).

American rule, science became more laboratorial and rigid.[27] Puerto Rican scientists of all persuasions continued to tout the importance of the laboratory under colonial-populism, but they also rediscovered ethnography as a tool that could help address the dual challenges of crime and stagnant colonialism. While the Spanish and US-Americans had contrasting rehabilitative logics, in both cases these emanated from the intersection of biopower and custodial-regulatory practices.[28] What rehabilitation looked like on the ground under colonial-populism, however, mirrored the socio-scientific outlook of Muñoz Marín and the PPD, who dominated Puerto Rican politics for several decades and articulated a "reformed colonialism" rather than reinforcing colonialism proper or insisting on independence.[29] The colonial-populist project of the PPD merged state-controlled development, modernization, and "democracy" without rupturing cultural and political bonds with Spain and the United States, respectively.[30]

A major site where Puerto Rican leaders hoped to manufacture civic foot soldiers for this project was the Río Piedras scientific corridor, a cluster of welfare institutions located on the outskirts of San Juan close to the University of Puerto Rico-Río Piedras (UPR-RP) campus, an agricultural experiment station, and a leper colony in Trujillo Alto.[31] Oso Blanco (Figure 6.1), modeled after Sing-Sing penitentiary in New York, opened in 1933 and was the capstone of a transinstitutional complex that included the prison, an insane asylum, and a tuberculosis hospital.[32] Each of these institutions reinforced one

---

[27] Medical experiments on convicts became commonplace in US prisons during this period and were also carried out in US imperial domains. Jon M. Harkness, "Research Behind Bars: A History of Nontherapeutic Research on American Prisoners," PhD dissertation, University of Wisconsin-Madison, 1996; *"Ethically Impossible": STD Research in Guatemala from 1946 to 1948* (Washington, DC: Presidential Commission for the Study of Bioethical Issues, 2011), 13–26.

[28] Michel Foucault, *Security, Territory, Population: Lectures at the Collège de France 1977-1978* (New York: Picador, 2009).

[29] Geoff C. Burrows, "The New Deal in Puerto Rico: Public Works, Public Health, and the Puerto Rico Reconstruction Administration, 1935–1955," PhD dissertation, Graduate Center, City University of New York, 2014.

[30] Emilio Pantojas García, "Puerto Rican Populism Revisited: The PPD during the 1940s," *Journal of Latin American Studies* 21, no. 3 (October 1989): 521–557; José Luis Méndez, *Las ciencias sociales y el proceso político puertorriqueño* (San Juan: Ediciones Puerto, 2005).

[31] University of Puerto Rico-Río Piedras Agricultural Experiment Station, *The Story of the Agricultural Experiment Station of the University of Puerto Rico* (Río Piedras: University of Puerto Rico, 1952); Julie H. Levison, "Beyond Quarantine: A History of Leprosy in Puerto Rico, 1898-1930s," *História, Ciências, Saúde-Manguinhos* 10, no. 1 (2003): 225–245.

[32] The Chicago-based architectural firm Bennett, Parsons & Frost devised the original plan for the complex. Bennett, Parsons & Frost co-spearheaded the "City Beautiful" movement, a reform philosophy of North American architecture and urban planning that

Figure 6.1 Front view of the Presidio Insular in Río Piedras, undated (but likely circa 1930s–1940s given the institution's name), photographed by M. E. Casanave.
Source: Colección Archivo Fotográfico El Mundo, Biblioteca Digital Puertorriqueña, Universidad de Puerto Rico.

another in a common epistemic fabric. A culture of rehabilitative corrections that triply pathologized inmates emerged on the grounds of the complex, where inmate bodies, minds, and behaviors served as raw material for a local yet cosmopolitan medical class whose research was well-disposed *and* self-aggrandizing.[33] In this context, incarcerated people conveyed their stories to penitentiary technocrats, while social scientists' narratives of inmates exposed their desires to resurrect the Puerto Rican "nation."

Knowing Precedes Rehabilitating

In the early twentieth century, academia helped advance and reinvent settler colonialism. Universities and scholars contributed to forging purposeful knowledge about colonies and drawing the boundaries of growing but racialized and exclusionary academies.[34] In the Western hemisphere, US-American scholars in the humanities and social sciences approached South America with fresh eyes to ingrain or reject prior generalizations and stereotypes of the subcontinent. Their research laid the groundwork for a new apparatus of

flourished in the 1890s and early 1900s with the intent of introducing beautification and monumental grandeur to cities around the world. William H. Wilson, *The City Beautiful Movement* (Baltimore: Johns Hopkins University Press, 1989); David Brody, *Visualizing American Empire: Orientalism and Imperialism in the Philippines* (Chicago: University of Chicago Press, 2010), chapter 6.

[33] Tuck and Yang, "R-Words," 226–227, 237, 244–245.

[34] Tamson Pietsch, *Empire of Scholars: Universities, Networks, and the British Academic World, 1850–1939* (Manchester: Manchester University Press, 2013); José Amador, *Medicine and Nation Building in the Americas, 1890–1940* (Nashville: Vanderbilt University Press, 2015); Ricardo D. Salvatore, *Disciplinary Conquest: U.S. Scholars in South America, 1900–1945* (Durham, NC: Duke University Press, 2016).

knowledge in the service of inter-American relations, imperial hemispheric hegemony, and informal empire.[35] Puerto Rico's flagship university (founded in Río Piedras in 1903) and Columbia University played key roles in informalizing the US Empire and building local capacity for education and the health and social sciences.[36] Columbia made inroads into Puerto Rico in the 1910s and 1920s, when its faculty documented folkloric traditions and established a tropical medicine school in San Juan. In 1915, the "father of American anthropology," Franz Boas, traveled to the Caribbean and participated in the *Scientific Survey of Porto Rico*.[37] He and the archeologist and linguist, University of California-Berkeley graduate John Alden Mason, together researched Puerto Rican Indigeneity, presumed to have been long decimated in the wake of Spanish colonial expansion in the sixteenth century. Later, Boas supervised Mason's research on rural peasant (*jíbaro*) and "negro" Puerto Ricans – the still living and visible portions of island identity by the mid twentieth century – efforts that "rescued" the equivalent of reams of highland oral folklore spanning poetry, sayings, songs, riddles, and folktales, and that documented Black vocabulary, religion, medicine, and customs.[38] Meanwhile, the tropical medicine school was inaugurated in September 1926 and operated as an imperial research outpost. It was celebrated as a private–public partnership and testament to North Americans' and Puerto Ricans' commitment to forging a shared America and equality in hemispheric relations. Asymmetrical relations between metropole and colony persisted unabated, however.[39]

[35] Salvatore, *Disciplinary Conquest*, 1–2.

[36] Marcial E. Ocasio Meléndez, *Río Piedras: ciudad universitaria, notas para su historia* (San Juan: Comité Historia de los Pueblos, 1985); Amador, *Medicine and Nation Building in the Americas*.

[37] Julio C. Figueroa Colón, "Introduction," *Annals of the New York Academy of Sciences* 776, no. 1 (June 1996): vii–viii; Ocasio, *Race and Nation in Puerto Rican Folklore*. On Boas's legacy, see George W. Stocking, Jr., *A Franz Boas Reader: The Shaping of American Anthropology, 1883–1911* (Chicago: University of Chicago Press, 1989 [1974]); Alan J. Barnard and Jonathan Spencer, eds., *Routledge Encyclopedia of Social and Cultural Anthropology*, 2nd ed. (London: Routledge, 2010), 88–91.

[38] Ocasio, *Race and Nation in Puerto Rican Folklore*. Mason went on to work in South America. See, for example, John Alden Mason, "The Languages of South American Indians," in *Handbook of South American Indians*, vol. 6, ed. Julian H. Steward (Washington, DC: Government Printing Office, 1950), 157–317.

[39] Courtney Johnson, "Understanding the American Empire: Colonialism, Latin Americanism, and Professional Social Science, 1898-1920," in *Colonial Crucible: Empire in the Making of the Modern American State*, eds. Alfred W. McCoy and Francisco A. Scarano (Madison: University of Wisconsin Press, 2009), 178; Amador, *Medicine and Nation Building in the Americas*, 136–137; Tuck and Yang, "R-Words," 245.

Columbia faculty again pushed the boundaries of Puerto Rican social science in the mid twentieth century. After World War II, UPR-RP's new Center for Social Research commissioned a large-scale study to determine how modern social science could be employed to examine and resolve Puerto Rico's social and economic problems. Anthropologist Julian Steward helmed the study, which was published in 1956. *The People of Puerto Rico* focused on how modernization affected local subcultures and underscored that health and welfare programs on the island had therapeutic and political significance.[40] While Puerto Rican authorities weighed the impact of social science on statecraft, penitentiary social scientists like Picart lent their expertise to raising convicts from mental and social death. In both cases, acts of knowing preceded acts of rehabilitation.

When Oso Blanco was inaugurated in May 1933, Puerto Rico's appointed governor, James Beverley, highlighted the prison's modernity. This meant a concrete structure with sanitary facilities, the efficient use of space inside and of land surrounding the prison, self-sufficient (and later profitable) institutional productivity, and the application of social sciences like criminology.[41] Modernity also signified professional subjectivities linked to notions of rationality and progress.[42] It connoted humanizing punishment, which manifested in the attempted procurement of regenerative treatment.[43] Colonial and local authorities alike envisioned Oso Blanco as a site of physical and mental rehabilitation. According to Attorney General Charles Winter, it was a place where convicts would be inculcated with healthy and moral habits through labor, education, and science. The point was to return "useful citizens" to their families, their communities, and society devoid of "criminal inclinations."[44]

---

[40] Julian H. Steward, *The People of Puerto Rico: A Study in Social Anthropology* (Urbana: University of Illinois Press, 1956), 482. Smaller-scale studies followed, including David Landy's *Tropical Childhood: Cultural Transmission and Learning in a Rural Puerto Rican Village* (Chapel Hill: University of North Carolina Press, 1959) and Sidney Mintz's *Worker in the Cane: A Puerto Rican Life History* (New Haven, CT: Yale University Press, 1960).

[41] This was the case in Latin American penitentiaries across the region in the nineteenth and early twentieth centuries. Carlos Aguirre and Ricardo D. Salvatore, eds., *The Birth of the Penitentiary in Latin America: Essays on Criminology, Prison Reform, and Social Control, 1830–1940* (Austin: University of Texas Press, 1996).

[42] Dorothy Ross, ed., *Modernist Impulses in the Human Sciences, 1870–1930* (Baltimore: Johns Hopkins University Press, 1994), 8–9.

[43] Fernando Picó, *El día menos pensado: historia de los presidiarios en Puerto Rico, 1793–1993* (Río Piedras: Ediciones Huracán, 1994), 29–30, 56; Carlos Aguirre, *The Criminals of Lima and Their Worlds: The Prison Experience, 1850–1935* (Durham, NC: Duke University Press, 2005), 1–2; Julia Rodríguez, *Civilizing Argentina: Science, Medicine, and the Modern State* (Chapel Hill: University of North Carolina Press, 2006), chapter 7.

[44] Departamento de Justicia, Oficina del Procurador General (Charles E. Winter), *Reglamento para el régimen y gobierno de la Penitenciaría de Puerto Rico en Río Piedras* (San Juan: Negociado de Materiales, Imprenta y Transporte, 1933), 5–6.

But a decade-plus would pass before the social sciences became fixtures at Oso Blanco.

Socio-scientific classificatory schemes, methods, and research in Puerto Rican corrections interfaced with intellectual currents emanating from the US mainland. In a paper published shortly after World War II by federal prisons official Frank Loveland, who around the same time was contracted by Puerto Rico's government to conduct a critical study about incarceration there and suggest improvements, he stressed the variability of rehabilitative ideas. Analyzing corrections in the periphery taught Loveland that rehabilitation was a multifaceted, cooperative process involving individuals and institutions. Without cooperation from both sides, constructive results could not be obtained.[45] Social scientists, critical ethnography, and psychometrics formed part of Loveland's arsenal to streamline the holistic improvement of incarcerated people.

By the mid-1940s, mental tests had become a vital instrument of prison classification by opening different pathways to work toward the institutional adjustment of inmates, their rehabilitation, and their eventual societal reintegration. With these goals in mind, the mental exams of the era sought to measure the intelligence, skills, and personalities of incarcerated populations.[46] In Puerto Rico and elsewhere, prison authorities believed that mental test results would clarify treatment paths for inmates. They also became synonymous with literal and symbolic health in Oso Blanco at a time when elite Puerto Ricans aimed to redefine the colonial pact with the US government amid exorbitant poverty, overpopulation, economic transformation, and political crisis.[47] Carceral health professionals associated with the PPD contributed to building a local constituency that bore evidence of this reformist undertaking. Racially diverse convicts were perceived as candidates in this vein. To an extent, this distinguished the Puerto Rican approach to carceral rehabilitation

---

[45] H. G. Moeller, ed., *The Selected Papers of Frank Loveland* (College Park: American Correctional Association, 1981), ix.

[46] Milton S. Gurvitz, "Psychometric Procedure in Penal and Correctional Institutions," in *Handbook of Correctional Psychology*, eds. Robert M. Lindner and Robert V. Seliger (New York: Philosophical Library, 1947), 58; Frank Loveland, *Classification in the Prison System* (Washington, DC: Bureau of Prisons, 1950).

[47] Pedro A. Caban, "Industrial Transformation and Labor Relations in Puerto Rico: From 'Operation Bootstrap' to the 1970s," *Journal of Latin American Studies* 21, no. 3 (October 1989): 559–591; Silvia Álvarez-Curbelo and María Elena Rodríguez-Castro, eds., *Del nacionalismo al populismo: cultura y política en Puerto Rico* (Río Piedras: Ediciones Huracán, 1993); Alberto Ortiz Díaz, "Pathologizing the *Jíbaro*: Mental and Social Health in Puerto Rico's *Oso Blanco* (1930s to 1950s)," *The Americas* 77, no. 3 (July 2020): 421–422, 434, 440; Pantojas García, "Puerto Rican Populism Revisited," 523; González, "Food for Every Mouth," 4, 8, 20–21, 25; Méndez, *Las ciencias sociales y el proceso político puertorriqueño*, 85–102, 149–202.

from the eugenic carcerality flourishing in the United States. Indeed, creole rehabilitation sought to (re)integrate the races anthropologists like Boas and Mason were so eager to disjoin earlier in the century. Still, penitentiary social scientists deployed rehabilitative techniques that revealed the caprices of scientific medicine and the spurious, philanthropic university-based research informing it.[48]

Psychometric Testing

Since the late 1800s, psychometric tools have served as technologies of power that create normative frameworks for thinking about racialized people.[49] They claim epistemological consistency and ontological universalism but can also be tools of juridical power and social control. The globalization of psychometric instruments in the twentieth century failed at universalizing "the human" yet contributed to colonial expansion. In mid-century Puerto Rico, neither Spanish nor American colonial frameworks were hegemonic. The creole nationalization of science and parallel pursuit of comprehensive human science under colonial populism subverted exclusively Spanish or American approaches to rehabilitation. Spanish, American, and creole flows of scientific knowledge production crashed into one another in the first half of the 1900s, with the latter becoming prominent at mid-century. Creole supremacy circumvented conventional exchange relations, reversed value, and upset predominant social structures and hierarchies associated with settler colonialism.[50] The colonial conceits that informed acceptable manifestations of modernity and state formation remained visible in the ways psychometric tools reproduced certain standards and tropes during the era of the PPD, but were also translated over.

The development and use of psychological tests and measures in the West and Puerto Rico spans more than a century.[51] In 1920, adaptations of the Pintner tests of Non-Verbal Abilities circulated on Puerto Rico's main island;

---

[48] Anne O'Brien, *Philanthropy and Settler Colonialism* (Basingstoke: Palgrave Macmillan, 2015).

[49] Joseph R. Buchanan, *Manual of Psychometry: The Dawn of a New Civilization* (Boston: Holman Brothers, 1885).

[50] Anderson, "Objectivity and its Discontents," 564.

[51] Roger Smith, *The Fontana History of the Human Sciences* (London: Fontana Press, 1997), 589–599; John Carson, "Mental Testing in the Early Twentieth Century: Internationalizing the Mental Testing Story," *History of Psychology* 17, no. 3 (August 2014): 249–255; Frances Boulon-Díaz, "A Brief History of Psychological Testing in Puerto Rico: Highlights, Achievements, Challenges, and the Future," in *Psychological Testing of Hispanics: Clinical, Cultural, and Intellectual Issues*, 2nd ed., ed. Kurt F. Geisinger (Washington, DC: American Psychological Association, 2015), 52.

the Stanford Achievement test then appeared, by 1925.[52] These were followed by other exams administered to public school students and incarcerated youths.[53] In 1927, for example, Attorney General George Butte, who was later active in the Philippines, reported that University of Puerto Rico psychologists completed an educational and mental survey of imprisoned young people in Mayagüez, using, among other tests, the Stanford Test of Ability, which had been previously used by the Commission from Columbia University in a study of public school children.[54] Clinical psychological services, on the other hand, were not routinely offered in psychiatric hospitals, facilities serving veterans and children, and prisons until the 1940s and later.[55] This slow expansion coincided with the maturation of psychology as an academic discipline emphasizing mental processes, introspection, behavior, and interpersonal relationships.[56]

Puerto Rican psychologists trained in the United States and elsewhere aspired to develop and perfect mental tests attuned to the local linguistic and cultural milieu. This was because cultural mistranslation weakened the reliability of these diagnostic tools. Therefore, it was difficult for Puerto Ricans to obtain the expected mean Intelligence Quotient (IQ) of 100, the prevailing Anglo standard. A clear bias emerged when speakers of languages other than English were tested within parameters designed for English speakers. Standardized culture-free tests could universally estimate the intellectual functioning and manual capabilities of different groups, in theory resulting in more equitable assessments of colonized people.[57]

[52] Irma Roca de Torres, "Perspectiva histórica sobre la medición psicológica en Puerto Rico," *Revista Puertorriqueña de Psicología* 19, no. 1 (2008): 11–48; Robert W. Rieber, ed., *Encyclopedia of Psychological Theories*, First Edition (New York: Springer, 2012), 797–798; Boulon-Díaz, "A Brief History of Psychological Testing in Puerto Rico," 51.

[53] Frances Boulon-Díaz and Irma Roca de Torres, "School Psychology in Puerto Rico," in *The Handbook of International School Psychology*, eds. Shane R. Jimerson, Thomas Oakland, and Peter Thomas Farrell (Thousand Oaks: Sage, 2007), 311–312.

[54] George C. Butte, *Report of the Attorney General of Porto Rico for the Fiscal Year Ending June 30, 1927* (San Juan: Bureau of Supplies, Printing, and Transportation, 1927), 17.

[55] Guillermo Bernal, "La psicología clínica en Puerto Rico," *Revista Puertorriqueña de Psicología* 17, no. 1 (2006): 353–364.

[56] Ana Isabel Álvarez, "La enseñanza de la psicología en la Universidad de Puerto Rico, Recinto de Río Piedras: 1903–1950," *Revista Puertorriqueña de Psicología* 9, no. 1 (1993): 13–29; Ellen Herman, *The Romance of American Psychology: Political Culture in the Age of Experts, 1940–1970* (Berkeley: University of California Press, 1995), 3–6; Boulon-Díaz, "A Brief History of Psychological Testing in Puerto Rico," 53–54.

[57] US Department of Health, Education, and Welfare, *Bulletin No. 12: Research Relating to Children* (Washington, DC: Government Printing Office, February 1960–July 1960), 21; Hussein Abdilahi Bulhan, *Frantz Fanon and the Psychology of Oppression* (New York: Plenum, 1985); Irma Roca de Torres, "Algunos precursores/as de la psicología en Puerto Rico: reseñas biográficas," *Revista Puertorriqueña de Psicología* 17, no. 1 (2006): 63–88; Boulon-Díaz, "A Brief History of Psychological Testing in Puerto Rico," 54.

Multiple tests were adapted in Puerto Rico between the 1930s and 1950s. These included the Wechsler Intelligence Scale, the Binet Intelligence Scale, and the Goodenough Draw-a-Person test.[58] Reliance on mental tests in Oso Blanco aligned with the growth of psychology in Puerto Rico at large. Prison psychologists connected to UPR-RP assessed the intelligence, cognitive and mechanical abilities, and personalities of convicts. Exam results reiterated to the Classification and Treatment Board that inmate minds harbored rich data that could and should be scientifically excavated, explained, and repurposed. In mapping prisoners in this way, penitentiary social scientists laid bare their own moral and behavioral preferences as well as those of Puerto Rico's criminal-legal system and government.[59] From their collective point of view, the end of a rehabilitated Puerto Rico "free" of colonial mismanagement justified the means of utilizing uneven psychometrics on prisoners to confirm their dysfunction *and* redemptive potential.

Pursuing Comprehensive Human Science

Oso Blanco endured scientific growing pains in its first decade of existence. Shortly after officially opening in 1933, the penitentiary lacked a psycho-pathological clinic to examine, diagnose, and treat convicts suffering from mental lesions.[60] The next year, in 1934, Attorney General Benjamin Horton encouraged Puerto Rican legislators to carve carceral mental health positions into the new fiscal year's budget. A psychiatrist and psychologist were "indispensable" for the study of prisoners' "defects" and treating the "mental disorders" either caused or influenced by the crime(s) they committed, Horton insisted.[61] As of 1935, the posts still had not been created.[62]

There is little to no mention of carceral mind science in Puerto Rican government justice publications between the mid-1930s and mid-1940s. By 1946, psychiatric and psychological services were routine in Oso Blanco and overlapped. Psychological testing fell under the umbrella of psychiatry in

---

[58] Boulon-Díaz, "A Brief History of Psychological Testing in Puerto Rico," 54.

[59] Jan E. Goldstein, "Toward an Empirical History of Moral Thinking: The Case of Racial Theory in Mid-Nineteenth-Century France," *The American Historical Review* 120, no. 1 (February 2015): 2.

[60] Charles E. Winter, *Report of the Attorney General of Puerto Rico for the Fiscal Year Ending June 30, 1933* (San Juan: Negociado de Materiales, Imprenta, y Transporte, 1934), 18.

[61] Benjamin J. Horton, *Report of the Attorney General of Puerto Rico for the Fiscal Year Ending June 30, 1934* (San Juan: Negociado de Materiales, Imprenta, y Transporte, 1934), 24.

[62] Benjamin J. Horton, *Report of the Attorney General of Puerto Rico for the Fiscal Year Ending June 30, 1935* (San Juan: Bureau of Supplies, Printing, and Transportation, 1935), 26.

terms of government reporting. Attorney General Enrique Campos del Toro, for instance, observed that "The number of new cases taken care of during the fiscal year by the psychiatrist at the Penitentiary were 138, of which the majority were cases of mental deficiency, psychopathic personality and psychoneurosis. Fifty-seven did not reveal apparent mental disturbances."[63] Mental tests helped prison authorities draw these conclusions. Governor Jesús Piñero relayed the following year, in 1947, that "Medical care to the indigent and public welfare activities in general increased throughout the Island, and a modest educational program [and census] on mental hygiene, with special emphasis on the problem of the feebleminded, was undertaken."[64] To be feebleminded was fused to promiscuity, criminality, and social dependence, equated "mental deficiency," and denoted a level of functioning just above "idiocy."[65] Puerto Rican penitentiary health professionals imbued the category with similar opprobrium in their estimations of incarcerated rural people in the mid twentieth century.[66]

Psychometric exams of convicts were administered on the UPR-RP campus and inside the penitentiary. Performing well or poorly on intelligence tests did not automate or preclude rehabilitation, though. For example, in September 1948 Picart assessed Enrique Carmona – a spoiled, "rebellious" seventeen-year-old wheat-colored (trigueño) prisoner from Toa Alta.[67] Carmona took the Wechsler-Bellevue test (Form I) on the UPR-RP campus and earned a complete score of 56, which meant he was "mentally retarded" or a "high moron."[68] However, Carmona had "psychological potential" that could bear fruit "under favorable environmental conditions."[69]

In contrast, in January 1949 Picart evaluated a thirty-one-year-old white "psychopathic" prisoner who grew up between San Juan and New York named Santiago Ocasio Soler.[70] Picart used the Wechsler-Bellevue to secure numerical values for the inmate's verbal and manual skills by having him respond to arithmetic problems, order blocks, and finish drawings. There was a

---

[63] Enrique Campos del Toro, *Report of the Attorney General of Puerto Rico for the Fiscal Year Ending June 30, 1946* (San Juan: División de Imprenta, 1947), 44.

[64] Jesús T. Piñero, *Forty-Seventh Annual Report of the Governor of Puerto Rico for the Fiscal Year 1946–1947* (San Juan: Service Office of the Government of Puerto Rico Printing Division, 1948), 31, 133.

[65] James W. Trent, *Inventing the Feeble Mind: A History of Mental Retardation in the United States* (Berkeley: University of California Press, 1995); Leila Zenderland, *Measuring Minds: Henry Goddard and the Intelligence Testing Movement* (New York: Cambridge University Press, 1998).

[66] Ortiz Díaz, "Pathologizing the *Jíbaro*," 429–430.

[67] Expediente del confinado Enrique Carmona, Caja 73, SJLBP, FDJ, AGPR.

[68] Ibid.

[69] Ibid.

[70] Expediente del confinado Santiago Ocasio Soler, Caja 61, SJLBP, FDJ, AGPR.

discrepancy between Ocasio Soler's verbal and nonverbal scores because he took longer than expected to complete the former and was too sure of himself on the latter. Yet, his "mental deterioration" was an "insignificant 7 [percent]," and overall, he scored a 118 (in the normal superior range), a number that exceeded the Anglo standard of 100.[71] Like Carmona, Ocasio Soler was deemed rehabilitable. In both cases, a psychologist of color (Picart) presided over administering and explaining their mental test results, an inversion of the presumed racial and scientific orders.

Oso Blanco Classification and Treatment practitioners had lucid exchanges about these and other cases and shared their research with one another. Mimicking developments in the United States to a degree, psychometric tools in Puerto Rico formed part of a culture of observation and evaluation that consolidated and expanded vertical forms of social control.[72] Notwithstanding the limits of many exams, psychologists like Picart deployed a variety of mental tests to determine on which minds different ones could be applied. His work, in conjunction with the efforts and interpretations of the Classification and Treatment Board, tended to essentialize "mentally deficient" prisoners. Convicts thus served as psychological specimens. But many social scientists also valued the "egalitarian" creed of the PPD and wanted the prisoners striving for rehabilitation to do so as well. As convicts showed signs of conforming progress, the liberatory effects of colonial-populist rehabilitation became more pronounced. Prisoners were corralled and studied in Oso Blanco, where social scientists viewed them as objects of research. Just as significantly, social scientists desired to comprehensively understand and civically redeem convicts, transforming the meanings of citizenship and socio-scientific knowledge production in a settler colonial context.[73]

Redeemable Prisoners

Rehabilitative corrections flourished in the mid twentieth century.[74] In Puerto Rico, Classification and Treatment officials investigated the lives of convicts, generated socioeconomic portraits of them, and organized health and social

---

[71] Ibid.

[72] Rebecca Schilling and Stephen T. Casper, "Of Psychometric Means: Starke R. Hathaway and the Popularization of the Minnesota Multiphasic Personality Inventory," *Science in Context* 28, no. 1 (March 2015): 77–98.

[73] Luis Negrón Fernández, *Report of the Attorney General to the Governor of Puerto Rico for the Fiscal Year Ended June 30, 1947* (San Juan: Real Hermanos, Inc., 1950), 30, 32.

[74] Volker Janssen, "Convict Labor, Civic Welfare: Rehabilitation in California's Prisons, 1941–1971," PhD dissertation, University of California-San Diego, 2005; Greg Eghigian, *The Corrigible and the Incorrigible: Science, Medicine, and the Convict in Twentieth-Century Germany* (Ann Arbor: University of Michigan Press, 2015); Ortiz Díaz, *Raising the Living Dead.*

science data they later channeled into treatment programs.[75] The director of Socio-Penal Services compiled findings and approved or modified prescribed plans. Treatment programs covered the medical, social, psychiatric, psychological, educational/vocational, and religious-spiritual aspects of rehabilitation. They usually concluded with a bottom line indicating whether individual prisoners could be rehabilitated at all. Crucially, incarcerated people had little to no say in whether they wholeheartedly consented to such practices, for at the time human subjects research was not legally micromanaged. Still, degrees of reciprocity were built into the human subject-researcher encounter, even if on highly unequal terms.

Classification and Treatment personnel utilized variable diction to convey inmate rehabilitative prospects. Proclaiming that prisoners were "rehabilitable" meant they were salvageable physically, mentally, socially, morally, and civically. Psychometric exams such as the Wechsler-Bellevue test, the Otis Mental Ability test, and the Rorschach Inkblot test either set the interpretive tone for Classification and Treatment Board efforts or built on already available ethnographies of convicts. In short shrift, however, board practitioners rejected exclusively considering IQ. Instead, they made a habit of contemplating exams intersectionally, meaning that Classification and Treatment experts believed multiple exams disclosed more together about individuals' rehabilitative prospects than apart. Although mental tests pointed to the intellectual, interactive, and mechanical "deficiencies" of inspected prisoners, they also served as a compass for the prison technocrats charged with crafting treatment programs. Regardless of inmate identities or the nature of their crime(s), they were generally considered eligible for rehabilitation – even if getting there would be an uphill climb. This does not mean that categories like race, class, or sexuality were irrelevant relics of a bygone era. Rather, health professionals defied and contradictorily engaged them to achieve the more pressing goals of self-government and constituency building.

The confluence of convict ethnographies, mental tests, and treatment programs repeatedly surface in the archival record. For example, in March 1950 prison authorities reviewed the case of a twenty-five-year-old white prisoner from Cayey serving time for mutilation named Onofre Rodríguez López. Classification and Treatment official Gloria Umpierre reported that this convict was well-educated, and that he appeared to be a "trustworthy," "serious,"

---

[75] Asymmetrical power relations and scientifically legitimated stereotypes are embedded in Classification and Treatment ethnographies and mental test results but so is the human science of rehabilitation. The modern archive is often reduced to a site of racialized, violent knowledge production given its genesis in colonial-imperial enterprise, but one can read into it counterintuitively as well, for archives produced under duress are also collectively inspired and defy reductiveness. The introduction to this book elaborates on these and other methodological nuances.

and an all-around "normal" person.[76] Perhaps most importantly, he showed repentance for his crime, possessed an excellent attitude, and was willing to adapt to what authorities demanded of him. He was a "good" case for rehabilitation. These conclusions would seem to suggest that Rodríguez López's whiteness dictated Umpierre's favorable assessment. However, his exam results divulge other interpretations.

Rodríguez López underwent protracted testing while incarcerated. He earned a high average score of 118 on the Otis test, above the Anglo standard of 100. More impressive was the fact that he answered questions quickly *and* correctly. Rodríguez López scored 39 points on the Bell Adjustment Inventory, which implied that he was "relatively well-balanced emotionally" compared to his immediate peers.[77] Yet, his social skills needed improvement, for he was an "isolated type."[78] The answers he gave to questions about interpersonal interactions (21 total) exposed his social shortcomings. In short, being white did not mean he had flawless social prowess. Psychologists also probed Rodríguez López's manual ability via the MacQuarrie test. He boasted average mechanical ability overall, performing well on the relational awareness, speed, and visual portions of the test, but underperformed on others. Prison officials recommended that the convict be given vocational work assignments to take advantage of the skills he had, and to cultivate and strengthen the ones he lacked. Interviews and religious services could further enrich his rehabilitation process, experts believed.[79]

As Rodríguez López's case suggests, mid-century Oso Blanco social scientists believed that determining intelligence had to be complemented by the measurement of nonintellectual characteristics, such as manual skills and/or personality traits.[80] The pattern is visible in cases of convicts of color as well. For example, in April 1950, Oso Blanco's Classification and Treatment Board studied a twenty-seven-year-old Mulatto prisoner from Guaynabo serving time for homicide named Antonio Hernández Alamo. The "always smiling" Hernández Alamo impressed board members as "trustworthy" and "humble" but also as "a bit ignorant" and not in full control of his emotions.[81] He earned

---

[76] Expediente del confinado Onofre Rodríguez López, Caja 195, SJLBP, FDJ, AGPR.

[77] Ibid.

[78] Ibid.

[79] Ibid.

[80] James D. A. Parker, "From the Intellectual to the Non-Intellectual Traits: A Historical Framework for the Development of American Personality Research," MA thesis, York University-Toronto, 1986; Kurt Danziger, *Constructing the Subject: Historical Origins of Psychological Research* (New York: Cambridge University Press, 1990), 158; Robert E. Gibby and Michael E. Zickar, "A History of the Early Days of Personality Testing in American Industry: An Obsession with Adjustment," *History of Psychology* 11, no. 3 (September 2008): 164–184.

[81] Expediente del confinado Antonio Hernández Alamo, Caja 475, SJLBP, FDJ, AGPR.

a 77 on the Otis test, a borderline deficient score. If exposed to interviews, religious guidance, challenging recreational activities like reading, and agricultural labor therapy, Hernández Alamo was a "good" case for rehabilitation. In fact, the prisoner teemed with "plenty of rehabilitable material."[82] His mixed-race background did not disqualify him.

Hernández Alamo's Otis test result was but one component of a more comprehensive psychology and broader human science that also depended on a social worker's evaluation of his personality. In his case and others, penitentiary social scientists bridged ethnography and psychometrics. A thirty-eight-year-old Black asthmatic inmate from Guayama serving time for homicide named Adolfo Ortiz Gutiérrez, who "feared the dark" and was tortured by nightmares about ghosts, is another case in point.[83] According to Classification and Treatment experts, who evaluated the convict in July 1950, he appeared to have "emotional problems," found "lying satisfying," and earned a 74 on an Otis test (borderline intellectual deficiency). These findings convinced them that they needed to administer personality tests to diagnose and treat Ortiz Gutiérrez more effectively. He had a "regular" chance to be rehabilitated. Interviews, religion, reading, and films could help bring him back from the mental and social brink.[84]

While white prisoners scoring in the normal range or exceeding it on a given intelligence test and mixed-race or Black prisoners scoring in the borderline deficient or inferior range can certainly be interpreted as settler colonial social science in action, that Oso Blanco social scientists experimented with combinations of tests in either racial scenario is suggestive of their awareness of the inequities baked into psychometrics and the need to evaluate incarcerated people on the basis of other shared criteria. Even if we assume prisoners of color were exclusively marginalized in this regard, a low intelligence score and rehabilitation were not mutually exclusive. This was the case for Salvador García Salamán, for instance, a twenty-one-year-old Black convict from Río Grande incarcerated at the Zarzal penal encampment in 1953 whose "below average intelligence" failed to raise serious concerns about his rehabilitative prospects.[85]

The opposite could also be true. For example, Picart administered a Wechsler-Bellevue test to a forty-six-year-old Black inmate from Santurce named Pedro Sánchez Alvarez in December 1948. The convict scored in the normal range (93). In his qualitative analysis, Picart noted that Sánchez Alvarez had satisfactory immediate and deep past memory, satisfactory mental concentration, and a "great ability to comprehend practical, real-life situations and to resolve situations involving arithmetic reasoning."[86]

---

[82] Ibid.
[83] Expediente del confinado Adolfo Ortiz Gutiérrez, Caja 475, SJLBP, FDJ, AGPR.
[84] Ibid.
[85] Expediente del confinado Salvador García Salamán, Caja 484, SJLBP, FDJ, AGPR.
[86] Expediente del confinado Pedro Sánchez Alvarez, Caja 101, SJLBP, FDJ, AGPR.

He expressed himself easily with ample vocabulary. The strengths and promise Picart saw in Sánchez Alvarez contrasted sharply with opinions of him shared by several of his immediate family members, who labeled him a "lost cause."[87] Here communal opprobrium did not flow from social scientists, but from Sánchez Alvarez's own brothers and other kin. For every García Salamán or Sánchez Alvarez, however, there was someone who incarnated socio-scientific stereotypes. A twenty-six-year-old Mulatto prisoner from San Juan named Ricardo Estrada Padilla, for example, combined "below average intelligence" and "criminal tendencies." These put his rehabilitation in jeopardy but never annulled it.[88] Similarly, health professionals understood a quinquagenarian Mulatto from Coamo named Sandalio Mateo Vázquez as a "suspicious hypocrite" with "below average intelligence," yet he was "rehabilitable."[89]

Prisoners' crimes, their backgrounds, and their (un)favorable personality traits and test scores did not automatically qualify them *for* or disqualify them *from* rehabilitation. Education, labor therapy, social orientation, religious services, and so on *could* transform convicts, although this was not guaranteed. While psychologists and social workers had occasion to belittle and racialize convicts, implicit in their mapping prisoners' worlds and articulating rehabilitative programming for them was the belief that inmates could be raised from living death. Social scientists and inmates together made the human sciences more human in a place (the prison) where dehumanization was and remains the expected norm. There, race largely functioned as a pivot on which creole science partially turned at a time when colonial-populists reimagined Puerto Rican national identity on race-neutral terms.

Conclusion

As Puerto Rican rehabilitative corrections hit their stride in the mid-1940s and 1950s (though not without challenges and shortcomings), Puerto Rico's government contracted federal consultants to conduct studies of the penal system. It was not until later in the 1950s and 1960s that justice officials finally implemented some of the recommendations put forward by the studies.[90] University of Puerto Rico-Río Piedras social workers drew from the studies to renew the promise of Puerto Rican rehabilitative corrections. In a critical analysis published in 1959, Rosa Celeste Marín, Awilda Paláu de López, and Gloria Barbosa de Chardón chronicled the work of contemporary university

---

[87] Ibid.
[88] Expediente del confinado Ricardo Estrada Padilla, Caja 484, SJLBP, FDJ, AGPR.
[89] Expediente del confinado Sandalio Mateo Vázquez, Caja 484, SJLBP, FDJ, AGPR.
[90] Rosa Celeste Marín, Awilda Paláu de López, and Gloria P. Barbosa de Chardón, *La efectividad de la rehabilitación de los delincuentes en Puerto Rico* (San Juan: Universidad de Puerto Rico, 1959), 96–97.

social science faculty in the prison system. Faculty assessed the personalities of maximum-security prisoners incarcerated in Oso Blanco and found that many of them tested as "mentally deficient" and were vulnerable to ongoing mental infirmity and recidivism.[91] To gather the data leading to these and interrelated findings, social scientists administered a diverse batch of psychometric tests: the Porteus Maze, Rorschach, Thematic Apperception, Draw-a-Person, and Bender-Visual Motor Gestalt tests. Psychologists also put exam results into conversation with ethnographic information about convicts. The collective data laid bare the psychological features of each inmate, as well as tendencies in their emotions, thinking, and behavioral and cognitive functioning.[92]

Discussion of prisoners' mental test results increasingly revolved around their dangerousness and propensity to mentally deteriorate when released from prison, however. A growing pessimism surrounding prisoner rehabilitation, flashes of which were apparent in the 1950s, gained momentum in Puerto Rico by the mid-1960s, evidenced by an uptick in studies that accepted the premise of criminal pathology yet lacked the rigor of previous generations to confront it.[93] This coincided with successive governments across Puerto Rico's party divide suffocating the rehabilitative ideal in the decades that followed.[94] No longer were social science tools and methods viewed as pathways toward convict redemption. Instead, they functioned as instruments of intense pathologization and othering, which aligned with how US researchers and authorities trafficked in them across groups and national borders earlier in the century.[95]

In the mid twentieth century, Puerto Rican social scientists exposed convicts to Spanish and American rehabilitative logics and practices. They transcended binaries (Spanish and US colonialism, Black and white, researcher and research subject) to foster a creole science and pugilistic nationalism that aspired to prove that prisoners could be rehabilitated and become colonial-populist citizens. This nationalism countered the American settler colonial image of Puerto Ricans as perpetual tutees. In a sense, Puerto Ricans pursued a decolonized science in the mold of their African peers, who as Erik Linstrum has argued, utilized psychology to disclose the limits of British imperial authority.[96]

---

[91] Ibid., 62–63.

[92] Ibid., 211–224. Maile Arvin's chapter in this volume examines Stanley Porteus's Maze Test and scientific work in greater detail than I do here.

[93] Manuel López Rey y Arrojo, Jaime Toro Calder, and Ceferina Cedeño Zavala, *Extensión, características y tendencias de la criminalidad en Puerto Rico, 1964–70* (Río Piedras: Editorial de la Universidad de Puerto Rico, 1975); Franco Ferracuti, Simon Dinitz, and Esperanza Acosta de Brenes, *Delinquents and Nondelinquents in the Puerto Rican Slum Culture* (Columbus: Ohio State University Press, 1975).

[94] Picó, *El día menos pensado*, 57, 73.

[95] Alexandra M. Stern, *Eugenic Nation: Faults and Frontiers of Better Breeding in Modern America* (Berkeley: University of California Press, 2005).

[96] Erik Linstrum, *Ruling Minds: Psychology in the British Empire* (Cambridge, MA: Harvard University Press, 2016).

While the mid-century social scientists involved in evaluating prisoners often cast them as deficient, returning incarcerated people of all colors to society as productive laborers, family providers, and citizens – indeed, as human capital – illustrated that the colonial-populist (and later commonwealth) state just might live up to its racial democracy rhetoric.[97] Socio-scientific research behind bars also showed that Afro-Puerto Rican health professionals like Picart could lay claim to scientific knowledge production, invest the epistemological riches in a more equitable domestic future without invalidating the relationship between metropole and colony, and unsettle assumptions about who got to possess research subjects and exercise health authority. Even though it was included as a category for describing incarcerated research subjects, race played a restrained role in inmate rehabilitation. The equity project within Puerto Rican corrections was paradoxical and failed, however, precisely because it depended on extracting pain and damage narratives from racially heterogeneous Puerto Ricans of perceived lesser status in the first place.

Colonial-populist Puerto Ricans' alternative deployment of social science contributed to Puerto Rico's modernization process but not in a way that automatically segued into rigid race-based exclusion.[98] Whereas anthropological ethnography is now cast as a valuable instrument in the struggle against systems of oppression, particularly in the Western academy, this was not always the case. Intelligence tests, for their part, are still being used for reactionary purposes. Social science literature linking race and intelligence continues to be published, and it appears the social science community that works on intelligence accepts this without challenge notwithstanding their recognition of the ethno-racial biases and controversies that have long tainted psychometrics.[99] The case of mid twentieth-century Puerto Rico, then, offers an inviting vantage point from which to understand how carceral human science, ironically, lent itself to the search for and realization of other realities, well in advance of our own preoccupation with reimagining the world.

---

[97] Michelle Murphy, *The Economization of Life* (Durham, NC: Duke University Press, 2017).

[98] As was the case in South Africa. Saul Dubow, *Scientific Racism in Modern South Africa* (New York: Cambridge University Press, 1995); William Beinart and Saul Dubow, *The Scientific Imagination in South Africa: 1700 to the Present* (New York: Cambridge University Press, 2021).

[99] Frederick T. L. Leong and Yong Sue Park, "Introduction," in Council of National Psychological Associations for the Advancement of Ethnic Minority Interests, *Testing and Assessment with Persons & Communities of Color* (Washington, DC: American Psychological Association, 2016), 1–2; Elliot Turiel, "Eugenics, Prejudice, and Psychological Research," *Human Development* 64, no. 3 (2020): 103–107.

7

The Imperial Logic of American Bioethics
Holding Science and History to Account

LAURA STARK

In 1974, the United States passed the National Research Act, which set the rules for the treatment of "human subjects" of research. The law pertained to both biomedical and social science research and it remains in place today, largely unchanged over fifty years, despite revisions in 2018 that nonetheless retained the basic structure and assumptions of the law. Those assumptions included a moral ontology organized around civic individualism and its safeguarding, as opposed to anticolonialism and its dismantling. In 1974, the immediate prompt for the law was the public revelation of the Tuskegee Syphilis Study: government scientists had been withholding a viable treatment for syphilis (penicillin) from people enrolled in the studies, who were low-income Black men in rural Alabama. The US government had been funding the Study for four decades; and scientists had been writing and reading about it in medical journals for just as long. Although the public exposure of medical suffering and abuse at the hands of the US government was the immediate prompt for the law's passage, the content of the rules – how procedural, government bioethics would work according to the law – had been two decades in the making within the National Institutes of Health.[1]

As a result, since 1970, a specific field known as modern American bioethics has dominated secular, English-language spaces of political power – the language of ethics as it is spoken in US domestic and foreign policy, international "medical diplomacy," global market regulation, and transnational corporations. It is the lingua franca of Euro-American science imperialism. It speaks louder and talks over the more context-informed, justice-based practices of science, making it easy to forget that this bullish and coercive variant of bioethics is a historical fluke.[2]

---

[1] Laura Stark, *Behind Closed Doors: IRBs and the Making of Ethical Research* (Chicago: University of Chicago Press, 2012).

[2] Jenny Reardon et al., "Science & Justice: The Trouble and the Promise," *Catalyst: Feminism, Theory, Technoscience* 1, no. 1 (September 8, 2015): 1–49; Jenny Reardon, "On the Emergence of Science and Justice," *Science, Technology, & Human Values* 38, no. 2 (2013): 176–200; Renee C. Fox and Judith P. Swazey, *Observing Bioethics* (New York:

Shortly after the law's passage in 1974, Carolyn Matthews, a white-settler free-spirit from Portland, Oregon, settled in her hometown and went back to college. In 1977, when she was in her late thirties and recently settled in Portland, she enrolled in a school that was designed for working adults. She could get course credit for her prior work experience, so she typed up her two decades of job experiences. She had worked a good deal in healthcare settings and been both a "human subject" of government medical research and a researcher of human subjects.

I met Carolyn after I put a description of my historical research in the Antioch College alumni magazine, and Carolyn got in touch with me. I was researching a program at the US National Institutes of Health (NIH), through which NIH "procured" healthy human civilians for medical experiments during the decades after World War II.[3] Across several conversations between 2016 and 2018, Carolyn relayed her life story. We talked by phone twice for official oral histories, we emailed updates about this project and our personal lives, and I visited her at her house in Portland, Oregon. Carolyn had been willing to tell me the story of her time at NIH's Clinical Center. But her NIH story extended into a longer, politically saturated narrative about bioethics – one that toppled the bookends of her NIH story and was impossible to ignore.

Carolyn shared with me the paper she had written in 1977 for course credit describing her job as a healthy human subject at the NIH Clinical Center (as well as her volunteer work as a lab technician there); she summarized the skills – and life experience – she gained as an x-ray technician in Arizona taking films of Akimel O'odham people; she listed her responsibilities as a research technician in Boston. It was her story – about her working life.

But woven into this 1977 story about her work life was another story – about her ethical awakening. This woven story compared her experiences as a human subject of NIH experiments to the experiences of the low-income hospital staff and patients whose organs she scanned in Boston after they received injections of radioactive tracers. The Akimel O'odham people, however, were absent. Whereas

Oxford University Press, 2008); Adriana Petryna, *When Experiments Travel: Clinical Trials and the Global Search for Human Subjects* (Princeton: Princeton University Press, 2009).

3 I am writing a book on the first healthy human subjects of NIH medical experiments from the time the agency's clinical research center opened, in 1954, until the death of the first "Normal" in 1980. The history of people's experiences and NIH's legal strategies to create what it called the Normal Volunteer Patient Program show that white bodies came to stand in for the "normal" body in postwar medicine – with ongoing effects. Carolyn was one of more than one hundred people with whom I created oral histories and archived photos, letters, and memorabilia from their time as "normal control" research subjects at the US National Institutes of Health. The collections are free and publicly accessible through the Vernacular Archive of Normal Volunteers. Laura Stark, *The Normals: A People's History* (Under Contract: University of Chicago Press), https://dataverse.harvard.edu/dataverse/vanv.

she inserted the Akimel O'odham people into the story of her working life, they were illegible as part of her bioethical understanding.[4]

This chapter tells Carolyn's story in two registers. It sets Carolyn's work experience prior to 1974 alongside her moral recounting of those experiences in her college portfolio – which she composed after the crystalizing moment of the Tuskegee revelations, which set the moral vocabulary and framework for research on people in terms of modern American bioethics. The point is not that Carolyn had a lapse in moral judgment in her practices or recall of her experiences. The premise of this chapter is that Carolyn perfectly articulated the logic of American modern bioethics. The insight of Carolyn's story is that the field of American bioethics operates with settler state presumptions. The question the chapter explores is how, specifically, the broad imperial logic of bioethics works – through what concepts, practices, and imperceptions.

The discourse of modern American bioethics is a geopolitical concern, and relationships across the Americas provide a special vantage on the field. Because of the coterminous geography, the history of science and ethics across the Americas points attention to the production of boundaries – to national borders, racial categories, citizenship status, and moral designations together through science.[5] For example, the same US government scientist who led the Tuskegee Syphilis Study from 1932 to 1972 also conducted related experiments in Guatemalan prisons during the 1940s, in which the research team intentionally infected incarcerated people with syphilis.[6] The production and enforcement of racial hierarchies within and between the US and Guatemala facilitated the research. In addition the research was predicated on logics of spatial containment and moral worth that justifies systems of incarceration and colonialism – within and between US, Latin American, and Native spaces (as also seen in Chapters 5 and 6). The study of science ethics across the

---

[4] There is large, excellent literature on the methods of oral history. As Spiegel explained in 2014, it nonetheless remains to be theorized "the materiality and reality of 'voices' from the past, without assuming the necessary truth of what they convey, at least in terms of the factuality of its content. In the end, however, what is at stake in not the epistemological question of 'truth' but an ethical response to the catastrophes of the last century." Gabrielle Spiegel, "The Future of the Past: History, Memory, and the Ethical Imperatives of Writing History," *Journal of the Philosophy of History* 8 (2014): 149–179.

[5] Megan Raby, "Science, the United States, and Latin America," in *The Routledge Handbook of Science and Empire*, ed. Andrew Goss (New York: Routledge, 2021), 264–274; Sandra Harding, "Latin American Decolonial Social Studies of Scientific Knowledge," *Science, Technology, & Human Values* 41, no. 6 (2016): 1063–1087; Eric V. Meeks, "Race and Identity across American Borders," *Latin American Research Review* 53, no. 3 (2018): 679–688; Eric V. Meeks, *Border Citizens: The Making of Indians, Mexicans, and Anglos in Arizona* (Austin: University of Texas Press, 2020).

[6] Susan M. Reverby, "'Normal Exposure' and Inoculation Syphilis: A PHS 'Tuskegee' Doctor in Guatemala, 1946–1948," *Journal of Policy History* 23, no. 1 (2011): 6–28; "*Ethically Impossible*": STD Research in Guatemala from 1946 to 1948.

Americas highlights how boundaries are strategically fabricated, not only *through* scientific efforts but also *for* science.

My intention as a white settler historian is to invoke the experiences of a fellow white settler knowledge maker – namely, Carolyn – in an imperfect effort to hold settler science (read: myself) to account.[7] My hope is to approximate Kim TallBear's technique of "studying up." While TallBear's standpoint, as a Sisseton Wahpeton Oyate scholar, in relation to white settler science is different from my own, the technique offers a way to study the (re) production of settler colonial structures within science and also to avoid co-opting and capitalizing on those injustices.[8]

The chapter follows Carolyn across three sites and over three decades. In 1962, Carolyn served as a healthy human subject at the NIH Clinical Center in Bethesda, Maryland, a time and place where Native people had a lively presence (second section). When she was not on study, Carolyn worked enthusiastically but without pay as a lab technician in the hospital. This volunteer work resulted in an offer to work for pay on an NIH research team collecting samples from a Native American tribe in Sacaton, Arizona, which she readily accepted (third section). After Sacaton, Carolyn worked as a scanning technician in Boston, Massachusetts (fourth section), before she returned after many years to Portland, Oregon, where she reflected on the ethical implications of her experiences in medicine (fifth section). The point of a critique of bioethics through the Americas is to strengthen existing alliances for justice-based science and to inform practices – in science, in history, and in transformative bioethics.

Bethesda, 1962: Carolyn as Research Subject

Carolyn enrolled in Antioch College in 1962 and arrived at the NIH Clinical Center three months later. The Clinical Center was the US government's main research hospital, located on what was called at the time NIH's "reservation" in Bethesda, Maryland. As part of its Congressional mandate, the Clinical Center could *not* admit people for treatment alone; everyone admitted to the hospital had to be a research subject (often as part of a treatment). For its part, Antioch

---

[7] Methodologically, scholars have improvised several anticolonial techniques for writing histories that highlight, then subvert, the structures of oppression built into many traditional archives, as well as the standards of professional history. See Marisa J. Fuentes, *Dispossessed Lives: Enslaved Women, Violence, and the Archive* (Philadelphia: University of Pennsylvania Press, 2016); Saidiya Hartman, *Wayward Lives, Beautiful Experiments: Intimate Histories of Social Upheaval* (New York: W. W. Norton, 2019); Kim TallBear, *Native American DNA: Tribal Belonging and the False Promise of Genetic Science* (Minneapolis: University of Minnesota Press, 2013).

[8] TallBear, *Native American DNA*; Eve Tuck, "Suspending Damage: A Letter to Communities," *Harvard Educational Review* 79, no. 3 (2009): 409–428.

College was one among a set of small colleges organized around a pragmatist pedagogy that prioritized "experiential learning." Every other quarter, for four years, students moved away from the tiny silvan town of Yellow Springs, Ohio, and took jobs anywhere they could imagine.[9] "I was very restless, even too restless for Antioch," Carolyn told me. "I just wanted to be on my own."

In the early 1960s, Antioch had a reputation for radical politics and drew students with a bent toward social activism. But Carolyn knew none of this when she was considering colleges. The hegemonic activism and what college histories called "militant intellectualism" was imperceptible from her high school in Portland, Oregon.[10] "It was a shock, and it was a good thing that I did not drop out right away."

"I was in a very conservative family, in a conservative town," Carolyn told me. Growing up, she was an only child and in 1951 her father's carpentry business went bankrupt. Her parents packed up their pickup truck with some clothes, the dog, their camping gear, and Carolyn. They let the bank have the house and drove east to the Rocky Mountains, stopping for a few weeks at a time for her father to do carpentry jobs and for Carolyn to go to school (Figure 7.1). When the weather turned cold, they crossed the Colorado border into Arizona and set up house for a few months in Phoenix. Carolyn's father worked, she went to primary school, and on weekends the family visited the national parks of the Sonora Desert (Figure 7.2).

For tribal members, the commodification of cultural authenticity offered one way of earning money out of the brutality of dispossession – turning the white American popular mythology into tourism dollars in the capitalist colonial structure that had long oppressed Indigenous groups. There was a way to look "Indian" to the white consumer eye that, in the postwar decades, reenacted a nineteenth-century fiction.[11]

While Carolyn and her parents were in Arizona, the latest US federal policy change related to Native Americans was emerging. In 1955, the Public Health Service, within the Department of Health, Education and Welfare (today's Health and Human Services), was handed responsibility for the Indian Health Service, formerly called the Division of Indian Health and located within the Department of the Interior's Bureau of Indian Affairs. The new Indian Health

---

[9] Burton R. Clark, *The Distinctive College: Antioch, Reed & Swarthmore* (Chicago: Aldine PubCo, 1970); Algo D. Henderson, *Antioch College: Its Design for Liberal Education* (New York/London: Harper & Brothers, 1946); Cary Nelson, "Antioch: An Education in the Real World," *The Chronicle of Higher Education* 53, no. 43 (June 29, 2007): B.5.

[10] Clark, *The Distinctive College*, 62.

[11] Philip Joseph Deloria, *Playing Indian*, Yale Historical Publications (New Haven, CT: Yale University Press, 1998); Philip Joseph Deloria, *Indians in Unexpected Places* (Lawrence: University Press of Kansas, 2004). Native people were also "supposed to" be poor under the white settler gaze. Alexandra Harmon, *Rich Indians: Native People and the Problem of Wealth in American History* (Chapel Hill: The University of North Carolina Press, 2010).

Figure 7.1   Carolyn Matthews, around age eight, and her father, circa 1951. Her dog Rip van Winkel (Rippy) is also in the photo. Roger Burmont Matthews stopped for food during a road trip to find work. The photo description on the back reads, "Chow time! On road/ between Boise Idaho & Salt Lake City." Photographer credit: Melba Cambridge Matthews. *Source:* Matthews Collection, VANV.

Service was a response to decades of federal cuts to Native clinics and reliance on private contractors, as well as state and local governments, to attend as they saw fit to the health needs of Native communities.[12] In Arizona as in other places, the poverty that caused poor health was not predetermined, but an

---

[12] Abraham B. Bergman et al., "A Political History of the Indian Health Service," *The Milbank Quarterly* 77, no. 4 (1999): 571–604; Betty Pfefferbaum et al., "Learning How to Heal: An Analysis of the History, Policy, and Framework of Indian Health Care," *American Indian Law Review* 20 (1995): 365. When Carolyn was in Arizona, a Cornell field hospital on the Navajo reservation was studying a new therapy for tuberculosis, while also trying to treat the disease. The study set up a wide net of surveillance in the name of public health. David Jones, "The Health Care Experiments at Many Farms: The Navajo, Tuberculosis, and the Limits of Modern Medicine, 1952-1962," *Bulletin of the History of Medicine* 76, no. 4 (2002): 749–790; see also Bergman, 583. Settler scientists had considered tuberculosis on federal reservations a key health problem since the late nineteenth century, attributing its prevalence to "race" rather than to the US state's past and continued discrimination. Christian W. McMillen, *Discovering Tuberculosis: A Global History, 1900 to the Present* (New Haven, CT: Yale University Press, 2015).

Figure 7.2   Carolyn Matthews in moccasins, Rippy, and her mother, circa 1951.
Photographer: Roger Burmont Matthews. Carolyn's father wrote a description on the back around 1951: "Grain Grinder. Tonto National Monument." The Tonto National Park is in the Upper Sonora Desert, near the ancestral home of Akimel O'odham people. *Source:* Matthews collection VANV.

expression of political structures working against many Native people's desires.[13]

After Carolyn's stay in Phoenix, the family drove back to Portland and built a new house. When Carolyn arrived at Antioch College at eighteen years old, she was a curly-haired aspiring anthropology major with braces on her teeth. One of Carolyn's first experiences as an Antioch student was working a co-op term at NIH. At the time, NIH administrators had "procurement contracts" with several colleges, a few labor unions, and the national organizations of two Anabaptist churches, to supply "normal control" human subjects for medical experiments.[14] In addition, the federal Bureau of Prisons flew or bussed twenty-five men to the Clinical Center every five weeks for most of the 1960s in an arrangement akin to convict labor leasing.[15]

She got free room and board, and a small "stipend" from NIH funneled through the college. When scientists were not using students in medical experiments they were allowed – encouraged – to work unpaid in "career placements" designed to keep the Normals busy, away from mischief or rumination, and advertised by NIH as a way to boost their resumes through (unwaged) work experience at a prestigious institution. Importantly, she also got course credit from Antioch and a chance to see Washington, DC in her downtime. To get these resources, however, she also had to give.

She arrived at the Clinical Center in early October and was assigned a bed in the ward on 8 West, having been allotted to NIH's Institute of Arthritis and Metabolic Diseases. Each of the institutes that comprised NIH was given space at the Clinical Center for their "bedside" research – their studies on whole people. Based on the studies they had planned, the scientists forecasted their need for Normals and every three months sent their order to the administrator for NIH's "Normal Volunteer Patient Program," the hinge between Antioch and NIH. Carolyn's body was projected into a study on insulin clearance. She started on Tuesday morning.

Her room had a private bathroom, which she shared only with her room-mate, a German Jewish grandmother from Brooklyn with thyroid disease, whom Carolyn adored. However, for the study, the nurses needed her to urinate, not inside the private bathroom, but in the open hospital room into a commode while they waited – and to do it every fifteen minutes. When

---

[13] Angela Garcia, *The Pastoral Clinic: Addiction and Dispossession along the Rio Grande* (Berkeley: University of California Press, 2010); Tuck, "Suspending Damage."

[14] Laura Stark, "Contracting Health: Procurement Contracts, Total Institutions, and Problem of Virtuous Suffering in Post-War Human Experiment," *Social History of Medicine* 31, no. 4 (2018): 818–846.

[15] Laura Stark and Nancy D. Campbell, "Stowaways in the History of Science: The Case of Simian Virus 40 and Clinical Research on Federal Prisoners at the US National Institutes of Health, 1960," *Studies in History and Philosophy of Biological and Biomedical Sciences* 48, Part B (December 2014): 218–230, https://doi.org/10.1016/j.shpsc.2014.07.011.

Carolyn and I met in 2016, I helped her get access to her NIH study record and she allowed me to see a copy, too. The study record includes a log of study procedures (researchers), social surveillance notes (nurses), and legal forms (administrators) for the autumn of 1962. It includes a note from Nurse Cushing the same day Carolyn started the study: "Unable to void @ prescribed times so test running irregularly."[16]

Her record does not include a consent form.[17] "Regarding informed consent: It's hard to tease out what I felt at the time, in 1962, from the perspective of 54 years later." Carolyn wrote me an email in the summer of 2016. "Although the NIH docs knew I had an interest in biology, I actually did not have much knowledge about it yet." She had two months of college course work at that point and told the doctor who admitted her that she was an anthropology major. "The docs may have credited me with a higher level of understanding than I deserved, and I wasn't assertive enough to say 'I do not understand'," Carolyn said. "I do know that I was very trusting of the whole thing, and it never occurred to me to question anything." She shared a sensibility with many white Americans of the early 1960s. The Cuban Missile Crisis took shape the week after she arrived, reinforcing public support for the sciences of national defense. A month later came the death of Eleanor Roosevelt, champion of social safety net programs as former First Lady and of international human rights as Chair of the United Nations Commission on Human Rights after World War II. Trust in authority – in government, in science, and in medicine – among middle-class white Americans would only unravel later in the 1960s. At the same time, sovereignty claims in the United States were being made ferociously by the American Indian Movement.[18]

While Carolyn was at the Clinical Center as a "normal" subject, there were also children from Native communities living in the research hospital as sick patients to study and treat. Irene was a thirteen-year-old Navajo girl who, in the summer of 1964, got a new roommate at the Clinical Center on the same floor where Carolyn had lived, the 8 West for insulin and diabetes. Irene's new roommate was a nineteen-year-old Normal from an Anabaptist college in Kansas who described her time with Irene in daily letters home to her boyfriend. "I had enjoyed being alone so much," the young woman wrote after Irene temporarily left, "but am glad she's back now since we still have not gotten the TV back (and I hope we never will)." Irene had a tracheotomy; she was shy and spoke little; the location of her family is unknown. Federal concern specifically with the health of children from Native communities is

---

[16] Carolyn Matthews Medical Record, "Nursing Notes," 10 AM, October 9, 1962.
[17] Carolyn Matthews Medical Record 1962. On the history of consent practices at the Clinical Center, see Stark, *Behind Closed Doors*.
[18] Elizabeth Rich, "'Remember Wounded Knee': AIM's Use of Metonymy in 21st Century Protest," *College Literature* 31, no. 3 (2004): 70–91.

a legacy of boarding school programs that removed Native children from their homes, cut them off from their families, and socialized them into white American habits, priorities, and networks.[19] Tuberculosis was a particular concern on reservations and in the total institutions of boarding schools especially. Irene's presence at the Clinical Center was likely an effect of the Bureau of Indian Affair's failures and Congress's reassignment of responsibility for Native health to the Public Health Service, which also subsumes NIH.

They developed a sweet intimacy. "Irene + I have been having a very good time together lately," the young woman wrote. "She acts so different around some people but not like a vegetable with me." The young woman was set to return to her Anabaptist college at the start of September. "She said she will miss me when I leave. I just hope I've been a good influence + have helped her see more in life than the TV set."[20] The following year, Irene was still living on the eighth floor of the Clinical Center, the young woman's boyfriend now living at the Clinical Center serving as a Normal himself. "I suddenly remembered you wanted me to look up Irene," he wrote to her the following year, "but she wasn't in." Irene was, however, still living on the same ward on 8 West.[21]

Another "normal control" Anabaptist young woman wrote to her grandparents about Native children at the Clinical Center. The young woman played with the children as part of her unpaid work assignment in the Clinical Center's recreation department, where she went during downtime from experiments. "[O]ne of my favorites is Alice [redacted], a 5 year old Am. Indian," she wrote to her grandparents, "I might have mentioned her before." Alice also lived on the endocrinology unit, 8 West. "Alice has a very rare condition, at least for a girl. Her blood does not clot," the Normal wrote. "I'm not positive, but I think she has to have transfusions something like every five days."[22] Like Irene, the nature of Alice's illness, the location of her family, or the condition of her assent are unclear.

The idea that it made sense to talk about "Am. Indian" as a group was a product of the US settler state. Until the American Revolution, settlers considered Indigenous people to be white, which was simultaneously a political and a biological statement. By the turn of the nineteenth century, however, white elites lumped various Native groups into the racial category of "red." When they were white, Native people were imagined as physically and mentally like white settlers, if socially different, and, therefore, capable of reform and worthy of assimilation. The contrast was with people whose families were

[19] David Wallace Adams, *Education for Extinction: American Indians and the Boarding School Experience, 1875–1928* (Lawrence: University Press of Kansas, 1995).

[20] Page 4 (MS page 10). August 1964 letter set, Reimer 2019, VANV.

[21] Keith Reimer to Susan (nee Stuckey) Reimer, June 1965 (page 19 of 89) and August 17, 1965 (page 80 of 89), Reimer letter set, VANV.

[22] Marnette (Bette) Hatchett to her grandparents. Marnette Hatchett Collection, VANV.

African, nearly all enslaved at the time. Ruling elites and citizens of a slave nation could better justify the institution by maintaining the strategic fantasy that any perceived physical differences between settlers (largely European descent) and the people they enslaved (largely African descent) indicated a physical incapability of adopting dispositions on which political rights rested.[23] When Native people became "red," they too were reimagined as biologically different from white settlers, politically intractable, and incapable of governance.[24] This recategorization justified explicit federal policies of termination starting in the early nineteenth century – including deportation, expulsion, and extermination.[25]

The political attitudes of the white settler state overlay a material need for territory – fields, mountains, water – and the resources they contained, as well as exigency of smooth travel that possession allowed. Thus, Native dispossession and scientific racism by the United States is always interdependent with Black subjugation.[26] After the Civil War, scientific racism, under the banner of social Darwinism, elaborated stage theories of society, including Lewis Henry Morgan's three-stage insult of savagery, barbarism, and civilization.[27] These stage theories were teleological, associating practices and people with a period in evolutionary time. The racial category of "red" was a political tool that built in the assumption of difference in social evolution and distance in

[23] Rana A. Hogarth, *Medicalizing Blackness: Making Racial Difference in the Atlantic World, 1780-1840* (Chapel Hill: The University of North Carolina Press, 2017).

[24] Alden T. Vaughan, "From White Man to Redskin: Changing Anglo-American Perceptions of the American Indian," *The American Historical Review* 87, no. 4 (1982): 917–953; Nancy Shoemaker, *A Strange Likeness: Becoming Red and White in Eighteenth-Century North America* (Oxford/New York: Oxford University Press, 2004). In his otherwise compelling, important, and no-doubt landmark study, Vaughan attributes the shift in settlers' visions of race to their ideas and attitudes, which he describes as "logical" extensions of settlers' experiences with Native groups. Vaughan admits the reasons for the shift in settlers' racial vision are obscure to him, and he is silent on material explanations, not least, settlers' forced relocation of Native people and seizure of Native lands, which the fabricated idea of racial difference helped justify.

[25] Claudio Saunt, *Unworthy Republic: The Dispossession of Native Americans and the Road to Indian Territory* (New York: W. W. Norton, 2020).

[26] Tiffany Lethabo King, *The Black Shoals: Offshore Formations of Black and Native Studies* (Durham, NC: Duke University Press, 2019).

[27] See especially chapter 6: Robert Bieder, *Science Encounters the Indian, 1820-1880: The Early Years of American Ethnology* (Norman: University of Oklahoma Press, 2003). Morgan introduced his three-stage (or "status") theory of the "Progress of Mankind" in the first chapter of Bieder, *Science Encounters the Indian, 1820-1880*; Lewis Henry Morgan, *Ancient Society Or, Researches in the Lines of Human Progress from Savagery, through Barbarism to Civilization* (Project Gutenberg, May 20, 2020), www.gutenberg.org/ebooks/45950; Yael Ben-Zvi, "Where Did Red Go?: Lewis Henry Morgan's Evolutionary Inheritance and U.S. Racial Imagination," *CR: The New Centennial Review* 7, no. 2 (2007): 201–229.

evolutionary time, which then prompted scientists to design studies that treated these assumptions as real.[28]

The US federal government introduced the category of "Indian" to the US Census in 1850 but the aim – and census-takers' activity – was to count only Native people who "renounced tribal rule" and "exercised the rights of a citizen." In the mid nineteenth century, the point was to track the settler-state goal of disappearance through "assimilation" and to count the number of people who needed to pay federal taxes particularly after the Indian Apportionment Act (1871) that parceled Native people's land for private ownership. The criteria for being "Indian" changed after the Dawes Severalty Act (1887), through which the US government took possession of Native land. After the Dawes Act passed Congress, people were required to register on tribal rolls (Dawes Rolls), which were based on ancestry. Thereafter, US census-takers were taught to count people as Native depending on their blood quota, not based on whether they renounced tribal rule (and were potential taxpayers). The addition, revision, and reintroduction of "Indian" into the census tracked the careening US policies toward Native groups.[29]

By 1962, the category of "Native American" lumped together the people that Carolyn had seen in Arizona as a child and the children from the Great Plains that lived in the Clinical Center in the 1960s, as well as many more groups – including Inuit people, Hawaiian islanders, and people who straddled the borders of settler states, like Mohawks (US–Canada) and Akimel O'odham (US–Mexico).[30] The creation of a bureaucratic category to capture a variety of Native groups suggested a coherent scientific racial grouping. It also suggested a homogeneity and commensurability, which belied a range of lifeways, lineages, and experiences under American Empire and global capitalism.

Although Carolyn was poor at being a "normal control" human subject, she was diligent in her career assignment as an unpaid lab technician. Dr. Jan Wolff was leader of the Clinical Endocrinology Branch, and supervisor of the

[28] Johannes Fabian, *Time and the Other: How Anthropology Makes Its Object* (New York: Columbia University Press, 1983); Michel-Rolph Trouillot, *Global Transformations: Anthropology and the Modern World* (New York: Palgrave Macmillan, 2004); Seth Garfield, *Indigenous Struggle at the Heart of Brazil: State Policy, Frontier Expansion, and the Xavante Indians, 1937–1988* (Durham, NC: Duke University Press, 2001).

[29] Kenneth Prewitt, *What Is Your Race?: The Census and Our Flawed Efforts to Classify Americans* (Princeton: Princeton University Press, 2013); Josh Pearl, "Native Americans and the Census," Journeys: Topics in Digital History, January 25, 2016, https://journeys.dartmouth.edu/censushistory/2016/01/25/native-americans-and-the-census/; Margaret M. Jobe, "Native Americans and the U.S. Census: A Brief Historical Survey," *Journal of Government Information* 30, no. 1 (2004): 66–80; Dan Bouk, *Democracy's Data: The Hidden Stories in the U.S. Census and How to Read Them* (New York: MCD, 2022).

[30] Audra Simpson, *Mohawk Interruptus: Political Life across the Borders of Settler States* (Durham, NC: Duke University Press, 2014).

young scientist who had enrolled Carolyn as a Normal. Together, they processed data for their endocrinology research in their laboratory.[31] Carolyn helped. How would people have perceived her then, more than fifty years ago, I asked in one conversation? "Oh, naïve," Carolyn told me, "Cooperative, except for not being able to pee on schedule, mostly cooperative."[32]

Hormones from the thyroid process iodine, and, troublingly, nuclear fallout sends out radioactive iodine. As of 1962, the United States was continuing a program of testing nuclear weapons in the ocean and upper atmosphere, as well as in the deserts of the American Southwest. Thus, the United States was funding both research on atomic science and research on the diseases that atomic science caused. Under its Atoms for Peace campaign, the US government paid the salary of Dr. Wolff, who was figuring out how to block the thyroid function in the event of a nuclear accident.[33] Another (paid) technician would stop at one of NIH's slaughter houses and bring thyroid glands from sheep, pigs, and other animals to the lab. Carolyn's job was to grind up the glands and prepare them for tests. She did not know what the researchers were trying to learn. "It had something to do with thyroid," she said. "It was very lofty and technical." She did not find her work interesting; she just wanted to do a good job.[34]

Interesting or not, she was happy. So it was easy to smile when the *Washington Post* photographer arrived at Dr. Wolff's lab (Figure 7.3). With Carolyn in a white lab coat, her situation was too delicious to resist: a human guinea pig doing research on other lab animals. The photographer snapped pictures of Carolyn rather than the scientists. Readers of the *Washington Post* were taught what it was like for her and other Normals to be subjects of NIH medical research. "Personal consent is essential," the article instructed. "No volunteer ever starts any test without first understanding its purpose, methods, duration, demands and inconveniences or discomforts." The people who served in medical research were portrayed as a type. "They are not daredevils, nor fools, nor even overly inspired idealists," the journalist wrote. "They're ordinary men and women, mostly in their twenties, who see a job

---

[31] Marvin C. Gershengorn, "History of the Clinical Endocrinology Branch of the National Institute of Diabetes and Digestive and Kidney Diseases: Impact on Understanding and Treatment of Diseases of the Thyroid Gland," *Thyroid* 22, no. 2 (February 2012): 109–111; Dewitt Stetten, ed., *NIH: An Account of Research in Its Laboratories and Clinics* (Orlando: Academic Press, 1984), 419.

[32] Matthews Oral History 2016, VANV.

[33] David V. Becker et al., "The Use of Iodine as a Thyroidal Blocking Agent in the Event of a Reactor Accident: Report of the Environmental Hazards Committee of the American Thyroid Association," *JAMA* 252, no. 5 (1984): 659–661; Kiyohiko Mabuchi and Arthur B. Schneider, "Do Nuclear Power Plants Increase the Risk of Thyroid Cancer?," *Nature Reviews Endocrinology* 10, no. 7 (2014): 385–387.

[34] Matthews Oral History 2016, VANV.

YELLOW SPRINGS, O., Feb. 1963 - Antioch College student Carolyn
Matthews is pictured on her co-op job doing laboratory work in
the Normal Volunteers program of National Institutes of Health
in Bethesda, Md. Carolyn is in the unusual position of both
dontaint research material - samples of her blood - and working
on it in lab tests. She assisted Dr. Jan Wolff in a study of the
circulation of anti-thyroid substances in the body.

Figure 7.3 Carolyn Matthews as "Normal Control," published in the *Washington Post*, 1963.

Source: Co-op Photo Collection, Olive Kettering Library, Antioch College, Yellow Springs, OH.

that needs to be done and offer to do it."[35] NIH strategically allowed access at the Clinical Center only to sympathetic journalists to manage its reputation.

In the evenings, Carolyn would get a pass from 8-West nursing station, trundle down to the first floor, and wait in the sweet autumn dusk for Jan Wolff's car. Wolff and the esteemed biochemist, Edith Wolff, who had married him, hired her as their babysitter. Word of Carolyn spread among the upper ranks, and she started babysitting for other scientists, too. Still, it came as a surprise when the doctor made a proposition. "[O]ne day when I was in my room on 8-West, another doctor who I'd never seen before came into my room – his name was Dr. William O'Brien – and said, 'Hey, I'm putting together a team to go to Arizona to take x-rays, to find the incidence of rheumatoid arthritis in this Indian population in Arizona. Would you consider coming?'"

Carolyn had never seen him before. He had just returned from a three-month stay at the Blackfeet Indian Reservation on the US border with Canada.[36]

"Me being this restless person I already told you about, I said, 'Are you kidding? I'd love to'."[37]

Sacaton, 1963: Carolyn as Research Technician

Carolyn arrived with the team in January 1963 and had ambitions of making an ethnological study of the Native community. It was a romance drawn from popular knowledge of the dominant school of cultural anthropology at the time, Claude Lévi-Strauss's *Structural Anthropology*.[38] Practitioners drew "plans" of space from a bird's-eye view, often Indigenous villages and tribal meeting places (Figure 7.4). Carolyn named the "dust, dirt, grit," and drew arrows to the world beyond the paper's edge as she imagined it. "The desert to infinity," they pointed. For structural anthropologists, it was unnecessary to learn of the world beyond the paper's edge because the "plans" of physical space corresponded to the social structure of a group, which itself could be "mapped" with ink and paper. Lévi-Strauss was credited with helping to dismantle the racist orthodoxy in mid-century Euro-American science. His target and that of many others was race science and eugenics, which claimed racial inequality was the natural consequence of biological difference, rather than the result of political oppression and discrimination, as the liberal academy agreed. In some camps, however, structural anthropology in general and Lévi-Strauss in particular were rebuked for replacing race science with a

---

[35] Patricia Griffith, "Conscientious Non-objectors to Important Medical Research," *Washington Post*, February 3, 1963.
[36] Anon., "Arthritis among the Blackfeet," *Modern Medicine* (February 4, 1963): 45–46.
[37] Matthews Oral History 2016.
[38] Claude Lévi-Strauss, *Structural Anthropology*, n.d.

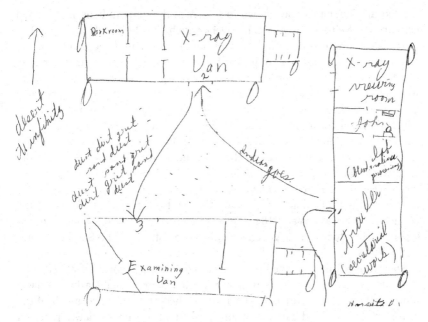

Figure 7.4 Carolyn Matthews's "plan" of field site, undated.
Source: Carolyn Matthews collection, VANV.

nonetheless essentialist and race-based concept of human difference. Essential racial difference was built into the visual grammar of structural anthropology. Its mapping method assumed that people with a different skin color also had within their bodies a fundamentally different social "structure" – where the white Euro-American experience was the unmarked category against which difference was compared. But Carolyn's amateurism left room for subversion. In drawing a plan of the NIH site, her untrained anthropology flipped a method and a viewpoint that scientists had originally developed on (and had used to discipline) Native groups back onto a scientific community.[39]

Soon Carolyn dropped her ethnographic ambitions as scores of people emerged out of the desert to infinity. That seemingly endless vacant sandscape was in fact striped with irrigation ditches and established villages.[40] People of

[39] Don D. Fowler, *A Laboratory for Anthropology: Science and Romanticism in the American Southwest, 1846–1930*, 1st ed. (Albuquerque: University of New Mexico Press, 2000); Nancy Parezo, ed., *Hidden Scholars: Women Anthropologists and the Native American Southwest* (Albuquerque: University of New Mexico Press, 1993).

[40] M. Kyle Woodson, *The Social Organization of Hohokam Irrigation in the Middle Gila River Valley, Arizona* (Sacaton, AZ: Gila River Indian Community, 2016); David

the Akimel O'odham tribe, known to settlers as the Pima Indians, had farmed the banks of Keli Akimel (the Gila River), which lent the people its name. The river also distinguished them from their kin, the Tohono O'odham (Papago), who lived to the south and migrated between desert and mountain, farming and hunting, depending on the season.[41] The Akimel O'odham, by contrast, were steady, resident agriculturalists. Until the late nineteenth century, they grew cotton, wheat, corn, beans, and melon for subsistence, often with surplus, which they traded or gave to neighboring tribes, to travelers en route to the Pacific coast's mineable mountains, or to administrators of the Mexican and American governments surveying the land and its resources. Mexico officially claimed the land after that colony itself gained independence from Spain, but the Mexican government still let the tribe govern itself as a matter of convenience.[42] Akimel O'odham land remained part of Mexico even after the Mexican–American War (1846–1848), when the United States claimed swaths of territory that it governs today as the states of the Southwest and Pacific Coast.[43]

The group's land remained part of Mexico – and its people effectively self-governing resident farmers – until the United States bought a bite of land south of the Gila River in 1854. Then the United States claimed much of the Akimel O'odham land as its own and pushed people into a newly parceled reservation, its linear borders drawn over a small section of their previous territory. The drive for railroads, mail service, and other accoutrements of national infrastructure hardened the US containment, as did US mining interests in the region.

In the 1950s, local scientists around Phoenix bemoaned the recent changes in the foods Akimel O'odham people ate. Where beans, tortillas, and chili

---

H. DeJong, *Stealing the Gila: The Pima Agricultural Economy and Water Deprivation, 1848–1921* (Tucson: University of Arizona Press, 2009).

[41] For an analysis of government-industry technology experiments on the Papago Reservation starting in the 19[80]s, see Jeremy Greene et al., "Innovation on the Reservation: Information Technology and Health Systems Research among the Papago Tribe of Arizona, 1965–1980," *Isis* 111, no. 3 (2020): 443–470.

[42] DeJong writes that "Mexican independence had little impact on the Pima . . . The arrival of Americans, however, did affect them. While Americans were prohibited from entering the country under Spanish rule, Mexican law was relaxed, and hundreds of American mountain men descended on the Gila River and its tributaries" (17). DeJong, *Stealing the Gila*.

[43] DeJong, *Stealing the Gila*; David H. DeJong, *Forced to Abandon Our Fields* (Salt Lake City: University of Utah Press, 2011); John P. Wilson, *Peoples of the Middle Gila: A Documentary History of the Pimas and Maricopas 1500s–1945* (Sacaton, AZ: Gila River Indian Community, 2014). At this moment, the Smithsonian Institution was created as an appendage of the US government and made responsible for collecting information on the land and people. See especially chapter 5, "The Great Surveys," in Fowler, *A Laboratory for Anthropology*.

peppers were staple for older folks, younger people bought packaged food at trading posts. "The Pima Indians are a southern Arizona tribe which has readily acclimatized itself to Western civilization through fairly close proximity to off-reservation urban communities," one local scientist explained in 1959. They were geographically near "civilization," the scientist said, and yet consumed "mostly non-perishable foods that are bought because of long distances traveled and lack of facilities for storage of perishable foods."[44] Scientific investigations left unremarked the relation between changes in the tribe's foodways and the United States' enclosure of the tribe in a reservation away from the Gila, the government's systematic denial of water for their irrigation systems that decimated their remaining farm lands, and federal policy on agriculture and economics favoring industrial capitalism. Left unremarked was the possibility that these changes in foodways were a political problem, not an outcome of nature (e.g., droughts), and therefore not natural and inevitable.[45]

When the new arrivals from NIH parked their trailers a short stretch from the US Public Health Service's Indian Hospital, they made their first discovery: they were far from the only scientists studying the local tribe (Figure 7.5.).[46] Inside the PHS Indian Hospital, a young clinician had been collecting information from Native medical records on a stack of index cards.[47] He was of the mind, as a colleague put it, that there was a "critical need for good health for Indian people if they were to take their rightful place in American Society."[48] The praise winced with settler assumptions and values. The clinician had first studied tribes in the area during the late 1950s under the auspices of the

---

[44] Frank G. Hesse, "A Dietary Study of the Pima Indian," *The American Journal of Clinical Nutrition* 7, no. 5 (September 1, 1959): 532–537.

[45] DeJong, *Stealing the Gila*. The US Indian Appropriation Act (1859) created reservations for "Pima" people. "Throughout the 1850s, the Pima continued to enjoy economic growth. While the Indians served notice that their land and resources were under their sovereign control, in the post-Civil War years the United States encouraged settlement of the territory, and by the end of the decade, the Pima stood on the precipice of far-reaching economic and political change. No longer did they control their own destiny, as the rapidity of change brought about by federal polices diminished Pima sovereignty and disadvantaged the Indians . . ." (56).

[46] On the NIH team's meeting of Dr. Maurice Sievers, see Stephanie Stegman, "Taking Control: Fifty Years of Diabetes in the American Southwest 1940–1990," PhD dissertation, Arizona State University, 2010, 88. Sievers would go on to coauthor many scientific articles with the NIH team, especially Dr. Peter Bennett. For analysis of the local scientific context and the NIH research team, as well as a comprehensive documentation of the 1963 field study on the ground, also see Stephanie Stegman, "Taking Control: Fifty Years of Diabetes in the American Southwest 1940–1990," PhD dissertation, Arizona State University.

[47] Maurice L. Sievers and James R. Marquis, "The Southwestern American Indian's Burden: Biliary Disease," *JAMA* 182, no. 5 (1962): 570–572.

[48] "Maurice Sievers: Obituary," *The Arizona Republic*, March 2020, www.azcentral.com/obituaries/par036229.

Figure 7.5 NIH trailers at data collection site, circa 1963.
Source: NIDDK, PowerPoint. Thanks to NIDDK Sacaton Branch.

National Cancer Institute, which was part of NIH. He interviewed and examined Navajo uranium miners who were digging radioactive materials.[49] Uranium is an essential building block of atomic weapons and nuclear states were the only buyers on uranium markets. The US government was spending money both to continue its nuclear weapons development program and to investigate the diseases that resulted from the program, especially cancer from exposure to uranium dust, nuclear waste materials, and fallout from test explosions.[50]

[49] Gabrielle Hecht, *Being Nuclear: Africans and the Global Uranium Trade* (Cambridge, MA: MIT Press, 2012). Through the study of French African uranium mines over the past six decades, Hecht examines the creation in the post–World War II period of a market for uranium, which, she shows, developed by creating risk for workers, not only producing a rhetoric of atomic threat for populations, raising questions of responsibility under global capitalism.

[50] The mines on the Navajo reservation were closed in 1986 by which point there were more than 500 mining sites run by private contractors of the Atomic Energy Commission and its incarnations. William L Chenoweth, "Navajo Indians Were Hired to Assist the U. S. Atomic Energy Commission in Locating Uranium Deposits," Arizona Geological Survey Contributed Report Series (US Department of Energy, September 2011); OECA US EPA,

By the time Carolyn arrived, she had gotten two weeks of official training at the NIH Clinical Center on how to use the equipment before flying to Phoenix with the team.[51] So with limited formal training, her working knowledge of making x-rays – how to arrange people's bodies like a portrait sitting, where to touch to tactfully drape the lead apron, when to swivel the machine like a carnival game – came from practice in conditions unlike those of the Clinical Center exam rooms.

Among other things, Carolyn was now standing inside a truck (Figure 7.6). When the door swung open it let in a burst of light and air – along with a person hoisting up into the vehicle. She took four pictures: neck, pelvis, hands, and feet. Then she stepped to the tail of the truck to let the films develop in the makeshift darkroom. At the end of January, she turned nineteen years old.

At first, the work was steady and manageable. Bernice, a member of the tribe, recorded each person's basic information in a mobile-home trailer when they arrived, then gave them a drink from the carton of Black Label beer. Bernice was hired as a temporary secretary for NIH, and her husband was hired too. He worked as a driver, one of the locals who knew the topography and motored around the reservation collecting tribe members for the study.[52]

"Case Summary: $600 Million Settlement to Clean up 94 Abandoned Uranium Mines on the Navajo Nation" (2017); Will Ford, "A Radioactive Legacy Haunts This Navajo Village, Which Fears a Fractured Future," *Washington Post*, January 19, 2020, www .washingtonpost.com/national/a-radioactive-legacy-haunts-this-navajo-village-which-fears-afractured-future/2020/01/18/84c6066e-37e0-11ea-9541-9107303481a4_story.html. See also DeJong, *Stealing the Gila*; Doug Brugge and Rob Goble, "The History of Uranium Mining and the Navajo People," *American Journal of Public Health* 92, no. 9 (September 2002): 1410–1419. See also Rafael Moure-Eraso, "Observational Studies as Human Experimentation: The Uranium Mining Experience in the Navajo Nation (1947–66)," *NEW SOLUTIONS: A Journal of Environmental and Occupational Health Policy* 9, no. 2 (1999): 163–178.

[51] NIH started offering regular training courses in x-ray technology. They took two years to complete. For example, "News from Personnel – X ray Technology Course," *NIH Record*, May 21, 1963: 2, https://nihrecord.nih.gov/sites/recordNIH/files/pdf/1963/NIH-Record-1963-05-21.pdf.

[52] This scene suggests how members of Native communities differentially adopted the vocabulary and skills of settler science – perhaps also adapting and co-opting its prestige in some circles for their own strategic purposes. My analysis is indebted to Gabriela Soto Laveaga's work documenting and showing how some Mexican peasants combine their own local knowledge of the (coveted) wild yam and scientists' chemical vocabulary (about progesterone) to produce themselves as local elites. The important insight is that Indigenous groups are internally stratified, that internal hierarchies are emergent (not preexisting and stable), and that people can use settler science as a resource to negotiate their own status in the broader context of citizenship. Gabriela Soto Laveaga, *Jungle Laboratories: Mexican Peasants, National Projects, and the Making of the Pill* (Durham, NC: Duke University Press, 2009). It would be interesting to develop further the suggestive parallel between Bernice in Sacaton and the legendary "Nurse Rivers" in the context of

Figure 7.6 Carolyn Matthews's NIH research team on location in 1963.
Photographer: Carolyn Matthews. Carolyn Matthews wrote a description on the back in 1963:
"X-ray van / Joel [Silverman] + Dr. [Thomas] Burch / (pineapple juice used / for glucose tolerance
test / is in Black Label beer / cartons)." *Source*: Carolyn Matthews collection, VANV.

When he pulled up to the site, he pointed his passengers to Bernice's trailer,
who then directed them to Carolyn's truck. He waited as they crossed the sand
to the last truck, where a member of the NIH team took their blood and saliva.
The drink from the beer carton had been pineapple juice, not alcohol, and the
needle in the vein allowed the researchers to test the level of sugar in their

the US Tuskegee Syphilis Studies. Susan M. Reverby, "Rethinking the Tuskegee Syphilis
Study. Nurse Rivers, Silence and the Meaning of Treatment," *Nursing History Review:
Official Journal of the American Association for the History of Nursing* 7 (1999): 3–28.

blood. Then he gathered his passengers to return them home. By late March, his trips were so frequent and his car so full, Carolyn was working ten-hour days.

In the evening, Carolyn and the rest of the NIH team drove to Chandler, the next town over from Sacaton and off the reservation. NIH had rented an old estate hotel, *La Hacienda*, with separate units for each of the families: Dr. William O'Brien, his wife, and two toddlers; Dr. Thomas Burch, head of the field unit; Dr. Peter Bennett and nurse Sally Bennett, his wife, both from the United Kingdom, and another young researcher also taking x-rays. Carolyn had a unit to herself as a family of one. Most nights, she cooked, cleaned herself, and promptly went to bed. On Sundays, the team cooked a communal meal in the central kitchen and ate together in the banquet hall.

At that time, there was a beguiling, and, to many researchers, highly suspect explanation circulating in the medical literature for a smattering of seemingly unrelated diseases: arthritis, diabetes, fever, and skin rashes.[53] These different conditions had a common cause, researchers argued, at least in some instances: an autoimmune disorder.[54] Before joining the NIH team, Bennett had done influential work in the UK to first argue that arthritis was often a sign of an autoimmune attack (rheumatoid arthritis), and not always the result of wear and tear on the joints (osteoarthritis). After his initial work, Bennett was keen to investigate whether genetics determined the autoimmune mal-function, but he conceded in a 1960 write-up: "No definite views on causation can be expressed on the evidence available."[55] Picking up the scent of a research question from across the Atlantic, NIH researchers got to work. In the fall of 1962, while Carolyn was serving as a normal control at the Clinical Center, O'Brien decamped for the Blackfeet Indian Reservation. He led a field team in the first of a two-phase study of the causes of autoimmune-based, rheumatoid arthritis. To help collect data from the Native community, NIH invited Bennett to Montana. Building on the work of the UK team, the NIH group wanted to find the "non-genetic etiology for

---

[53] In the late 1920s, a scientist in Norway was led to the seemingly improbable conclusion that diabetes might be caused by an antibody response (assumed to be an infection). Edvard Gundersen, "Is Diabetes of Infectious Origin?," *The Journal of Infectious Diseases* 41, no. 3 (1927): 197–202.

[54] The possibility of a common cause of disparate conditions also meant that diseases previously categorized together (diabetes, arthritis) were split and reshuffled. Some kinds of arthritis were now more akin to some kinds of diabetes than to diseases that shared the same name because they were both caused, endocrinologists claimed, by this auto-immune malfunction. Arleen Tuchman, *Diabetes: A History of Race and Disease* (New Haven, CT: Yale University Press, 2020).

[55] J. S. Lawrence and P. H. Bennett, "Benign Polyarthritis," *Annals of the Rheumatic Diseases* 19, no. 1 (March 1960): 20–30.

these conditions," namely, rheumatoid factor (an indicator of autoimmune disease). When O'Brien had come to Carolyn's hospital room at the Clinical Center, he had been a stranger to her because he had been in Montana gathering x-rays and blood from Native community members.

In addition to genetics, another possible cause of arthritis was climate. The NIH team designed their field survey in Sacaton to allow them to test the hypothesis that climate (hot/dry versus cold/moist) predicted a particular form of arthritis better than genetics ("heredity"). The logic of comparison articulates the racialized assumptions of NIH scientists. There are plenty of hot dry places in the United States. But scientists wanted to control for another key variable: race. By this logic, scientists needed, then, to pick people (a "stable population") in a hot dry place that were related to people thousands of miles away on the US–Canada border. There is little that is intuitive about the idea that a group of people in the Sonoran Desert were a "stable population" given migration and relationships across the US–Mexico border; there was little reason to imagine they were more closely related to people in Montana than to residents of Phoenix, where Akimel O'odham people moved, married, and worked. Through their logic of comparison it is possible to see how scientists imagined this political difference in how the US government treated groups as a biological reality.[56]

By April, the team had brought nearly all of the adults living on the reservation – one thousand in all – to the trucks for an exam, x-rays, saliva swab, and blood draw.[57] The team packed up the trailers and their families, said goodbye to Bernice and the other local workers, and went back to Bethesda. Carolyn returned to school.

Boston, 1964: Carolyn as Research Technician

In the spring of 1964, Carolyn decamped for another co-op assignment: in Boston, where she worked as a live-in cook for a communal home for the

---

[56] "Race" is a meaningful category in the modern United States, but that category has multiple meanings – political, biological, administrative, and more. As Steven Epstein writes, historically, the idea that "the categories of political mobilization and state administration also functioned as the categories of biomedical differentiation was given crucial support by federal health officials." As a result, "the question of why the categories of political mobilization and administration should also be viewed as the categories of greatest biomedical relevance was effectively bypassed." Steven Epstein, *Inclusion: The Politics of Difference in Medical Research* (Chicago: University of Chicago Press, 2007).

[57] H. Bennett and T. A. Burch, "The Distribution of Rheumatoid Factor and Rheumatoid Arthritis in the Families of Blackfeet and Pima Indians," *Arthritis and Rheumatism* 11, no. 4 (1968): 546–553. The team examined 86 percent of adults on the "well-defined reservation communities" (551).

Quaker church.[58] At the end of the quarter, she decided to stay. She had gotten out of sync with her Antioch cohort when she skipped a beat to go directly from Bethesda to Sacaton. ("I was never one of the tribe," she said.) She also never fit the mold of the politically radical Antioch student. And the pacifist politics of her Quaker housemates had started to compel. US imperialism bore down in Southeast Asia and nuclear weapons enhanced its menace. Within American borders, the practice of nonviolent resistance of the mainstream Black Civil Rights movement aligned with the goals and philosophy of the Quaker church. In the early 1960s, only a few groups connected US imperial projects, on the one hand, and the Black Civil Rights movement, on the other hand, in a way that articulated the movement as a response to American imperialism via chattel slavery and its extensions.

As a result, it was possible for middle-class white Americans, like Carolyn, to miss imperial projects close to home. Like the Black Civil Rights movement, the Red Power movement mobilized in the late 1950s and 1960s.[59] Unlike the Black Civil Rights movement, however, the demands of Red Power were not for civil rights defined by liberal individualism. Native American sovereignty movements, contemporary with the nonviolent Civil Rights movement, demanded recognition of their full political authority over land and lives that the US claimed or presumed to control.[60]

With her NIH credentials, despite now being a college drop-out, Carolyn was hired as a scanning technician in the Radiation Department at Peter Bent Brigham Hospital, next to the Harvard Medical School. This steady job, one she had found herself not pointed at in a college binder, felt very good in concept. She wore a lab coat, fetched doctors to inject people with radioisotopes, and scanned their bellies with a new machine the department was testing. At the end of the week, she got her own paycheck.

In practice, the work felt considerably worse. Instead of sending radioactive atomic particles into the bodies of Akimel O'odham people (her job as an x-

---

[58] Carolyn lived and worked at the Beacon Hill Friends House, which still operates, at 6 Chestnut Street, Boston, MA. Oral History 2017. Carolyn also lived for a year in a small apartment with Sandy Calloway Ferguson, who had been a Normal at the US National Institutes of Health through the Antioch co-op program and also dropped out of Antioch. Sandy had severe reactions to the experimental drugs she was given at NIH. She had to leave NIH and take a term off from Antioch to recover at home in upstate New York. Ferguson Oral History 2019; Woods Oral History 2018.

[59] Rich, "Remember Wounded Knee."

[60] King, *The Black Shoals*. The US government's prior gestures toward Native sovereignty nonetheless held that Indian nations were not independent but "domestic dependent nations" according to the Marshall Trilogy. As Richard Scott Lyons writes, "It's a paradox: sovereignty produced by colonization." Scott Richard Lyons, "Actually Existing Indian Nations: Modernity, Diversity, and the Future of Native American Studies," *The American Indian Quarterly* 35, no. 3 (2011): 294–312.

ray technician with NIH), she was using a new machine that absorbed atomic waves from people's temporarily radioactive bodies. Carolyn worked under Dr. James Potchen, who got his medical training during the Eisenhower administration and the federal "Atoms for Peace" campaign, through which the Atomic Energy Commission urged researchers to find uses for atomic science beyond its use for weapons.[61] Under the Atoms for Peace program, the federal government made and mailed radioactive versions of molecules to scientists around the country, including the radioactive iodine used in Jan Wolff's lab at the NIH Clinical Center, where Carolyn had worked without pay when she was not serving in experiments. Radiologists like Potchen knew that when the radioactive version of a molecule, called radioisotope tracer, is put into the body (in a drink, or through an injection), it moves through the body just like a regular molecule, but it emits radiation. If researchers had a machine that could detect radiation as well as human tissues (x-rays are only good for bones), they could see how "normal" bodies work and where sick bodies are having problems. During his first year of residency, Potchen designed such a machine. Today he is credited with making the first radionucleotide image in the esteemed Harvard hospital system.[62]

To get clinicians beyond Boston to use his scanner – and, through the instrument, to extend the reach of his influence – he needed to show its usefulness. He corralled the experimental results of his scanner to get funding from the Atomic Energy Commission to continue testing the machine and to establish its precision.[63] He hired Carolyn to do the scans.

The work felt ethically wrong to Carolyn, and by Potchen's own admission it was. "That first image was produced at night somewhat surreptitiously," Potchen said, "because all clinical radioisotopes at the Brigham were in the domains of the departments of surgery, endocrinology, or hematology. Radiology did not have permission from the 'clinical chiefs' to be using isotopes on patients."[64] The Atomic Energy Commission had control over

---

[61] "Potchen, E James," in *American Men & Women of Science: A Biographical Directory of Today's Leaders in Physical, Biological, and Related Sciences*, 23rd ed., vol. 5 (Detroit, MI: Gale, 2007), 1191.

[62] E. James Potchen, "Reflections on the Early Years of Nuclear Medicine," *Radiology* 214, no. 3 (2000): 624.

[63] Potchen presented on new scanning techniques at a conference in Shaker Heights, OH, October 30–31, 1964, based in part on work that Carolyn had done for him. United States Atomic Energy Commission, "Isotopes and Radiation Technology," 2, no. 2 (Winter 1964–1965): 192–193. See also Teresa J. C. Welch, *Fundamentals of the Tracer Method* (Philadelphia, PA: Saunders, 1972); James E. Potchen and Alexander Gottschalk, *Diagnostic Nuclear Medicine*, Golden's Diagnostic Radiology Series; Section 20 (Baltimore, MD: Williams & Wilkins, 1976).

[64] Potchen told his story of derring-do and it became a favorite lore of the field of radiology. Potchen, "Reflections on the Early Years of Nuclear Medicine."

isotope distribution, and sent isotopes to medical departments for their exclusive use.

It was unclear what patients and healthy people knew about the injections they got and about the scans. Carolyn herself did not know much, nor did she ask. Sometimes she scanned patients for a diagnosis or to ready them for surgery. Other times she scanned the healthy hospital orderlies and janitors, her fellow low-skill laboring friends. Every few days, she carried the radioactive waste down the hall to an open storage room in the hospital.[65]

Years later, Potchen reflected on the ethics of his experiments in nuclear medicine to refine the scanner: "I am still not convinced that the early unregulated years inflicted any harm to patients." Ethically, the interest was whether scientists thought the procedure could harm, rather than whether a patient or a community had a right simply to know they were part of an experiment and to have the power to decide. The concern was with the result rather than with the ethics of the action. "Such an approach would be unthinkable today, and would require many years to reach the same conclusions." It is worth noting that Potchen's practices were not uniformly endorsed, even then. Carolyn lasted a year.

Portland, 1977: Carolyn as Story-Worker

In the thirteen years after Carolyn dropped out of Antioch, she moved more than a dozen times. After her year as a Boston lab technician, she moved to Switzerland as a nanny; later, she briefly took classes at Columbia University, where she watched the women's movement pulse into popular awareness on the streets of New York City. She kept her Boston connection to the Quaker church and, though she was not a religious believer, she took a job with the American Friends Service Committee in 1968 in San Francisco. Within months of Carolyn's arrival, an activist group, Indians of All Tribes, landed on Alcatraz Island in the San Francisco Bay. Just a mile offshore from the city's North Beach, the island was by 1969 a legendarily brutal maximum-security federal prison. The activists of Indians of All Tribes aligned with the Red Power movement. They organized to occupy Alcatraz to protest the US government's policies of termination of Native communities. They carried weapons and stayed for nineteen months. Carolyn's activist sensibilities were, however, elsewhere.

San Francisco was the context for what she came to see as her "political awakening." In that city and that year, the war in Vietnam was the dominant target of protest and the summer of love its emblem. Her work and social life

---

[65] At the time, the federal government did not require that people label radioactive material when it was being moved. United States Atomic Energy Commission, "Isotopes and Radiation Technology," 192–193.

was oriented around her Christian pacifist Friends, so it was the US proxy war on Red China, not the Red Power movement, which tuned her attention. She was involved in the Movement for a New Society, which opposed violence abroad and supported Black Civil Rights at home. Yet it had less to say about war as a strategy of continental American Empire or about Native sovereignty claims. In 1973, two years after the Alcatraz occupation ended, Native groups operating under the American Indian movement stood off with federal agents at Wounded Knee, the site of the US federal massacre of Lakota people in 1890.[66]

Through another antiwar job with the Quaker church, Carolyn moved back to Boston in 1974. The same year, the US Congress passed the National Research Act in the wake of public exposure of the Tuskegee Syphilis Studies, the four-decades long racist and exploitative study of African-American men funded by the US government. Federal health agencies, led by the National Institutes of Health, had been quietly working to write policies to guide research on "human subjects" since the mid-1960s.[67] When an enterprising journalist at the *Washington Star* finally followed up in 1972 on a tip about the Public Health Service-sponsored research on poor Black men in Tuskegee, Alabama, the agency had already drafted a set of policies for the US Congress to adopt as law. There were other options proposed, including a centralized federal ethics review office (to encourage consistent decision-making and transparency), rather than innumerable local Institutional Review Boards, modeled on the NIH Clinical Center's Clinical Review Committee. But Congress lined up behind the plan from the NIH proposal, to deflect legal responsibility away from the federal government and toward local research institutions.[68]

Thus, US federal protections for human subjects research – and popular knowledge of those protections (e.g., through the mass media) – bore the mark of the specific historical event that prompted the protections: the Tuskegee Syphilis Study. In the present day, American ethics training, as well as mass-market accounts of the importance of the field of bioethics, begins with Tuskegee.[69]

The essential feature of Tuskegee in its public iterations was that the human subjects racially identified as Black. They also lived in a rural area, earned less

---

[66] Dee Brown and Hampton Sides, *Bury My Heart at Wounded Knee: An Indian History of the American West* (New York: Holt Paperbacks, 2007).
[67] Stark, *Behind Closed Doors*.
[68] Ibid.
[69] John Hyde Evans, *The History and Future of Bioethics: A Sociological View* (Oxford: Oxford University Press, 2014); John H. Evans, *Playing God?: Human Genetic Engineering and the Rationalization of Public Bioethical Debate* (Chicago: University Of Chicago Press, 2002); Fox and Swazey, *Observing Bioethics*; Reardon et al., "Science & Justice."

money than they needed, and experienced structural racism, as well as casual prejudice and interpersonal discrimination. Yet, American law and ethics about human research was coded as a set of practices especially "protecting" people who were Black – which was important but also an insufficient recognition of the structures of injustice at play. The 1974 National Research Act, in addition to requiring regulation, also mandated that a committee of experts write ethical guidelines that would translate the regulations into useable language and a moral framework. This mandate resulted in the Belmont Report, following a four-year effort in the mid-1970s, through which a team of scientists, theologians, and ethicists gathered background papers and wrote a several-hundred-page manual. This influential report and its background papers discussed Black Americans in research, but failed to mention other racial-identity groups or the political dispossession that connects them.

Carolyn was in her early thirties when she moved back to Portland to be near her aged parents and the landscape that had raised her. In 1977, as the Belmont Commission completed its work and five years after the Tuskegee Syphilis Study had hit the headlines, Carolyn went back to school at Portland's Marylhurst Education Center (now Marylhurst University). With the hopes she might earn some course credit, Carolyn wrote up her past decade of transient work experience.

From the vantage point of 1977, she first explained her previous work at the NIH Clinical Center:

> Normal Voluntary Control (human guinea pig)
> National Institutes of Health
> Bethesda, Maryland
> Oct.-Dec. 1962
>
> I was an object of medical research, "employed" as a normal voluntary control. I lived in the Clinical Center, availing myself at all times for medical / experimental study. I understood very little of the research that was being done on me – partly because I was not informed of much, and partly because what was told to me was too complicated for me to understand. (I have since read parts of two articles [Carolyn's college advisor and friend] loaned me: "Accountability in Health Care" and "A Political Perspective on Accountability & Research on Humans," which confirm many of my principled questions on the subject, developed since my experience at the NIH.

Her "principled questions" were organized around whether she, as a human subject, had been on the receiving end of an adequate informed consent process. The write-up documents that the scholarly literature emerging in the late 1970s, as the field of American bioethics consolidated, informed Carolyn's bioethical vocabulary and analysis.

Carolyn also analyzed her earlier job as a research technician in Boston. Among other things, her responsibilities included:

– Arrange for patient's departure (hospital orderly, taxi), making appropriate and sensitive response to patient's anxious questions (such as, 'you should ask you [sic] doctor, since I do not make diagnoses – he'll be looking at this test along with the other's you have had . . .')

– [Participate] from time to time [in department"s] "research cases" to determine how fast a compound would concentrate in a given area of the body and how fast it would dissipate. Sometime these coincided with clinical cases, but often the "volunteers" were solicited. In my opinion, these people were not informed adequately what the procedure really was and what the possible risks were. In fact, my supervisor deliberately disguised what he knew to be certain facts of such a procedure in order to get volunteers, taking advantage of the fact that most of these volunteers were from the uneducated, lower echelons of the hospital hierarchy – orderlies and supply room workers.

Carolyn explained from the vantage of 1977 what she learned from the job in Boston. "This was one of my early exposures to medical deception," she wrote. "I did not at this time, link this up with my having been an ill-informed subject of medical experimentation at NIH." Carolyn affirmed that by 1977 she did not connect her earlier experience at NIH as a "human guinea pig" with that of her research subjects in Boston. Carolyn's 1977 write-up shows how she reframed her past with tools of the field of American bioethics that emerged in this decade.

Carolyn also thought back to her work in Sacaton with the Akimel O'odham people. What is worth noting is what was left unstated under the frame of American bioethics:

X-ray technician
NIH field survey
Pima Indian Reservation
Sacaton, Arizona
Jan.-March, 1963

I fell into this job accidentally, while at the NIH. I heard about a survey team that was going to Arizona to determine the incidence of rheumatoid arthritis in an Indian tribe. They were looking for a flunkie to take x-rays, and invited me along. So I went on an "own-plans" job with Antioch, and arranged to return to campus one quarter later.

I was quickly trained to take and manually develop x-ray pictures of hands, feet, neck and pelvis on 1000 Indians in a portable unit right on the reservation. An Indian tribe was used because it was considered a relatively stable population; the family members do not move around much, so hereditary factors could be considered. Also, the results were to be compared with a previous survey done in Montana (on the Blackfoot [sic]

Indians), the idea being to compare the effect of a hot-dry climate with a cold-dry climate on the incidence of arthritis. Interestingly, no significant difference was discovered. Plans were being made to go to Guam (hot-wet climate).

I had hopes of being able to do some kind of anthropological study with the Pima Indians for some Antioch credit, but that soon proved impractical as we were working about ten hours each day.

Unlike her accounts of her experiences in Boston and Bethesda, Carolyn did not articulate her experiences in Sacaton in terms of questionable consent or possible coercion. Unlike the other two medical research experiences, she did not see a similarity between herself at the Clinical Center, her human subjects in Boston, and her research subjects in Sacaton – the Akimel O'odham people. In her 1977 report to Marylhurst, Carolyn described each of the three work experiences in chronological order. To draw the ethical connection between Boston and Bethesda as she did, Carolyn had to leapfrog her work with Native research subjects.

The puzzle is how it was possible for a person like Carolyn to fail to draw a connection. She was a deeply compassionate and educated person: she was informed about the ethics of research on people, had been employed in anti-imperial work for social justice, and was lucid in her memory of her part in research on the Akimel O'odham people ("Pima Indians"). Based on her 1977 account of her earlier experiences in Boston, Bethesda, and Sacaton, I want to suggest that Carolyn enunciated the formally sanctioned American bioethical frame. She is a best-case scenario not merely of what was allowable but what was perceptible under the dominant bioethical frame – which is a frame that set the structure of bioethical thinking that continues to dominate white American sensibilities today. To support justice through the work of history, it is important to precisely identify how and why the research setting that was most overtly characterized by racialist imperial power dynamics could be illegible within such a solid frame of ethical analysis. Carolyn is an indicator of what American bioethics made morally imaginable and legible as a matter of ethics.

Conclusion

Carolyn's story testifies to the embeddedness of modern American bioethics in a logic of colonialism. Dominant strains of modern science have invented and stabilized purportedly natural categories, such as race, which settler states and other global actors have used to naturalize difference, universalize place-specific relations, and justify political rule. No doubt, territorial expansion, forced labor, and violence are tactics shared beyond settler states; Indigenous groups have used these tactics, too. What is distinctive of Euro-American settler states, however, is their enlistment of formal science in the service of setter logics and practices.

At the same time that settler states have operated through science, in the twentieth century, the field of bioethics has safeguarded the moral authority of science.[70] Instead of operating as place-specific, responsive practices of valuation, American bioethics reproduced the ontologies of the human sciences, life sciences, and clinical sciences that it claimed to hold to account. Bioethics functioned to uphold the settler logics of the sciences it claimed to regulate because it was a product of those sciences – designed to allow them to continue racialized, extractive practices without serious question or reform.[71] The implication is that research with people, even in the most formally ethical conditions among well-intended people, can enact this colonial logic.[72] To the extent that professional history has uncritically absorbed the discourse of modern American bioethics, the field – even in its best efforts – can rearticulate a colonial moral logic.

Carolyn's story materializes the perfect functioning of bioethics' settler logic. It reveals that the formally sanctioned and globally dominant American bioethical frame actively makes imperial situations illegible as such. In doing so, it defuses the settler past that structures the possibilities of action in a given situation, and the possibilities of scientific research in the first place. The imperial power dynamics of the Sacaton research situation was illegible in Carolyn's moral imagination, even from the vantage of 1977. With Carolyn as a guide, it is possible to recognize that the dominant American bioethical frame is marked by its origins in the US federal government – as a historically specific outgrowth of the Tuskegee exposé, which allowed elite actors to narrow ethics concerns, to emphasize racial discrimination as a preeminent issue, to organize ethics of race around a Black–white binary, and to exclude consideration of settler colonialism.[73]

This points to the framing of bioethics as an issue of safeguarding civic individualism, rather than dismantling anticolonial logics. It also suggests the bioethical frame is organized around expectations of citizenship and the demand that the state observe individual civil rights. Thus, the possibility that sovereignty is at stake, rather than civil rights, is not part of the officially sanctioned American bioethical frame. An imagination of the United States as

---

[70] Stark, *Behind Closed Doors*.
[71] Laura Stark, "Reservations," *Isis* 113, no. 1 (March 2022): 128–136.
[72] Tess Lanzarotta, "Ethics in Retrospect: Biomedical Research, Colonial Violence, and Iñupiat Sovereignty in the Alaskan Arctic," *Social Studies of Science* 50, no. 5 (October 1, 2020): 778–801; Joanna Radin, "'Digital Natives': How Medical and Indigenous Histories Matter for Big Data," *Osiris* 32, no. 1 (September 1, 2017): 43–64.
[73] Aileen Moreton-Robinson, *The White Possessive: Property, Power, and Indigenous Sovereignty* (Minneapolis: University of Minnesota Press, 2015).

an empire is inactive. Projects on research justice, such as CARE, as well as new critical histories of human experimentation, offer strategies reengaging research as an ethical relational practice.[74] In addition to promises that have not yet been met, there are solidarities yet to activate between movements for Native sovereignty and against anti-Black racism, both grounded in the bioethics of the US state.

[74] Cristina Mejia Visperas, *Skin Theory: Visual Culture and the Postwar Prison Laboratory* (New York: New York University Press, 2022).

PART III

Governance, Politics, and Self-Determination

8

Investigating Cuauhtémoc's Bones

Politics, Truth, and Mestizo Nationalism in Mexico

KARIN ALEJANDRA ROSEMBLATT

This chapter examines a very public controversy in which people of Native ancestry from a rural village in Mexico engaged the human and natural sciences and Mexican nationalism. On February 2, 1949, in the village of Ixcateopan (also rendered as Ichcateopan), in the state of Guerrero, a villager named Salvador Rodríguez Juárez found some papers hidden behind a shrine to the Virgin of Asunción in his home. Among the papers were an eighteenth-century book with some marginalia and an eight-page booklet in a leather cover. Both were signed by the colonial-era Franciscan friar Toribio de Benavente, known as Motolinía. They appeared to indicate that the last Mexica emperor, Cuauhtémoc, was a Native of Ixcateopan and had been buried in the village church.[1] Rodríguez Juárez consulted the village priest about what to do with the papers, and the priest disclosed the astounding discovery at a Sunday mass.[2] An official from a nearby town who attended the mass told a landowner from his village who worked as a stringer for the major Mexican daily El Universal.[3] The news spread quickly.

The politically imperiled governor of Guerrero, General Baltasar Leyva Mancilla, immediately set up a commission of local experts to investigate. But the committee members had limited expertise, and the governor soon sought help from the Instituto Nacional de Historia e Antropología (INAH, National Institute of History and Anthropology), headquartered in Mexico

---

[1] The story is well-documented, most recently in Paul Gillingham's compelling and deeply researched book, *Cuauhtémoc's Bones: Forging National Identity in Modern Mexico* (Albuquerque: University of New Mexico Press, 2011). See also Lyman L. Johnson, "Digging up Cuauhtémoc," in *Death, Dismemberment, and Memory: Body Politics in Latin America*, ed. Lyman L. Johnson (Albuquerque: University of New Mexico Press, 2004); Salvador Rueda, "De conspiradores y mitógrafos: Entre el mito, la historia y el hecho estético," *Historias* 39 (October 1997–March 1998): 17–26.

[2] My account here follows Gillingham, *Cuauhtémoc's Bones*, chapter 2, unless otherwise noted. A different version of how the story reached Mexico City is found in Ángel Torres y González, *La tumba de Cuauhtémoc: Un reportaje histórico* (Mexico City: Nacionalidad, 1950).

[3] Alejandra Moreno Toscano, *Los hallazgos de Ichcateopan, 1949–1951* (Mexico City: UNAM, 1980), 11.

City. The INAH commissioned Eulalia Guzmán, who had spent several years in Europe locating and transcribing Mexican codices and had written a manuscript on Hernán Cortés's *Relaciones*.[4] Guzmán knew colonial-era documentation well.[5]

Once in Ixcateopan, Guzmán would have realized at once that something was amiss with the documents. Motolinía could not have penned marginalia in a book published more than two hundred years after his death. Guzmán nonetheless went ahead with her investigations, returning to the village several times between February and September, and accumulating additional evidence in support of the authenticity of the burial. Still unsure about the truth of the matter, she believed that archaeological explorations in the church could provide definitive confirmation or refutation,[6] and she asked the INAH to appoint an archaeologist. When the commissioned archaeologist was waylaid, the governor decided the excavation should begin anyway. After a few days of digging under the church altar and past three sets of floors, on September 26 villagers uncovered a pile of stones reminiscent of a pre-Columbian ritual mound. Below, there was a stone slab. Finding a gap in the stone, the diggers hit softer mud. A pick went through it, releasing a foul cuprous odor. Lifting some stone, the crew found a dagger and a copper plaque. The plaque was engraved with the numbers 1,525 and 1,529 separated by a cross and, below, "Rey, é, S, Coatemo." Lifting the plaque, the assembled villagers saw a cranium with some beads in it. They unearthed more bones. Cuauhtémoc's remains had been found!

The church bells rang. Guzmán brought the copper items out of the church, displayed them to the assembled public, and delivered an impromptu speech praising Cuauhtémoc. The discovery set off national celebrations. Schoolchildren from around the country placed handfuls of soil from their villages at the feet of the statue of Cuauhtémoc in Mexico City. Over seven

---

[4] Eulalia Guzmán, comp., *Relaciones de Hernán Cortés a Carlos V sobre la invasión de Anáhuac* (Mexico City: Instituto Nacional de Estudios Históricos de las Revoluciones de México, 2019), originally published in 1958.

[5] According to Gillingham, *Cuauhtémoc's Bones*, 51, the governor requested Guzmán as "the scholar most likely to come up with the right result." Gillingham adduces that because Guzmán was from a provincial town, she was well-suited for fieldwork in a place like Ixcateopan. (In reality, Guzmán had left her Zacatecas village at the age of eight.) On Guzmán's early life, see Beatriz Barba de Piña Chan, "Eulalia Guzmán Barrón," in *La antropología en México*, v. 10, *Panorama histórico: Los protagonistas*, eds. Carlos García Mora and Lina Odena Güemes (Mexico City: Instituto Nacional de Antropología e Historia, 1988), 255.

[6] Eulalia Guzmán to the director of *El Universal*, August 15, 1949, exp. 23 serie proyecto Ichcateopan, subserie correspondencia, Eulalia Guzmán Archive, Museo Nacional de Antropología (hereafter AEG). Unless otherwise noted, all archival citations are from the caja 1, subserie correspondencia of AEG.

thousand people traveled to Ixcateopan on Columbus Day 1949. Members of Congress speechified. Ixcateopenses rallied.[7]

The Professional and Political Stakes

But did the remains belong to Cuauhtémoc? In the ensuing months and years, experts examined the documents, the church, the burial site, and the village's oral traditions, using the tools of physical anthropology and history as well as architecture, chemistry, and archaeology. One group of scientists, led by Guzmán, concluded that it was at least possible, perhaps probable, that Cuauhtémoc was buried in Ixcateopan. Two other government-sponsored commissions came to the opposite conclusion.

This chapter, like the chapters by Rosanna Dent and Eve Buckley in this book, shows how the views, methods, and moral stances evinced by the experts in both camps emerged through interactions taking place across local, regional, national, and transnational scales – in villages, state agencies, and the transnational scientific community. The experts who investigated the Ixcateopan discovery dialogued with each other but also with Ixcateopense commoners and village officials, public opinion, US experts, national leaders, and regional elites like Governor Leyva Mancilla. The noted muralist Diego Rivera weighed in on the Ixcateopan controversy, as did President Miguel Alemán (1946–1952). Because the controversy took place during the years immediately following World War II, it became entangled in discussions regarding Mexico's place in a Cold War world order that placed a premium on science. Intellectuals on both sides of the dispute deployed a science they deemed cosmopolitan, but they viewed the relation of science to Indigeneity and nationalism in different – and very gendered and racialized – ways.

For all the participants in the debate, the nature of Indigenous contributions to the Mexican nation was at stake, as was the role of science in authorizing narratives regarding Indigenous and Mexican history. Villagers did not reject science or expertise; they deployed it strategically and in combination with local views. The members of the two official commissions repeatedly voiced their adherence to scientific protocols as well, while characterizing Guzmán and her allies as irrational and branding them ideologically motivated *indófilos* (Indophiles). Members of the official commissions dismissed Ixcateopenses as likewise guided by fervor rather than evidence. Situating themselves as proponents of a cosmopolitan and detached science, these experts eschewed inevitably affect-laden relationships with the Ixcateopenses and propounded a Mestizo identity that embraced, they said, Indigeneity and Spanishness

---

[7] Gillingham, *Cuauhtémoc's Bones*, 60–63; Moreno Toscano, *Los hallazgos*, 11–13; "Tierra de todas las zonas indígenas en el monumento a Cuauhtémoc," *Excélsior*, October 15, 1949, in ibid., 108–109.

equally.[8] Guzmán and her collaborators responded by saying that the official investigations were incomplete and superficial. The authenticity of the burial and the documents, they affirmed, were too easily dismissed by experts who diminished the importance of Indigenous ancestry to the mixed, *Mestizo* Mexican nation. Guzmán's opponents denied these accusations, proclaiming their respect for Cuauhtémoc and the Mexica Empire. Members of the first INAH commission laid a wreath at the statue of Cuauhtémoc in Mexico City before beginning their work.[9]

Gender played an important role in the controversy. Because the main investigator was a woman working within an otherwise almost exclusively male scientific community, and because the virility of Cuauhtémoc was so often invoked in the debate, the episode laid bare the ways in which participants related gender differences to both scientific authority and Indigeneity – and of both science and Indigeneity to nationalism. As we shall see, Guzmán was caricatured as embodying a feminine, quasi-religious hysteria and at the same time portrayed as both unnaturally masculine and unnatural in her claim to scientific authority. Wigberto Jiménez Moreno, secretary of the second official investigation – dubbed the Gran Comisión because it included Mexico's most well-respected scholars – wrote that the pro-Guzmán investigative commission was "more robust, intransigent, aggressive, and dangerous than its counterpart." And villagers' hostility toward the committee was, he said, due to their cultish adherence to Guzmán: "The discoverer's harangues animated them, creating a tendency opposed to any kind of serene attitude, and they called any hint of skepticism a sacrilege."[10] By contrast, the secretary of the Gran Comisión, Arturo Arnáiz y Freg, praised Mexico's "anthropological and historical disciplines" for their "maturity" and lauded commission members for not "letting themselves be lulled [*dejarse alucinar*] by deficient testimonies, documents plagued with anachronisms, or the mystical inclinations of enthusiastic but misguided groups."[11] For the official scientific establishment, then, neither Indigenous villagers nor Guzmán – a woman – could speak authoritatively for the nation. Nor could they fruitfully deploy science and rationality.

Subsequent scholarship by Salvador Rueda, Lyman Johnson, Paul Gillingham, and others has echoed these characterizations. Gillingham calls the reports of the indófilos "partisan," and "resting on unsupported assertion."

---

[8] Wigberto Jiménez Moreno, "Los hallazgos de Ichcateopan," *Historia Mexicana* 12, no. 2 (October–December 1962): 161–181; México, Comisión Investigadora de los Descubrimientos de Ichcateopan, *Los hallazgos de Ichcateopan: Actas y dictámenes de la Comisión Investigadora* (Mexico City, 1962), xii–xiii.

[9] Comisión Investigadora, *Los hallazgos*, xv; Jiménez Moreno, "Los hallazgos," 170.

[10] Jiménez Moreno, "Los hallazgos," 163, 175–176.

[11] Comisión Investigadora, *Los hallazgos*, xiii.

"Discussion," he says, "was largely replaced with shrill assertion." He further characterizes villagers' attitudes as "defensive hostility" laced with "aggression and resentment," and he repeats without critique the contemporary opinion that "to have taken Ixcateopan seriously was 'another of Doña Eulalia's insanities'."[12]

In sum, as this chapter shows, those who rejected the authenticity of the burial – in the 1950s and since – have subtly deployed notions of femininity to discredit the burial, Guzmán, and the Ixcateopan villagers. Guzmán and her allies countered by deploying science and casting doubt on the impartiality of scientists who did not denounce the Spanish conquest. Indeed, Guzmán was not as unscientific as her opponents claimed, and her opponents, despite their claims of neutrality, often offered poorly substantiated arguments. Unsurprisingly, Guzmán's opponents also held biases that shaped their investigations, including a bias against women in science.

My goal in showing Guzmán's allegiance to science and her opponents' biases is not to claim that Guzmán was correct. Recent scholarship has shown convincingly that she was not.[13] But by highlighting Guzmán's unacknowledged rigor and her opponents' scientific missteps, this chapter unveils how a mix of ideology and proof colored *all* the experts' opinions, along with their views of Indigenous Mexico and its history. Scholars' relationships to Ixcateopenses mattered too, as did the gendered relationships they forged among themselves.

Guzmán herself had an ambivalent position. Her ambitions no doubt drove her too, and at times she evinced deep suspicion of Salvador Rodríguez Juárez and other villagers. Yet she also probed their oral traditions and rituals with care. At times, her political proclivities along with her suspicion of her academic opponents blinded her to contrary evidence. Yet she showed greater understanding of ideological biases than her opponents.

The Ixcateopenses had ambitions and plans as well. Like the Terra Indígena Pimentel Barbosa villagers examined by Rosanna Dent in this book, they engaged science and scientists to pursue their aspirations and reduce harm. For Ixcateopenses, that meant deploying a pro-Indigenous nationalism and forging alliances with scientific allies who deployed a similar national narrative. For everyone involved nationalism was a guiding and limiting framework, a terrain of struggle.

The Historiographical and Theoretical Stakes

This chapter historicizes experts' moral stances, while stressing the competing contemporary moralities *within* the scientific community, born of distinct

---

[12] Gillingham, *Cuauhtémoc's Bones*, 67, 68, 71, 76–78.
[13] Ibid., chapter 7.

alliances.[14] The mores of investigators were conditioned in part by the material rewards and forms of recognition available to intellectual elites. Yet scientists' views of, and experiences in, Ixcateopan mattered as well. The allegiance to scientific proof that Guzmán and her allies demonstrated was tempered by their knowledge of how it could be manipulated. Perhaps more than their opponents, the so-called indófilos were attuned as well to what was *not* proven or *not* known, a fact missed by subsequent scholarship that has deployed hindsight (and the findings of a new scientific investigation in 1976).[15]

This essay also builds on Latin Americanist studies of the *testimonio* and Indigenous historical narratives. Those studies have probed how we make judgments regarding historical truths and historical narratives and how those may change over time. For instance, explorations of the collaboration between Rigoberta Menchú and Elizabeth Burgos Debray have suggested that readers of a testimony will inevitably view, and judge, Menchú's testimonio differently now than when it was first published during a brutal civil war. If distortions in Menchú's narrative were meant to save lives, how should that affect how we view those distortions? Does the relationship of narrative truth figure differently for Menchú because she is Maya? What level of cynicism, credulity, or contextual knowledge do we expect of distinct audiences?[16] Our own moral judgments may rest on accessing multiple, current and past, points of view.

Discussions of Menchú's testimonio have also raised questions about how narratives are authored. Menchú's account, like the accounts regarding Cuauhtémoc, involved conflictual and power-laden collaboration between Indigenous and non-Indigenous people. And both invoked political projects that were at once local, national, and transnational. In the Cuauhtémoc episode, villagers of Native ancestry came to work, somewhat reluctantly, with the group of outsiders led by Guzmán. In that regard, the controversy – like the testimonio genre – qualifies notions of Indigenous refusal.[17] Villagers

---

[14] In stressing the distinct moralities that exist in the past, I complicate Jan E. Goldstein, "Toward an Empirical History of Moral Thinking: The Case of Racial Theory in Mid-Nineteenth-Century France," *The American Historical Review* 120, no. 1 (February 1, 2015): 1–27.

[15] José Gómez Robleda, "Anexo al acta anterior: Carta del Dr. Gómez Robleda," in Comisión Investigadora, *Los hallazgos*, 381–383. I draw here on Robert Proctor and Londa L. Schiebinger, eds., *Agnotology: The Making and Unmaking of Ignorance* (Stanford: Stanford University Press, 2008). On the later investigation, Alicia Olivera de Bonfil, "Los restos de Cuauhtémoc y la política de los años setenta," in *Los archivos de la memoria* (Mexico City: Instituto Nacional de Antropología e Historia, 1999), 181–190.

[16] Rigoberta Menchú and Elisabeth Burgos-Debray, *I, Rigoberta Menchú: An Indian Woman in Guatemala* (London: Verso, 1993); Arturo Arias, ed., *The Rigoberta Menchú Controversy* (Minneapolis: University of Minnesota Press, 2001); Greg Grandin, *Who Is Rigoberta Menchú?* (London: Verso, 2011).

[17] Audra Simpson, "On Ethnographic Refusal: Indigeneity, 'Voice' and Colonial Citizenship," *Junctures: The Journal for Thematic Dialogue* no. 9 (2007): 67–80.

embraced Guzmán while refusing, as we shall see, to give *gringo* archaeologists access to their secrets.

In addition, the episode prompted what we might view as an affect-laden *scientific* refusal by Guzmán and her intellectual allies. Guzmán visited Ixcateopan repeatedly, and her opponents did not. This mattered. Power and affect shifted as words and objects circulated and as people traveled from the field to the city and back – or stayed put. Florencia Mallon has likewise noted that while the subjects of testimonios may have significant power at the moment they tell their stories, they inevitably lose control as their stories move outward.[18] If Cuauhtémoc's bones, like the kuru examined by Warwick Anderson, are good to think with, in the Ixcateopan controversy, specific actors' control over narratives varied as stories moved outward from the village and into regional, national, and international political debate. Yet neither the local histories nor the objects were lost. Today, what used to be the village church houses a shrine to Cuauhtémoc and a display of the remains. Ixcateopan's yearly mardi gras festivities celebrate Cuauhtémoc.[19]

The existence of these local histories suggests that the nation-state's use of symbols like Cuauhtémoc might be thought of, at least in certain circumstances, not just as appropriation but as something else, perhaps a form of imperfect recognition but also signs of active persistence, resurgence, or political savvy on the part of Native people.[20] Following Helen Verran, might we acknowledge that peoples can act together without necessarily thinking or being in the same ways?[21]

In what follows, I first provide additional details regarding the development of the scientific investigation and the surrounding public controversy. I underscore the relations between villagers and the experts, while remaining attuned to disparate opinions within each of these groups. I then discuss the scientific and political context in which these debates developed. A fictional dialogue I discovered in the archive of a member of the Gran Comisión provides a window onto questions of gender as it relates to science and Mexican nationalism.

---

[18] Florencia Mallon, "Editor's Introduction," in Rosa Isolde Reuque Paillalef, *When a Flower Is Reborn: The Life and Times of a Mapuche Feminist* (Durham, NC: Duke University Press, 2002), 1–34.

[19] Anne W. Johnson, "El poder de los huesos: Peregrinaje e identidad en Ixcateopan de Cuauhtémoc, Guerrero," *Anales de Antropología* 48, no. 2 (June 2014): 119–149.

[20] A nod here to Eve Tuck and K. Wayne Yang, "R-Words: Refusing Research," in *Humanizing Research: Decolonizing Qualitative Inquiry with Youth and Communities*, eds. Django Paris and Maisha T. Winn (Thousand Oaks, CA: SAGE, 2014), 223–248.

[21] Helen Verran, "A Postcolonial Moment in Science Studies: Alternative Firing Regimes of Environmental Scientists and Aboriginal Landowners," *Social Studies of Science* 32, no. 5–6 (2002): 729–762. See also Marisol de la Cadena, *Earth Beings: Ecologies of Practice across Andean Worlds* (Durham, NC: Duke University Press, 2015).

Eulalia Guzmán and the Ixcateopenses

Guzmán was skeptical when she first arrived in Ixcateopan, but this changed over the coming days and months. Shortly after her arrival, Guzmán visited the village *momoxtli*, a ritual mound that seemed to be the ruins of a castle. There, she found pottery that she identified as Aztec. She also heard of a ritual during Carnival in which villagers danced and pantomimed carrying a body that they took down from a tree. She learned of oral traditions that supported the account in the documents. Although Guzmán did not immediately reveal this detail to her superiors in the INAH, Rodríguez Juárez had in confidence shared with her additional documents, including an "Instruction" written by Rodríguez Juárez's grandfather, Florencio Juárez. According to this new document, the book with the marginalia and the papers bound in leather had been copied from an older set of documents when the latter were in poor condition.[22]

If Guzmán was suspicious, so too were villagers. Initially, only three village residents agreed to share with Guzmán versions of an oral tradition regarding Cuauhtémoc's burial. Most villagers blamed village leaders for revealing their long-held secret and for the unwanted attention the village was receiving. Village officials, however, recognized that the publicity might help the village gain access to resources. On the day Guzmán was leaving the town, they asked her to address a town meeting to see if she could convince more villagers to come forward with their stories. It worked. Eleven elders subsequently shared their own versions of oral traditions they had heard from their parents and grandparents.[23]

In 1949, as today, the details of Cuauhtémoc's life and death were sketchy, and the evidence Guzmán collected in Ixcateopan filled in gaps in prevailing accounts. At the time of the Spanish conquest, Ixcateopan was a largely Chontal-speaking village in an area conquered by the Triple Alliance of Mexico Tenochtitlan–Texcoco–Tlacopan. According to the Ixcateopan documents, local rituals, and oral histories, Cuauhtémoc's grandmother was a member of the Ixcateopan nobility who married a Texcocoan man sent to rule over the region around Ixcateopan. Cuauhtémoc's mother, a figure about which existing accounts said almost nothing, was imprisoned by the Mexica, along with her father, as a result of disputes over taxation, and they were taken together to Tenochtitlan, the seat of Mexica power. There, Cuauhtémoc's mother met his father, who was the son of the Mexica *tlatoani*, or ruler.

---

[22] Eulalia Guzmán a Humberto Colín, March 2, 1949, exp. 5. Parts of Guzmán's *informe* to the INAH are attached to this letter; Eulalia Guzmán a Salvador Rodríguez Juárez, March 14, 1949, exp. 8; César Lizardi Ramos, "Eulalia Guzmán revisa las actas de Ichcateopan," *Excelsior*, February 20, 1949 in Moreno Toscano, *Los hallazgos*, 77.

[23] Eulalia Guzmán to Humberto Colín, March 2, 1949, exp. 5.

Their son, Cuauhtémoc, returned to his mother's homeland, departing for Mexico's central valley when called to defend the Mexica capital from the arriving Spaniards.[24]

According to Bernal Díaz del Castillo's well-known chronicle, in the aftermath of the Spanish defeat of the Mexica, Cortés imprisoned Cuauhtémoc and took him on a trip to subdue rebellious Spaniards at Hibueras. On that trip, Cortés learned that the young man was planning to kill Spanish members of the expedition, and Cortés had the Mexica leader assassinated, leaving Cuauhtémoc's corpse hanging from a tree. Díaz del Castillo, who followed Cortés as he continued his trip, could not say anything about what happened to Cuauhtémoc's body afterwards. Here, the local Ixcateopan accounts filled in. Indigenous members of Cortés's expedition escaped and returned to the site of Cuauhtémoc's assassination, wrapped his body in cloth, transported him hundreds of miles to his hometown, and buried him at the site of his family's castle. Motolinía, who later visited the village, had him reburied at the site where the church was subsequently built. Because Motolinía was afraid the villagers might be persecuted by the Inquisition, he told them to keep quiet about the burial.[25]

Returning to Mexico City, Guzmán looked in archives for information regarding the Ixcateopan region. Denizens of Ixcateopan and the surrounding region began to write her with tips, local lore, and even rumors of an unknown codex held in a nearby village. She learned that long before Rodríguez Juárez proffered his documents there had been stories in the region connecting Ixcateopan and Cuauhtémoc. The Ixcateopan oral tradition, it seemed, abounded in neighboring villages.[26] Guzmán also visited Ixcateopan at least twice more between February and August. On one of these trips, Rodríguez Juárez, under pressure from village authorities, produced additional documents named in the Instruction, including, most intriguingly, two pieces of paper sewn shut with writing in invisible ink.[27] In August, Guzmán opened the document with invisible ink before assembled villagers and applied gentle heat, revealing some hard-to-read words. Initial examinations of the paper and

---

[24] See Rueda, "De conspiradores"; José Gómez Robleda, *Dictamen acerca de la autenticidad del descubrimiento de la tumba de Cuauhtémoc en Ixcateopan* (Mexico City: Secretaría de Educación Pública, 1952), 17–18.

[25] Gómez Robleda, *Dictamen*, 17–18. The original documents are published in translation in Gillingham, *Cuauhtémoc's Bones*, 227–235.

[26] Rodríguez Juárez to Guzmán, June 13, 1949, exp. 14; Alicia Olivera de Bonfil, *La tradición oral sobre Cuauhtémoc* (Mexico City: Instituto de Investigaciones Históricas UNAM, 1980), 94 on prior documentation of an oral tradition.

[27] Rodríguez Juárez to Guzmán, March 25, 1949, exp. 9. Two sheets are mentioned in Héctor Pérez Martínez et al., *La supervivencia de Cuauhtémoc: Hallazgo de los restos del héroe* (Mexico City: Ediciones "Criminalia," 1951), 27.

ink indicated that they were produced during the early colonial era, but Guzmán determined that further chemical analysis was needed.[28]

At this point, Guzmán was still unsure about the veracity of the story, but felt that physical evidence would help confirm or refute it. She began thinking about digging in the church after the rains had passed, in November or December. But villagers were keen on digging, and the governor of Guerrero was in a rush to unearth Cuauhtémoc before the hundredth anniversary of the state of Guerrero in October. Guzmán asked the INAH to send an archaeologist, but when the archaeologist was delayed, villagers and the governor decided to begin the excavation. Despite some efforts on the part of Guzmán, archaeological procedures regarding visual and written documentation of the excavation process were not followed. There were no drawings, and few photographs or notes.[29]

Gillingham adduces that villagers were excited about the stature they might gain. But village leaders had another reason for wanting to begin the excavation: a fear of looting. In letters to Guzmán, members of the village council told her that they had received letters from anthropologists wanting to excavate. Villagers were aware that if things did not move forward quickly, local riches might end up in the hands of US collectors. Guzmán was clearly attuned to villagers' suspicions. She reassured the village that INAH experts would respect the rights of the village. "We take extreme precautions," she wrote Rodríguez Juárez, "so as not to stir people up too soon, or awaken their ambition of finding valuable or historical material for personal gain." Guzmán also cited concerns about the villagers' sensibilities when INAH director Ignacio Marquina offered to commission archaeologist Pedro Armillas, who was a Spanish emigré. Guzmán told Marquina not to send Armillas because the Ixcateopenses would not trust a foreigner. If the town was to reveal its secrets, it wanted to ensure that collaborators would be on their side. Guzmán offered those assurances.[30]

To be sure, village residents were not always in agreement. The village appears to have retained divisions between descendants of earlier Chontal inhabitants and those descended from the Indigenous conquering forces of the Triple Alliance. Many of the village leaders, including Rodríguez Juárez, descended from those conquerors, and village commoners mistrusted their motivations. Rodríguez Juárez's grandfather Florentino was literate, a mason who enriched himself by wresting lands from other villagers.[31] Another division – which perhaps built

---

[28] Guzmán to INAH, August 15, 1949, exp. 21.

[29] Gillingham, *Cuauhtémoc's Bones*, 58.

[30] On the United States, see Rodríguez Juárez to Guzmán, February 25, 1949, exp. 9; Rodríguez Juárez to Guzmán, June 14, 1949, exp. 14. On Guzmán's assurances, see Guzmán to Salvador Rodríguez Juárez, July 8, 1949, exp. 14.

[31] Gillingham, *Cuauhtémoc's Bones*, 136–140; Olivera de Bonfil, *Tradición oral*, underscores the fact that both Rodríguez Juárez and his grandfather were part of a village elite.

from the former rift – arose during the dig itself when the excavation team found that the church altar did not have a foundation and would need to be dismantled to avoid it from falling. A group of villagers opposed the removal of the altar and protested loudly outside the sanctuary. They desisted only when the governor offered to rebuild the altar and fix the church.[32]

A team of village residents did the excavating. After six days, on September 26, they unearthed the tomb. Two days later, INAH archeologist Jorge Acosta finally arrived, and the following day, the noted archeologist Alfonso Caso and INAH director Ignacio Marquina reached Ixcateopan.[33] Caso had excavated the fabulous Zapotec ruins at Monte Albán, assisted by Guzmán and others. One of his first questions was to ask where the jewels were. It seemed improbable that a noble like Cuauhtémoc would have such a poor gravesite.[34] Marquina was even more dubious. Though he had commissioned Guzmán, he had not expected her investigations to yield results. In September, he told Guzmán he did not think they would find anything in the church, and he made public declarations to that effect the day before the discovery. Once in the town, he made inquiries about whether anyone had entered the church on the night before the burial was unearthed. He presumably suspected that someone had planted the bones there that evening. Later, he and Caso visited Rodríguez Juárez, and, as Rodríguez Juárez later told Guzmán, warned Rodríguez Juárez not to trust Guzmán.[35]

For her part, Guzmán was still not fully convinced that the burial site belonged to Cuauhtémoc. She trusted the oral testimonies she had collected but thought that Cuauhtémoc might be buried elsewhere in the village. Perhaps, she thought, the townspeople had intentionally pointed to the wrong site so as to keep outsiders from finding it. Or they might have moved the body in the intervening 420 years. Guzmán continued to look for a second, more majestic, burial. She made some preliminary excavations at the momoxtli and at another chapel in the town. Like Caso and Marquina, she kept silent about her doubts. The crowd celebrating around the church would not have appreciated their suspicions. Caso publicly validated the find, and he refused to badmouth Guzmán publicly, calling her a friend.[36]

[32] Gillingham, *Cuauhtémoc's Bones*, 59–60.

[33] Armando Rivas Torres, "Caso y Marquina confirmaron todo. Sí pertenecieron a Cuauhtémoc los restos encontrados en Ichcateopan," *Excélsior*, October 1, 1949, in Moreno Toscano, *Los hallazgos*, 98.

[34] Evidence of Caso's concerns can be found in Moisés Mendoza, *Rey y señor Cuauhtémoc: El hallazgo de Ixcateopan* (Mexico City: Compañía Importadora y Distribuidora de Ediciones, 1951), chapter 34; Gillingham, *Cuauhtémoc's Bones*, 69.

[35] Guzmán, Relación de la visita de Jorge Acosta, Alfonso Caso y Ignacio Marquina, September 26, 1949, exp. 46; Guzmán to Alfonso Caso, October 6, 1949, exp. 83.

[36] Guzmán, Relación de la visita; Guzmán to Alfonso Caso, October 6, 1949, exp. 83. See also Gillingham, *Cuauhtémoc's Bones*, 262 n62. On Caso and Marquina, Rivas Torres,

Soon after, Marquina appointed the first INAH commission. Its investigation was brief. The villagers were distrustful and possessive. When military leaders tried to transport the bones to Mexico City for safekeeping, village leaders told them no.[37] When the physical anthropologists Eusebio Dávalos and Javier Romero visited Ixcateopan on October 6, villagers prohibited them from taking the bones outside the church to examine them in the sunlight. The INAH investigators spent no more than eight hours in the town. Nonetheless, even this cursory examination convinced them the bones did not belong to the nobleman. For one thing, they quickly determined the bones belonged to at least five different individuals, one of them a child. There were two left femurs. The cranium apparently belonged to a woman.[38] Less than three weeks after the initial discovery of the bones, on October 14, the INAH commission delivered its verdict to the Minister of Education. Only the archaeologist who investigated the building of the church abstained from signing, citing a lack of sufficient evidence. The Minister of Education, aware that the verdict would not meet with the approval of a good part of the Mexican populace, consulted with President Alemán before releasing the results five days later.[39]

Meanwhile, Guzmán had begun assembling her own set of experts to examine the evidence. When she had opened the document with the invisible ink – she was still working on behalf of INAH – she had requested the help of a chemist from the Banco de México. Then, just days before the bones were uncovered, the criminologist Alfonso Quiroz Cuarón, who headed the Bank's investigations department, wrote to offer his assistance.[40] Quiroz Cuarón and his investigations team presumably knew how to spot fakes, including forged signatures. Together, Quiroz Cuarón and Guzmán found additional experts. Historian Luis Chávez Orozco agreed to weigh in on the history and the historical documentation. Chemists examined the oxidation on the copper plaque to determine its age, and they looked at its chemical composition to see where it might have been mined. A paleographer dated the handwriting on the various documents. Architect and art historian Alejandro von Wuthenau

---

"Caso y Marquina confirmaron todo," *Excélsior*, October 1, 1949 in Moreno Toscano, *Los hallazgos*, 98.

[37] Moreno Toscano, *Los hallazgos*, 11–12.

[38] "Dictamen de los antropólogos físicos: Eusebio Dávalos y Javier Romero," in "El hallazgo de Ichcateopan: Dictamen que rinde la comisión designada por acuerdo del c. Secretario de Educación Pública, en relación con las investigaciones y exploración realizadas en Ichcateopan, Guerrero," *Revista Mexicana de Estudios Antropológicos* 11 (1950): 205–217; Rueda, "De conspiradores"; Mendoza, *Rey y señor*, chapter 37.

[39] "El hallazgo de Ichcateopan: Dictamen que rinde la comisión designada por acuerdo del c. Secretario de Educación Pública," 197–295. Archaeologist Carlos Margáin did not sign the report, alleging there was not enough evidence on which to base conclusions. Jiménez Moreno, "Los hallazgos," 170.

[40] Quiroz Cuarón to Guzmán, September 22, 1949, exp. 41.

worked to understand when the church had been built and when the altar had been erected so as to pinpoint when a burial in that location might have occurred. This group would eventually produce a series of reports.[41]

Faced with Guzmán and Quiroz Cuarón's efforts and public repudiation of the INAH verdict, the Minister of Education determined to set up a new commission. Guzmán asked Minister Gual Vidal to delay this new review until her experts had rendered their verdicts, but after consulting President Alemán, the Minister forged ahead. The Gran Comisión met for the first time on January 6, 1950.[42] Members of the Gran Comisión were assigned to report on distinct issues, depending on their areas of expertise. There were many technical issues to resolve, but the minutes of their deliberations suggest that commission members were convinced from the start that the find was fraudulent, and that they saw their mission as easy: proving that fraud. They were also under extraordinary political pressures from national government officials who wanted them to confirm that the bones belonged to Cuauhtémoc, and some members were called before high government officials. This political maneuvering simply convinced them of the need to assert their scientific neutrality and independence.[43]

The Banco de México and Gran Comisión teams had many questions to address: Regarding the documents: Did the handwriting and spelling correspond to the sixteenth century? The forms of expression? The paper and ink? Regarding the plaque: Did the spelling of Cuauhtémoc as Coatemo correspond to the sixteenth century? Would people back then have rendered the date with a comma between the thousands and the hundreds columns? Would they have referred to Cuauhtémoc as "Rey"? Would they have written the conjunction "é" with an accent? Or would they have used the letter "e" as a conjunction at all? Was the copper from the region? How old was it? What did the oxidation reveal? What about the handwriting? Regarding the human remains: How old were they? (An effort was made to send them to the United States for carbon dating.[44]) Could they belong to a young man of Cuauhtémoc's supposed age? Regarding the burial in the church: When would it have been possible to

---

[41] Eulalia Guzmán, *Ichcateopan, la tumba de Cuauhtémoc: Héroe supremo de la historia de México. Tradición oral, documentos, los dictámenes negativos, los concluyentes estudios químicos, antropológicos, históricos, matemáticos, anatómicos, paleográficos* (Mexico City: Aconcagua, 1973); Comisión Investigadora, *Los hallazgos*.

[42] Comisión Investigadora, *Los hallazgos*, xii.

[43] Acta de la sesión celebrada el 22 de septiembre, caja 25, Fondos Documentales Alfonso Caso, Instituto de Investigaciones Antropológicas, UNAM, published in Comisión Investigadora, *Los hallazgos*, 110–119.

[44] Gómez Robleda, *Dictamen*; Rafael Molina to Dr. F.W. Libby of the University of Chicago, to Dr. Kalewo Rankama, October 20, 1949, exp. 112; Alfonso Quiroz, José Gómez Robleda and Liborio Martínez, Investigación y estudio de los restos óseos de Cuauhtémoc (borrador definitivo), February 28, 1950, exp. 39, caja 3 subserie investigaciones, AEG.

undertake such a burial? Specifically, when was the church built and when was the altar installed? And what of the Ixcateopan story? Was it possible that Cuauhtémoc was of Tlatelocan heritage? Who was his mother? Could he have been raised in Ixcateopan? Could he have been transported from the presumed site of his death, about one thousand miles away? Could Motolinía have been in Ixcateopan around the time of Cuauhtémoc's death?[45]

Both the INAH commission and the Gran Comisión focused on the human remains, which both official commissions readily determined to have been arranged in the form of a human skeleton by someone with little anatomical expertise. For the experts, and for the broader public, the physical evidence was paramount. The original documentary discovery had made headlines, but it received substantially less attention than the excavation.[46] Guzmán and her allies focused on physical evidence too, but for them the physical evidence included the plaque and the paper as well as the human remains. They also paid considerable attention to the historical narratives provided by the documents and oral histories. In so doing, they manifested an interest in the local histories that made up Mexico's larger national history and in the timeworn papers held in Ixcateopan, as in so many Mexican villages.

Mestizo Nationalism in the Cold War

Both the rival groups of experts used science, and foreign scientific expertise, in their efforts to establish the truth regarding the burial, but both groups were partial – influenced by the broader political contexts in which they worked; by existing narratives regarding Mexico's past; by ideas about the relation of science to affect; and by group members' relationships to Indigeneity in general and Ixcateopan in particular. The group assembled by Guzmán and Quiroz Cuarón argued that their opponents' investigations were superficial. With the exception of Caso, Gómez Robleda, and the chemist Rafael Illescas Frisbee, none of the commission members ever visited Ixcateopan. Their science would, in contrast to Guzmán's, not be swayed by their relation to Ixcateopenses. In subtle and not so subtle ways, members of the Gran Comisión dismissed local histories.[47]

For instance, the Gran Comisión used the 1946 book of US historian Gilbert Joseph Carraghan to discredit the oral evidence Guzmán had collected.

---

[45] Comisión Investigadora, *Los hallazgos*; Gómez Robleda, *Dictamen*; Eulalia Guzmán et al., *Pruebas y dictámenes sobre la autenticidad de los restos de Cuauhtémoc*, rev. 1962 ed. (Mexico City, 1962).

[46] Moreno Toscano, *Los hallazgos*, 10.

[47] On whether the Gran Comisión should visit Ixcateopan, see ibid., 97–98, 356; Jiménez Moreno, "Los hallazgos," 175–176. See also Eulalia Guzmán, "Nuevas pruebas científicas," in Pérez Martínez et al., *Supervivencia*, 202–203.

According to Carraghan, oral traditions were valid only if historians could document uninterrupted transmission and two unrelated, independent series of transmittals. He also required that the tradition refer to an important event, and be widely accepted across time. The Gran Comisión then added multiple additional criteria, which, its members recognized, most traditions could not meet. Members of the Banco group countered that in fact the Ixcateopan traditions met many if not all of these conditions.[48]

Here, as in other instances, members of Guzmán's group did not reject science but instead tried to make it compatible with local viewpoints. And, they were likely less convinced than their opponents made them out to be. A close reading of their reports shows that a number of them viewed the authenticity of the burial as *plausible or probable* rather than fully proven. Given their plausibility, and the Guzmán group's desire to affirm Mexico's Indigenous heritage and local histories, its members went along with a version of events that satisfied populist politicians and the broader Mexican public. As Gómez Robleda argued, "Insurance companies have very lucrative businesses based exclusively on probable truths."[49]

As the debate regarding Cuauhtémoc was taking place, discussion of Spanish colonial and US imperial power was at a high point, with each growing out of and reinforcing the other. The Cuauhtémoc controversy should therefore be understood, as past scholarship has noted, in the context of proximate events that touched on Mexico's relationship to its Indigenous past, its Spanish heritage, and its place in a Cold War world that placed a premium on science. Guzmán, Gómez Robleda, Chávez Orozco, and Diego Rivera were all part of the left-wing opposition to the ruling party that rallied around Vicente Lombardo Toledano's Partido Popular, founded in 1948.[50] For this group, Cuauhtémoc symbolized the fight against colonial oppression, a fight that had been continued by the Mexican patriots of the wars of independence, in the 1846–1848 war against the United States, and in the war against the French invasion. Ironically, this view jibed with that of a ruling party that deployed a populism that elevated the nation's glorious Indigenous past.

One important event that framed the controversy took place in 1946, when Cortés's corpse was exhumed – for the eighth time. The conquistador's body had been hidden from the public since the independence era on the theory that Mexicans so despised Cortés that they might destroy his remains. In 1945, two foreign historians, the Cuban Manuel Moreno Fraginals and the Spanish exile Fernando Baeza Martos, produced a document that signaled the spot where

---

[48] Gómez Robleda, *Dictamen*, 27–31; Gilbert Joseph Carraghan and Jean Delanglez, *A Guide to Historical Method* (New York: Fordham University Press, 1957), cited in ibid., 27–28.

[49] Gómez Robleda, *Dictamen*, 29.

[50] Jiménez Moreno, "Los hallazgos," 164–165.

conservative politician Lucas Alamán had reburied Cortés in 1836. Moreno Fraginals and Baeza Martos showed the document to the Mexican historian Francisco de la Maza, and de la Maza looked for a set of impartial arbiters to judge its veracity. After deciding the document was likely real, de la Maza and his collaborators decided to avoid bureaucratic hurdles by eschewing official sponsorship and searching for the body themselves. The found it almost immediately in the spot indicated.[51]

The burial was then examined by experts. Historian Silvio Zavala, acting on behalf of the INAH, confirmed the documentary evidence by comparing de la Maza's document with a copy that Alamán had deposited with Spanish authorities. Physical anthropologist Eusebio Dávalos examined the human remains and judged them to be authentic. Cortés was 1.58 meters, Dávalos noted, and diminished in size by his advanced age. His bones were diseased, and he had rickets, but his deformities had not been caused by an infection.[52] Based on photographs of the remains, Quiroz Cuáron rebutted Dávalos's claim that infection was not the cause of the bone anomalies. To the consternation of Hispanophiles, Quiroz Cuarón claimed that Cortés was syphilitic.[53] Diego Rivera, who had already decided he would paint an elderly, stooped Cortés in his Palacio Nacional mural, latched on to Quiroz Cuarón's conclusions.[54]

Guzmán shared with Quiroz Cuarón and Rivera a desire to demystify the conquistador. Before undertaking the Cuauhtémoc investigation, she had been working on a book based on the conquistador's letters to the king of Spain. The book argued that Cortés's letters were politically motivated exaggerations with little relation to what actually happened. The INAH had refused to publish her book, and prominent anthropologists appear to have sabotaged her efforts to get it published elsewhere.[55]

[51] Francisco de la Maza, "Los restos de Hernán Cortés," *Cuadernos Americanos* no. 2 (April 1947): 165; Salvador Rueda, "El descuido de los héroes. Apuntes sobre historiografía marginal," *Historias*, no. 75 (2010): 63–80.

[52] Rueda, "El descuido"; de la Maza, "Los restos"; Salvador Rueda Smithers, "Don Silvio Zavala y la piel del historiador. Apuntes sobre historiografía marginal," *Historia Mexicana* 65, no. 2 (October 1, 2015): 819.

[53] Alfonso Quiroz Cuarón, "Estudio de los restos de Hernán Cortés descubiertos en la iglesia de Jesús Nazareno, anexa al hospital de la Convención de México, en noviembre de 1946," in Guzmán, comp., *Relaciones de Hernán Cortés*, 967–995. Quiroz Cuarón's study is dated January 5, 1949; it was published in Guzmán's book in 1958. Gillingham, *Cuauhtémoc's Bones*, 52 implies that Guzmán affirmed Quiroz Cuarón's statements about Cortés's physical condition but his sources refer to her 1958 book. Statements she made on this topic in 1949–1950 seem to have been relatively cautious. See, for instance, her interview in Torres y González, *La tumba de Cuauhtémoc*, 58–61.

[54] Rueda Smithers, "Don Silvio Zavala," 828–829.

[55] Gillingham, *Cuauhtémoc's Bones*, 52; Torres y González, *La tumba de Cuauhtémoc*, 39, 58–59.

In the ensuing public debate over Cortés, many Mexicans, Guzmán included, openly objected to the reverence shown the conquistador, and the Cortés discovery reignited discussion about the respective contributions of Spaniards and Native Americans to Mexico, with conservative Hispanists arrayed against progressive indigenistas.[56] In this context, members of the anti-Guzmán group portrayed themselves as level-headed, middle-of-the-road thinkers, who recognized the quintessentially Mestizo nature of their nation. Wigberto Jiménez Moreno, who served as secretary of the Gran Comisión, wrote in 1960:

> Cortés y Cuauhtémoc, as symbols that personify those two cultural traditions that seemed irreconcilable, were imbued with a terrible affective charge capable of clouding for those with less serene minds the concept of a Mexican nation that – seen from biological, psychological, cultural and social angles – had emerged, basically, from Mestizaje and transculturation. Those of us who preached the necessity of accepting the indissoluble Hispanoindian fusion and recognizing the positive value of both patrimonies, saw ourselves being repudiated by an exalted Indophile, Hispanophobe current . . .[57]

Again and again, Jiménez Moreno and other commission members stated that regardless of their determinations regarding the remains, they recognized Cuauhtémoc as a brave national precursor, but their science was separate from their affect. Zavala wrote that it was "necessary to separate clearly the admiration and respect we Mexicans feel for the figure of Cuauhtémoc from the purely scientific problem that entails establishing the authenticity of the discovery of the Ixcateopan remains."[58]

Jiménez Moreno's characterization of the indófilos was not necessarily accurate. According to the story circulating in Ixcateopan, the Spaniard Motolinía had presumably ordered the village to keep silent about Cuauhtémoc's burial in order to protect them. The Ixcateopenses thus underscored reconciliation between Spaniards and indígenas, not Hispanophobia. A fictionalized dialogue that I attribute to Zavala made a similar point, while suggesting that Guzmán's allies refused to recognize the positive Spanish role. In the dialogue, one friend says to another:

> The Western world surrounds the presumed discovery of the indigenous hero . . . It is a Spanish and Christian friar who collects the remains and buries them lovingly. That is why the Discovery makes the indophiles so happy, because it seems to give them their supreme hero but it also

---

[56] Moreno Toscano, Los hallazgos, 12.
[57] Jiménez Moreno, "Los hallazgos," 163.
[58] Rueda, "De conspiradores," 24–25, quotes Zavala. See also Jiménez Moreno, "Los hallazgos," 171; Rueda Smithers, "Don Silvio Zavala," 828–829.

dignifies the hispanists because it concedes that one of their friars carried out that generous act ... And I do not want to be a pain but I have to mention the burial of that great defender of the pagan world in a Catholic Church ... and what about the praises and insults exchanged in Spanish ...
The health of our historic conscience depends of the assimilation of that indigenismo and that hispanicismo. We must move beyond that duality and toward a broader synthesis ... Curiously, the life of Mexico marches forward and moves beyond the harshness of the encounter while [our] historical consciousness, which is the arena of controversy, opinion, and sentiment – lags behind reality.[59]

To be modern and fully Mestizo required Mexicans to reject popular historical memory and instead adhere to a science associated with the industrialized West.

In the late 1940s, the approaching centennial of the US–Mexico war made Mexicans particularly sensitive to their country's position relative to the United States. In fact, shortly after Cortés's remains had been found, nationalism flared when, in March 1947, Harry Truman visited Mexico on the first official visit by a US president to the country. On his visit, Truman stressed the need to continue good neighborly relations, and on March 4, he paid tribute to the "*niños héroes*" (child heroes) who had perished while defending Mexico City during the US invasion in 1847. By the end of that month, the remains of the niños had been dug up in Chapultepec Park. A quick investigation confirmed their authenticity. INAH scientists subsequently examined them and declared that the age of the deceased corresponded to that of the martyred child-cadets. As icing on the cake, the Mexican Congress declared the remains to be authentic. The message was clear. Even the US president valued Mexican bravery and nationalism.[60]

In this context, experts on both sides of the Cuauhtémoc conflict sought to buttress their science by showing that it stood up to US scrutiny. For instance, Guzmán's team had good chemical evidence that the copper in the plaque was not of recent vintage. To confirm this finding, it sought out US experts who could date the metal. Caso, a member of the Gran Comisión, also wrote to US experts regarding the plaque, but his goal was to undermine the Banco group's dating. During one of the Gran Comisión's sessions, physician José Gómez Robleda (its only member aligned with Guzmán's team) and Caso engaged in

---

[59] "Diálogo sobre Cuauhtémoc," caja 25, Fondos Documentales Alfonso Caso, Biblioteca Juan Comas, Instituto de Investigaciones Antropológicas, Universidad Nacional Autónoma de México. I attribute authorship to Zavala because of similar wording found in "Dictamen del Doctor Silvio Zavala sobre los manuscritos e inscripción del hallazgo de Ichcateopan," *Revista Mexicana de Estudios Antropológicos* 11 (1950): 290–295.

[60] Jiménez Moreno, "Los hallazgos," 165–166.

a long and testy debate regarding whether water could have dripped into the burial site. If the plaque had been in contact with water, as Caso argued, it would have rusted more quickly and was therefore newer than the Banco group claimed. Gómez Robleda denied that the gravesite could have been infiltrated by water.

But even as they argued about the conditions at the burial site, all participants recognized that ultimately a solid conclusion regarding the plaque's age could never be drawn from the evidence of oxidation. There were simply too many variables.[61] Yet both groups persisted in this debate and continued to invoke foreign experts. In so doing, they showed they could dialogue with US and European colleagues and that their cosmopolitan science could dignify the Mexican nation. In another case, physical anthropologists Dávalos and Romero touted their use of internationally recognized scientific methods to examine the human remains. Whether these foreign formulas actually shed light on the controversy was, for them, secondary: "To determine the height," they wrote, "we have used Manouvrier's table or Pearson's ten formulas although the results have to be considered relative given that we are talking about indigenous remains."[62]

The connections between Spanish and US imperialism – and their relation to both nationalism and science – were made clear in Zavala's fictionalized dialogue, in which two friends, one an immigrant to Mexico from Spain, the other a Mexican, talked during a visit to Cuernavaca. In the dialogue's opening paragraph, we learn that the friends were bored of visits to Cortés's Cuernavaca palace, a landmark that ironically housed a Diego Rivera mural commissioned by US ambassador Dwight Morrow. Yet they were at home among the many tourists, "perhaps because of habit, or since they know that it is because of the love of their dollars that comfortable hotels have been built. And we must not forget that the 'girls' [in English in original] are soft on the eyes ..." The friends asserted their masculine superiority over the wealthy 'girls,' and their affinity for US-style consumption just as they manifested their indifference to the symbols of US power, Spanish colonialism, and of the left-wing Mexican nationalism that overwrote that colonialism, all instantiated in the palace.[63]

The fictional friends debated the merits of myth and science. "Mario," the author told readers, was "cosmopolitan." The Spaniard roamed the world not for fun but to "understand mankind" (comprender los hombres). The Mexican "Aníbal," was also a world traveler, but over the years had once again turned his gaze to Mexico, which he now saw more clearly and in greater detail.

---

[61] Comisión Investigadora, Los hallazgos, 17–43.
[62] "Dictamen de los antropólogos físicos," 221.
[63] "Diálogo sobre Cuauhtémoc." The quotations that follow on pp. 229–232 are all from this document, unless otherwise noted.

By explaining the controversy to Mario, Aníbal displayed Mexican science to the world. As Mario noted, "A Mexican has written that we must defend Mexico's scientific prestige since the verdict [of the commission] will be known in circles frequented by foreign experts." Referencing perhaps a Cold War distinction between the superior democratic West and the totalitarian Soviet bloc, the dialogue ended with Aníbal asserting that the broad-ranging media coverage of the controversy was made possible by Mexico's freedom of the press. He urged his friend to write something about Cuauhtémoc to "repay the hospitality we have shown you." The word of the foreigner would validate that of the Mexican.

Like the expert reports emitted by all the investigating commissions, which went to great lengths to establish their scientific nature, Zavala's dialogue contrasted the friends' contemplative "tranquil spirits" to the "burning debates" in the press. At the core of Zavala's argument, and that of others who questioned the authenticity of the burial, was the notion that masculine rationality and science should triumph over feminine emotion. Jiménez Moreno made similar arguments, denouncing the "thunderous ... shrieking press" and the excesses of a public that labeled them "traitors" who should be shot. (Rivera had famously claimed that those who denied the authenticity of the discovery should be shot in the back.) The members of the Gran Comisión, Jiménez Moreno noted, worked without any remuneration and on top it had to endure insults for "upholding the jurisdiction of scientific investigations without twisting the truth to suit patriotic motives." Jiménez Moreno further castigated indófilos "who – inspired at times by a racist attitude that negated the positive contributions of Mestizaje and Hispanic-Indigenous transcultura-tion and that therefore destroyed the roots that created and nourish our nationality – have in the end abandoned legitimate and well-grounded patri-otic sentiment."[64] Guzmán and her allies were thus unscientific pro-Indigenous "racists" who refused to recognize Mexico's mixed heritage or adhere to cosmopolitan scientific standards.

The fervor of Guzman's camp was often equated with a religious fervor that, given the anticlericalism of the postrevolutionary state, was un-Mexican. Jiménez Moreno claimed that the "popular work" of author and former governor of Campeche Héctor Pérez Martínez, who favored the Guzmán camp, presented the Mexican past to the public with "*dramatismo,*" providing what was "perhaps a Bible for those who with great passion agitated [*mili-taron*] in favor of the authenticity of the Ichcateopan discovery."[65] In Zavala's dialogue, Mario compared the Cuauhtémoc "myth" to that of the Virgin of

---

[64] Jiménez Moreno, "Los hallazgos," 177; Moreno Toscano, *Los hallazgos,* 15; Rueda Smithers, "Don Silvio Zavala," 829; Johnson, "Digging Up Cuauhtémoc," 221.

[65] Jimenez Moreno, "Los hallazgos," 165.

Guadalupe, affirming that the Cuauhtémoc "cult" had "priests like Miss Guzmán and Diego Rivera, surrounded by their acolytes."

Since the allegiance of Guzmán and her allies to a revolution that advocated rationality against religion was unquestionable, their detractors had to labor to prove that the alleged indófilos were irrational. As Aníbal noted in Zavala's dialogue, the quasi-religious zeal with which Guzmán and her allies defended Cuauhtémoc was ironic given that they were also supporters of Mexico's official "socialist" education, which advocated "rationalism" and even banned religion from the schools.[66] Aníbal explained away this support of science by reminding Mario that Mexicans liked to "*razonar el mito guadalupano*" (reason out the Guadalupan myth). "Our myth makers," Aníbal explained, "do not openly display themselves as such. Instead, they look for the support of scientific veneer. And, so, they imitate the testimonial tradition of the Church, for which miracles must be 'proven' before they are officially accepted."

Guzmán and her allies tried to do what for their opponents was unimaginable: uphold local history, nationalism, and feminine belief along with cosmopolitan masculine science. As a single woman and intellectual, and a feminist moreover, Guzmán personified this unnatural mixing. In contradistinction, her opponents claimed to propound a harmonious Mestizaje. Yet they deployed a language of gender that relied on hierarchy and distinction rather than mixing or equalizing. Their vision of Mestizaje was, in its adherence to science, Westernized or whitened, and masculine. Caso, Jiménez Moreno, and Zavala peppered their assessments of Guzmán with praise. Zavala, for instance, characterized Guzmán as having an "indisputable persistence and unquestionable integrity."[67] But this praise was overshadowed by less favorable assessments. Guzmán's aggressive harangues (buttressed with scientific posturing) did not seem fitting for a woman.

At the same time, the Mestizaje promoted by these mainstream scientists depended on the masculine Indigenous presence of the "Aztec warrior" who embodied and engendered a virile Mestizo Mexico. For that reason, members of the Gran Comisión repeatedly pointed to a second unacceptable gender discrepancy: the cranium found in Ixcateopan, they said, belonged to a

---

[66] Shortly before the Cuauhtémoc episode erupted, in 1946, Article Three of the Mexican Constitution had been reformed, doing away with dramatic restrictions on instruction imparted by religious institutions. The revised article nonetheless still backed instruction "based on the results of scientific progress [and that] will do battle against ignorance and its effects: servitude, fanaticism, and prejudice ..." It stipulated that education should stimulate in the student "love of Patria and a sense of international solidarity, arrived at with independence and justice." Hernán Cortés Medina and Manuel Ortiz C., *Instituciones jurídico-políticas de México* (Mexico City: Ediciones Cicerón, 1950), 5. See Jiménez Moreno, "Los hallazgos," 162.

[67] "Dictamen del Doctor Silvio Zavala," cited in Rueda Smithers, "Don Silvio Zavala," 25.

woman. As Aníbal admitted: "I am bothered that the defenders of the remains may crown the most virile hero of our history with the cranium of a woman. In regards to this aspect, I am indeed interested in rigorous anthropological judgment." Elsewhere, too, this fact was mentioned as the definitive proof of the inauthenticity of the discovery. Physical anthropologists Dávalos and Romero wrote, "The cranium, which is the key piece [pieza capital] is a woman's. It is inconceivable that anyone would want to represent a hero that has figured as a symbol of the virility of the Indigenous Aztec with a cranium of the female sex."[68] Here was one more reason to reject the conclusions of their opponents. The nationalism of Guzmán and her supporters was based on an Indigenous heritage that was judged, scientifically, to be in fact feminine. The remains from the village of Ixcateopan could not be the fount of Mexico's Mestizo national identity. If Mexico had an Indigenous heritage, it would need to be linked to something grander and more manly – more imperial and Aztec, less conquered and Chontal.

In their report for the first INAH commission, Dávalos and Romero cautioned that in examining the cranium and other bones "to the point possible we have abstained from crossing the limits that science imposes in this particular case."[69] "We recognize," they later added, "that any physical feature varies and that masculine characteristics are not always clearly displayed in specimens of this sex, and we can say the same of feminine characteristics." Yet this did not stop them from suggesting that it was "inconceivable" that the virile Cuauhtémoc should have a woman's cranium and that the masculine and feminine crania from the Aztec period were "totally different."[70] The anti-Guzmán camp portrayed themselves as careful scientists. But their assessments were not always sober and neutral.

Affect, Evidence, and Truth

As I have suggested, recent historical accounts have repeated the contrast between an exalted Guzmán and her rational opponents. To cite one example, Lyman Johnson writes that when pressed by regional authorities to begin the dig against her will, Guzmán's "political ambitions . . . made successful resistance to these pressures unlikely." Once the dig began, "Guzmán was unable to exercise full control at the site, with military and civilian officials and even members of the Rodríguez Juárez family directing workers at times." Johnson further refers to her "lack of professional discipline." After the first commission report, "a frontal assault on her achievement," Guzmán – who, Johnson reminds us, originally trained as a schoolteacher rather than as a historian or

---

[68] "Dictamen de los antropólogos físicos," 209–210.
[69] Ibid., 225.
[70] Ibid., 217.

archeologist – "signaled clearly both her independence from INAH superiors and her political sophistication." Those INAH superiors did not, it should be noted, have more anthropological training, since they were part of a generation that came of age before anthropology was established as a professional activity.[71] One problem with this argument is that, as contemporary observers had to admit, Guzmán and her followers sought to be as scientific in their proofs as their foes. In at least some ways, they were equally if not more rigorous.

That did not ensure they came up with the right result. The findings of the pro- and anti-commissions are voluminous and complex. Without deep expertise in paleography, physical anthropology, history, chemistry, and other disciplines, it is difficult to understand the arguments of each camp, much less evaluate them. But even with those caveats, it is obvious that spurious arguments could be found in both groups, along with more robust proofs. And often truth was impossible to determine. Just a few of the easier to understand examples: The Gran Comisión claimed that the handwriting on the copper plaque was from the nineteenth century and that it was similar to that of the documents. The Guzmán group countered that chiseling into copper was different from writing on a sheet of paper. The Comisión argued that writing the date with a comma was not common in the colonial era. The Guzmán group offered copious examples of periods and other signs used between words and minimized the difference between a chiseled comma and a period. The Guzmán group affirmed that the paper on which the documents were written was from the colonial era. Their opponents countered that it was easy to tear colonial-era paper from old books or folios. The Guzmán group argued that the remains had to have been buried before the altar was constructed; that the flooring over the burial was intact, evidence that the grave had not been tampered with; that the oxide on the plaque and its positioning were not consistent with a later burial.

The most convincing argument offered by the official investigators was the forensic evidence. The skeleton made up by Liborio Martínez of the Banco de México group was put together incorrectly, they said, an assertion to which even Gómez Robleda, a member of the opposing group, had to assent. The human remains belonged to different people, and it looked as if someone inexperienced – someone who could not tell a left from a right femur – had tried to reconstruct a complete skeleton. There was clearly an old burial ground in the church which could have provided ample materiel for a forger. The most convincing evidence offered by Guzmán's group – what had likely convinced Guzmán herself – was the oral evidence and folklore. It pointed clearly to a burial and was able to signal the place where the remains had been

[71] Johnson, "Digging Up Cuauhtémoc," 210, 220.

found. Cuauhtémoc himself might be buried somewhere else in the village, as Guzmán suggested. Or the remains might belong to another colonial-era personage. But the oral evidence and the widespread rituals were hard to dismiss. The anti-Guzmán camp was able to do so only through superficial procedures and appeals to foreign expertise. To do so, they cited a book by a foreigner in order to claim that the oral evidence was unreliable because it had not been confirmed by a written source.[72]

Careful sleuthing has revealed that the village schoolteacher found a sealed room in the back of the home that had belonged to Florentino Juárez and that Salvador later sold to the municipality. It was full of charred bones and old books. It was in essence a factory for making heroes. Gillingham and Mexican scholars offer a convincing account that says that Florentino planted the bones while the church steeple was being rebuilt. His goal was to earn favor in Mexico City in order to buttress Ixcateopan's position in a land dispute with a neighboring village. He then spread rumors throughout the area, planting the roots of the oral tradition. Most of the bearers of the oral traditions had been peons working for Juárez and his close associates.[73] The oral tradition, or at least aspects of it, was a real tradition, but it was a nineteenth-century tradition.

However, the truth of the matter is not the only issue of import. Rather, my argument is that the portrayal of Guzmán as shrill, opportunist, politically motivated, and unscientific was in itself politically motivated as well as affect-laden. And it was politically motivated by a particular vision of Mexico, one that favored a whitened, cosmopolitan, masculine identity and was unconcerned with the needs or histories of villages like Ixcateopan.

Guzmán was a highly trained specialist, whose work was criticized in ways that others' was not. Her investigations and those of her collaborators were not engaged with as seriously as they might have been. As Guzmán herself pointed out, the bones of the niños heroes were declared real with only minor questions raised. Uneasiness with Guzmán's gender, her feminism, and her self-assurance colored her opponents' assessments of her, as did suspicions of both Indigenous peoples and the left-wing pro-Indigenous nationalism promoted by Guzmán. A nation that was both virile and Mestizo – and that would be accepted in a world order that valued scientific proof and technological advances – could not be represented by the Indigenous woman buried in the Ixcateopan church or by a left-wing nationalist, and feminist, like Eulalia Guzmán. Nor could that nation take villagers' opinions as proof. The members of the Gran Comisión did not have the time to do further research regarding Ixcateopan. They were not interested in doing so.

---

[72] Pérez Martínez et al., *Supervivencia*; Comisión Investigadora, *Los hallazgos*.
[73] Olivera de Bonfil, *Tradición oral*.

If we reject how Guzmán's opponents characterized her, how do we reflect critically on her actions and historicize the morality of the scientists involved? Can we think of Guzmán's work as active scientific refusal? Or an unconscious refusal? On one level, Guzmán does seem to have had an awareness that the bones did not belong to Cuauhtémoc. Perhaps then, her investigations sought to give a scientific veneer to the villagers' beliefs. On the other hand, the Banco group seems to have been sincere in its efforts to prove that the discovery was authentic. It is perhaps most likely that Guzmán and her followers were unsure about the veracity of the find. Pleading with members of the Gran Comisión, Gómez Robleda argued not so much that the burial was authentic but that, if there was any uncertainty, the commission should refrain from passing judgment. Was this humility? An attempt to save face? Perhaps a bit of both? Given what Guzmán and her allies saw as the political motivations of their opponents, they felt justified in insisting that the burial was real.

Standards of scientific proof could, and did, vary, and the volume of evidence made it more difficult to reach a clear result. In many ways, what we know today to be the real story is more fantastic than the assertions of Guzmán and her crew. Even more important, Guzmán and her supporters rejected the characterizations thrust on them and refused to counterpose truth and politics. The members of the Gran Comisión refused to see themselves as defying the imperatives of a nationalism that embraced its Indigenous peoples and local histories.[74]

We should also ask to what extent Guzmán's insistence was a response to a scientific feud, an attempt to buttress her own ambitions. Certainly, that was how her opponents portrayed her, and more recent scholarship persists in denouncing her desire for recognition. One might think of the dueling visions of Mestizaje at play as part of an elite dispute that silenced popular voices. I have instead suggested that Guzmán's experiences in Ixcateopan may have *affected* her. And, at least in some ways, she was responding to her village allies. The question about the nature of Mexican Mestizaje that shaped the dispute between scientists was one Ixcateopenses, at least some of them, cared about. Their position within the Mexican polity and their ability to extract resources depended on the presumed value of Indigeneity to the nation. And hegemonic views regarding Indigeneity no doubt helped shape their views about who they were and about their past. Some of them at least sought to claim a glorious imperial Indigenous past.

In Ixcateopan, the Cuauhtémoc myth persists today. In 1976, Guerrerenses asked that the case be reopened and a new set of investigations was conducted, leading to a good many publications. The burial and the oral traditions were

---

[74] José Gómez Robleda to Comisión Investigadora, February 7, 1951, Anexo a Acta 38, in Comisión Investigadora, *Actas y dictámenes*, 382.

judged by these experts, again, to be fraudulent. In the process the history of the village and the region were further documented. Still, Ixcateopan's reputation as Cuauhtémoc's hometown remains. Indeed, the village now celebrates his birthday with dancing and singing. The alternative view, so satisfying to truth-seeking historians, also persists. Along with the new investigation, a new series of archeological excavations began in 1976. They are ongoing.

9

Unequal Encounters
Debating Resource Scarcity, Population, and Hunger in the Early Cold War*

EVE E. BUCKLEY

Introduction

In the early 1950s, Brazilian physician and nutritionist Josué de Castro published a book, entitled *The Geography of Hunger* in its English translation, which emerged from decades-long observation of chronic starvation in his native city of Recife, Brazil. The book, which received reviews both laudatory and scathing in the global press and academic journals, was a direct response to a 1948 publication by American conservationist William Vogt called *The Road to Survival*, which warned of a looming global population crisis. Vogt's writing, influenced by mid twentieth-century American eugenicist thinking, was unapologetic about the need to curb reproduction in what would soon come to be termed the "third world."[1] He adopted the concept of "carrying capacity" (originally used in reference to steam ships) as a purportedly object-ive measure of how many humans and other species a specified region can sustain, and he used this seemingly straightforward mathematical equation to justify recommending a range of population control measures.

Vogt viewed people in less industrialized countries as abstractions, based on ecological and demographic data but limited direct experience. De Castro's response to Vogt's assertions stemmed from personal encounters, since child-hood, with the sort of people whose reproduction Vogt feared. The Brazilian intellectual countered Vogt's call for population control with an emotional appeal to readers, insisting that they confront the geopolitical injustices that make the global poor chronically hungry. Until his death in 1973, de Castro demanded in a variety of forums that the problem of hunger be addressed by reforming political and economic structures that failed to prioritize food

---

* Archival sources cited in this article are from the Josué de Castro Archive, Fundação Joaquim Nabuco, Recife.
[1] Initially by Alfred Sauvy in "Trois mondes, une planète," *L'Observateur* 118 (August 14, 1952): 14.

cultivation and access. This essay examines the work of Vogt and de Castro with an emphasis on their contrasting forms of encounter with the subjects of their analyses and on the different affective rhetorical strategies that each employed to persuade their readers.

In the past two decades, several historians have written histories of twentieth-century family planning. Matthew Connelly critiques population control in several contexts as examples of the pitfalls of social engineering, highlighting the racism and classism inherent in numerous family planning campaigns. Thomas Robertson examines population concerns as one element of twentieth-century debates that pitted economic growth against environmental conservation, particularly in the United States. Alison Bashford considers immigration and birth control promotion as opposing strategies for managing global population growth. Emily Merchant illustrates how family planning advocacy emerged in conjunction with a new approach to eugenics, both movements understood by their supporters as progressive routes to economic uplift for the poor. She portrays demographers as more tempered advocates for population reduction than biologist Paul Ehrlich and other popularizers of "zero population growth."[2] Other scholars have provided insightful analyses of demography as a social-scientific discipline with ideological commitments shaped by its initial funders and institutional locations.[3] This essay foregrounds the effort by one briefly prominent intellectual from the Global South to oppose the emerging, US-led consensus that population control was central to economic development, particularly in the world's poorest regions.

Both the American conservationist and the Brazilian scientist attuned to structural inequalities were politically progressive within their specific social contexts. Yet Vogt was willing to objectify and dismiss the individual aspirations of people in places far removed from his own experience in order to protect what he cherished as a North American "standard of living." De Castro imagined the global poor as familiar, due to his own experiences among

---

[2] Matthew Connelly, *Fatal Misconception: The Struggle to Control World Population* (Cambridge, MA: Harvard University Press, 2008); Thomas Robertson, *The Malthusian Moment: Global Population Growth and the Birth of American Environmentalism* (New Brunswick: Rutgers University Press, 2012); Alison Bashford, *Global Population: History, Geopolitics, and Life on Earth* (New York: Columbia University Press, 2014); and Emily Merchant, *Building the Population Bomb* (Oxford: Oxford University Press, 2021).

[3] Susan Greenhalgh, "Social Construction of Population Science: An Intellectual, Institutional and Political History of Twentieth-Century Demography," *Comparative Studies in Society and History* 38, no. 1 (1996): 26–66; Edmund Ramsden, "Confronting the Stigma of Eugenics: Genetics, Demography and the Problems of Population," *Social Studies of Science* 39, no. 6 (2009): 853–884; and Carole McCann, *Figuring the Population Bomb: Gender and Demography in the Mid-Twentieth Century* (Seattle: University of Washington Press, 2017).

marginalized Brazilians, and vigorously opposed the relatively affluent North American's agenda as an imperialist imposition on the autonomy of less empowered people. De Castro feared that prioritizing population control as the solution to entrenched poverty would undermine other movements for economic transformation, such as agrarian reform in rural Latin America.

Debating Population during the Early Cold War

In her 2014 history of population studies Bashford notes that while "global food, hunger and population were key elements in the [Cold War] discourse of anti-Communism ... [they were also key elements of] postwar anticolonialism."[4] That latter framing, which emphasized the alignment between endemic hunger and areas of former colonial rule, was articulated in the postwar writing of intellectuals from the newly designated "Third World." Their analyses of resource scarcity rejected the rising preoccupation with "overpopulation" emerging from industrialized nations, particularly the United States. Rather than reflecting an overabundance of people, these scholars argued, famines and social unrest were the result of economic injustices that directed inadequate resources (particularly food) to the world's poor. Development efforts, therefore, should focus not on reducing the numbers of poor people but on elevating their food consumption and economic productivity through economic restructuring, both within and among nations. In a 1954 book entitled *Hungry People and Empty Lands* Indian social scientist Sripati Chandrasekhar promoted the emigration of his country's poor to less populated regions of the Earth as a way to better align human population with food supply.[5] As another prominent intellectual from one of the regions targeted for population reduction in the 1950s, de Castro argued vigorously against the solidifying US-based orthodoxy that blamed poverty on overpopulation.

Many nationalists from the Global South were wary of population control measures because of the eugenic intellectual roots of American and British fertility control advocates such as Margaret Sanger. In a foreword to Sanger's 1922 book promoting birth control, *The Pivot of Civilization*, author H. G. Wells – Sanger's friend and lover – summarized her message as follows:

> The New Civilization is saying to the Old ... We cannot go on giving you health, freedom, enlargement, limitless wealth, if all our gifts to you are to be swamped by an indiscriminate torrent of progeny. We want fewer and better children who can be reared up to their full possibilities in unencumbered homes, and we cannot make the social life and the world-peace

---

[4] Bashford, *Global Population*, 269.
[5] Sripati Chandrasekhar, *Hungry People and Empty Lands: Population Problems and International Tensions* (London: George Allen & Unwin, 1954).

we are determined to make, with the ill-bred, ill-trained swarms of inferior citizens that you inflict upon us.[6]

Critics of such views from the Global South portrayed foreign-sponsored birth control promotion as a form of imperialism that blamed poverty on the over-fecundity of the poor rather than on global economic relationships that fed citizens of powerful nations at the expense of those in the periphery. Examples like Puerto Rico, in which per capita income tripled over the 1940s at the same time that the population was growing (due to decreased infant mortality), threw US-influenced demographic orthodoxy into question.[7]

In order to combat accusations of imperialist motivation, new organizations focused on global population reduction – like the International Planned Parenthood Federation, established in 1952 – adopted the term "family planning" to connote a focus on the best interests of parents and children rather than the more insidious "control" of birth rates. Nevertheless, when Cornell University sociologist J. Mayone Stycos analyzed Latin American press coverage of population issues in the 1960s and interviewed scholars across the region, he concluded that most Latin Americans viewed "overpopulation" as an imperialist myth used to justify reducing birth rates in the Third World. Those who did believe in a looming demographic crisis thought it should be solved by promoting industrial economic growth, which – according to dem-ographers' own models – would lead naturally to reduced family size.[8]

In the past decade, numerous scholars have traced the multi-institutional focus on population control that was pursued by philanthropic foundations and government agencies in the decades following World War II. Connelly highlights the technocratic modernization ideologies that underlay such efforts.[9] Robertson traces the merging of concerns about ecological limits to growth with both eugenics and Cold War geopolitical concerns.[10] Many Cold Warriors viewed American-style suburban domesticity as the exemplar of capitalist modernity. They upheld small, nuclear-family households as essen-tial to industrial development driven by increased *per capita* consumption of

---

[6] Quoted in Bashford, *Global Population*, 233. Historian Piers Hale asserts that Wells owed his Malthusian views to Thomas Huxley; see *Political Descent: Malthus, Mutualism, and the Politics of Evolution in Victorian England* (Chicago: University of Chicago Press, 2014), 219, 272.

[7] Laura Briggs, *Reproducing Empire: Race, Sex, Science, and U.S. Imperialism in Puerto Rico* (Berkeley: University of California Press, 2002), 15.

[8] Karina Felitti, "La 'Explosión Demográfica' y la Planificación Familiar a Debate: Instituciones, Discusiones y Propuestas del Centro y la Periferia," *Revista Escuela de História* 7, no. 2 (2008): 1–16.

[9] Connelly, *Fatal Misconceptions*.

[10] Robertson, *Malthusian Moment*; Paul Ehrlich, *Population Bomb* (New York: Ballantine Books, 1968).

manufactured goods. Large families were antithetical to such progress, draining resources at both the household and national levels. As one historian argues, by the early 1950s, "[a]n intellectual orthodoxy concerning the importance of the relationship between national economic development and [managed] population growth solidified among social scientists, economic planners, and political leaders in the West and in those nations that looked predominantly to the liberal democracies of the West" for assistance and emulation.[11] This new orthodoxy was based on demographic transition theory, developed by Frank Notestein at Princeton University. In 1952, several philanthropic foundations, including Rockefeller, Ford, and Milbank Memorial, jointly formed the Population Council, allowing them to distance their names from controversial population control efforts; Notestein became head of the council in 1959. Reducing fertility in poor nations became a focus of many international development agencies during the 1950s and 1960s, motivated by demographically based theories of economic development and by Cold War concern that the impoverished masses of Europe's former colonies would be vulnerable to Marxist political movements.

The consolidation of overpopulation as a way of conceptualizing global resource scarcity is an example of intellectual imperialism grounded in social and natural sciences (notably demography and ecology), with significant consequences for globally marginalized people when concerns about population growth were translated into policy – most tragically when such policies were carried out by authoritarian regimes, or by political classes that sought to reduce the influence of minority ethnic groups. Even in less draconian cases (those that did not involve coerced sterilization or abortion), it is a significant intrusion into people's most intimate life decisions to assert that their reproductive choices will have negative consequences for national welfare and the wider human community.[12]

American Conservationist William Vogt and the *Road to Survival*

Vogt's 1948 book *Road to Survival* argued that curbing population growth was an urgent priority for developing nations. The American conservationist based his analysis of human population limits on what he termed carrying capacity,

---

[11] Simon Szreter, "The Idea of Demographic Transition and the Study of Fertility Change: A Critical Intellectual History," *Population and Development Review* 19, no. 4 (1993): 659.

[12] This analysis is informed by Warwick Anderson's examination of imperialist science, knowledge, and policy and Jan Goldstein's advice to consider the moral field in which particular historical actors – including those deeply committed to curbing human population growth during the 1950s – operated. Warwick Anderson, "Objectivity and Its Discontents," *Social Studies of Science* 43 (4) 2013, 557–576; Jan Goldstein, "Toward an Empirical History of Moral Thinking: The Case of Racial Theory in Mid-Nineteenth-Century France," *American Historical Review* 120, no. 1 (2015): 1–27.

expressed as a simple equation: C = B:E, where C is carrying capacity, B is biotic potential to sustain animals, and E is environmental resistance to this potential.[13] Later ecologists, notably Eugene Odum, would question the usefulness of this means of estimating a region's capacity to sustain life, due to the number of variables in dynamic interaction that impact the maximum population a particular land area can sustain.[14] Soil fertility was a central focus of Vogt's analysis, reflecting concerns about soil depletion that arose during the 1930s as a result of the US Dust Bowl, along with evidence of severe erosion in Australia and South Africa.[15] In Vogt's view, the problem of overpopulation was not only an ecological concern. He believed that an intensifying struggle for *lebensraum* (literally room to live, a term deployed by the Nazis to justify Germany's territorial expansion) underlay the terrible violence witnessed in Europe and Asia during the first half of the twentieth century. Without a curb on human population increase, Vogt predicted, such violence would intensify.

Vogt's analysis acknowledged that overconsumption and waste in the industrialized world contributed to resource scarcity elsewhere. He also noted the rapacious impact of capitalism on natural resources. Uniting these views, he asserted that "Ecological health for the world, requires, above all ... 1. That renewable resources be used to produce as much wealth as possible on a sustained-yield basis ... and 2. We must adjust our demand to the supply, either by accepting less per capita (lowering our living standards) or by maintaining less [*sic*] people." However, Vogt quickly dismissed the possibility of reducing consumption in the industrialized world by any significant measure: "Since our civilization cannot survive a *drastic* lowering of standards, we cannot escape the need for [global] population cuts," he concluded.[16] And although Vogt saw capitalism as partly to blame for global poverty, he took the eugenicist's view that the poor are somewhat responsible for their own misfortune. Vogt tentatively recommended financial "bonuses" for men (in particular) who voluntarily underwent surgical sterilization. "Since such a bonus would appeal primarily to the world's shiftless," he postulated:

> it would probably have a favorable selective influence. From the point of view of society, it would certainly be preferable to pay permanently indigent individuals, many of whom would be physically and psychologically marginal, $50 or $100 rather than support their hordes of offspring

[13] William Vogt, *The Road to Survival* (New York: W. Sloane Association, 1948), 16.
[14] Nathan F. Sayre, "The Genesis, History, and Limits of Carrying Capacity," *Annals of the American Association of Geographers* 98, no. 1 (2008): 127.
[15] William Beinart, "Soil Erosion, Conservationism and Ideas about Development: A Southern African Exploration, 1900–1960," *Journal of Southern African Studies* 11 (1984): 52–83.
[16] Vogt, *Road to Survival*, 265.

that, by both genetic and social inheritance, would tend to perpetuate their fecklessness.[17]

Vogt was not reticent about the struggle for survival between people in wealthier and poorer countries that carrying capacity limitations would eventually necessitate, and he did not disguise his bias against populations he viewed as overly fecund. Why should the United States, Canada, Australia, and Brazil "open their doors to Moslems, Sikhs, Hindus (and their sacred cows)," he asked rhetorically, in a section on immigration as one possible solution to food scarcity, "to reduce the pressure caused by untrammeled copulation. Our living standard must be dragged down, to raise that of the backward billion of Asia."[18] With reference to a UN FAO proposal for economic development in Greece, Vogt noted: "At no point in the entire report is there any suggestion that a positive effort be made to reduce the breeding of the Greeks ... Such neglect [of this possible solution] would disqualify a wildlife manager in our most backward states!"[19] Vogt warned his American readers that "Since Greece seems to have planted its hand firmly in the American dinner pail, the question [of population growth in that war-ravaged country] is of more than academic interest to the American taxpayer."[20] Regarding President Truman's "Point Four" program of foreign aid following World War II (which Vogt opposed and de Castro praised), Vogt advocated freedom to access reliable contraception as a precondition for receiving American assistance. "Quite as important as the Four Freedoms ... is a Fifth Freedom – from excessive numbers of children," he quipped.[21]

Throughout *Road to Survival*, Vogt criticized twentieth-century public health programs for having exacerbated the mounting problem of overpopulation. "Was there any kindness in keeping people from dying of malaria so that they could die more slowly of starvation?" he asked in the introduction.[22] Vogt's critique pointed to Latin America's medical sanitarians as among the well-meaning progressives responsible for unsustainable population growth. His discussion of that world region noted that "drinking water has been improved in many cities to such an extent that intestinal diseases, the most effective factor limiting populations, have dropped sharply."[23] This observation was certain to raise the ire of public health professionals, including de Castro, for whom their nations' early twentieth-century sanitarians were

[17] Ibid., 282.
[18] Ibid., 228.
[19] Ibid., 206.
[20] Ibid.
[21] Ibid., 211.
[22] Ibid., 13.
[23] Ibid., 164.

heroes. Vogt was evidently aware that his cavalier attitude toward infant mortality could be jarring. He followed the observation that "One of the greatest national assets of Chile, perhaps the greatest asset, is its high death rate" by adding, "This is a shocking statement. Nevertheless, if one does not believe there is a virtue in having more people live ever more miserably, destroying their country with increasing rapidity, the conclusion is inescapable."[24] Unabashed, later in the volume Vogt chastised British colonial administrators in India for contributing "to making famines ineffectual, by building irrigation works, providing means of food storage, and importing food during periods of starvation" rather than allowing disease and famine to limit Indian population growth as they had for generations.[25] Vogt's book repeatedly depicts the global poor as wallowing in misery. In his view it would be better not to have been born than to live as they did.

Vogt's dismal depiction of hunger and poverty in many regions of the world was influential in shaping overpopulation discourse and related philanthropic efforts over the following decade.[26] His warnings about the dangers of overpopulation became the basis for Paul Ehrlich's widely read book *The Population Bomb*, published in 1968, two months before Vogt's death. An entomologist by training, Ehrlich became a member of Stanford University's faculty in 1959, where he interacted with sociologist Kingsley Davis, ecologist Garrett Hardin, and others who believed firmly in looming demographic catastrophe. Critics of Ehrlich's 1968 book have noted that he attributed to population biology and ecology problems (like the global proliferation of urban slums) that were more directly attributable to industrialization, increasingly concentrated land ownership, and resulting rural–urban migration.[27] During the 1960s, Ehrlich cited India as an example of the population crisis that threatened all humanity, but by the following decade he had begun to focus attention on American consumerism as a significant engine of environmental crisis.[28] Ironically, this brought him closer to the viewpoint of Josué de Castro. Historian Thomas Robertson argues that this

---

[24] Ibid., 186.

[25] Ibid., 226. Notably, British economist John Maynard Keynes opposed immunization efforts in colonial India because he thought population growth would impede economic progress. Michelle Murphy, *The Economization of Life* (Durham, NC: Duke University Press, 2017), 21.

[26] Robertson, *Malthusian Moment*, 56.

[27] Charles Mann, *The Wizard and the Prophet* (New York: Knopf, 2018), 402.

[28] This shift in Ehrlich's position was probably influenced by his debates with ecologist Barry Commoner, whose 1971 book *The Closing Circle: Nature, Man, and Technology* (New York: Random House, 1971) "insisted that the consumption patterns of industrial society, not the birth rate, lay behind the ecological crisis of the planet." Erika Lorraine Milam, *Creatures of Cain* (Princeton: Princeton University Press, 2019), 186. Commoner debated both Ehrlich and ecologist Garrett Hardin.

shift made Ehrlich's position less politically viable than his prior attacks on the overfecundity of the third-world poor.[29]

In a biography of Vogt that juxtaposes the ecologist's gloomy predictions about humanity's future with those of the more optimistic agronomist Norman Borlaug (often termed the father of the Green Revolution), author Charles Mann emphasizes aspects of Vogt's life that may have contributed to his pessimistic outlook.[30] Vogt's father abandoned their small family shortly after his son's unplanned birth. During a lonely childhood on Long Island the young Vogt became an avid amateur ornithologist, and as an adult the rapid urbanization around his childhood haunts alarmed him. In 1934, Vogt became editor of the Audubon Society journal *Bird-Lore*; through that position he met Aldo Leopold and other prominent naturalists. In 1939, on the recommendation of Robert Cushman Murphy of the American Museum of Natural History, Vogt traveled to guano islands off the Peruvian coast to advise Peru's government about seabird population decline. While there he witnessed a famine among the birds caused by diminished plankton supply during an El Niño period. Vogt interpreted this in Malthusian terms, as a natural cycle that kept the bird population from exploding.

Upon his return to the United States, Vogt was hired by Nelson Rockefeller's Office of Inter-American Affairs, and from 1943 to 1949 he served as head of the Pan-American Union's (later the Organization of American States's) Conservation Section, in which capacity he toured a number of Latin American countries. In 1948, Vogt created the Conservation Foundation in partnership with Aldo Leopold and Fairfield Osborn, whose father had written a preface to Madison Grant's eugenicist 1916 tome *The Passing of the Great Race*; 1948 was also the year in which Vogt and Osborn each published books warning of disastrous environmental decline as a result of human population growth.[31] Vogt continued this argument in subsequent publications, such as a 1949 *Harper's* magazine article that portrayed Mexico as teetering on the edge of ecological collapse due to accelerating human population growth. Historian Nick Cullather notes that Mexico's population density at the time was lower than that of the United States and speculates that Vogt confused urbanization, caused primarily by

---

[29] Thomas Robertson, "Revisiting the Early 1970s Commoner-Ehrlich Debate about Population and Environment: Dueling Critiques of Production and Consumption in a Global Age," in *A World of Populations: Transnational Perspectives on Demography in the Twentieth Century*, eds. Heinrich Hartmann and Corinna Unger (New York: Berghann Books, 2014), 116 and 120.

[30] Mann, *The Wizard and the Prophet* in which Borlaug is cast as the wizard in contrast to pessimistic prophet Vogt.

[31] Fairfield Osborn, *Our Plundered Planet* (London: Faber and Faber, 1948).

internal migration, with overall population increase.[32] That internal migration was the result of both agricultural modernization, spurred in part by the US-promoted Green Revolution, and industrialization.

Also in 1948, with help from Julian Huxley and UNESCO, Vogt and several colleagues formed the International Union for the Protection of Nature. During its first conference the following year the IUPN criticized President Truman's Point Four program for aid to developing countries, which advocated increased productivity and consumption worldwide. Due to such outspoken opposition to the economic development agenda promoted by many government officials, Vogt was asked to resign from the Pan-American Union. In 1950, he accepted Fulbright and Guggenheim grants to travel with his second wife, Marjorie, to Scandinavia, to investigate successful population control measures. In 1951, he was hired by Margaret Sanger to head the Planned Parenthood Federation of America where he remained for a decade. Vogt's criticism of economic policies that prioritized growth without heeding environmental costs made him a pariah in some circles during that critical Cold War decade and strained his professional relationships. In 1967, Vogt's third wife died, and he killed himself a year later. None of his marriages produced children.

Vogt's descriptions of the obstacles to global security and prosperity posed by unmanaged population expansion in Asia, Latin America, and elsewhere were ruthless. From his perspective, as an American who prioritized conservation to maintain existing standards of living, this stance undoubtedly seemed eminently defensible, merely embracing hard-nosed solutions to postwar global challenges. Viewed from a very different geographic and political orientation, Vogt's recommendations were appalling. De Castro countered Vogt's dramatic vignettes of impending catastrophe with an optimistic vision of improved global nutrition fueling boundless human ingenuity. Both authors hoped to persuade readers through emotional appeals about the values at stake in this debate.

Brazilian Physician Josué de Castro and the *Geography of Hunger*

Josué de Castro grew up in the coastal capital city of Recife, Pernambuco, in northeast Brazil (Figure 9.1). There a centuries-old sugar-exporting sector historically dependent on enslaved labor was in decline, and de Castro witnessed the deep poverty of many in the region. His childhood home was adjacent to mangroves where families constructed makeshift shacks to forge a meager living from crabmeat as their primary sustenance. Some of these

---

[32] Nick Cullather, *The Hungry World: America's Cold War Battle against Poverty in Asia* (Cambridge, MA: Harvard University Press, 2010), 65.

Figure 9.1 Portrait of Josué de Castro, undated.
Source: Acervo Fundação Joaquim Nabuco. Recife-PE, Brazil.

desperate souls were escaping droughts in the semiarid interior *sertão*, as de
Castro's own father had done; others faced the endemic hunger that plagued
many inhabitants of Brazil's coastal sugarcane zone. "Right next to [our]
house," de Castro recalled in a memoir, "started the tightly packed area of
the hovels – straw and mud huts, piled one on top of the other in a network
of alleys in desperate anarchy. The houses penetrated the water, the tide
invaded them. The branches of the river overtook the street and the mire
overwhelmed everything."[33] As a child, de Castro listened to stories of
servitude told by two formerly enslaved men who worked for his father.
Over time he came to understand that the crabs on which residents of the
mangroves subsisted provided a more reliable source of nutrition than what
many people in northeast Brazil enjoyed – and that the "drama" of constant
hunger that colored the daily existence, and even the metaphorical language,
of his childhood playmates played out with slight variation in communities
around the world.[34] The physician had witnessed first-hand the human

---

[33] Josue de Castro, *Of Men and Crabs*, trans. Sue Hertelendy (New York: Vanguard Press,
1970), xvii–xviii.
[34] Ana Maria de Castro, ed., *Fome: Um tema proibido*, 3rd ed. (Recife: CONDPE/CEPE,
1996), 28–29.

misery of slow starvation. He disagreed vehemently with Vogt as to its fundamental causes and solutions.

De Castro's professional training took him to medical school in Rio de Janeiro during the late 1920s and back to Recife, where he graduated from the city's new faculty of medicine in 1932. Soon after, he received funding from the Pernambuco (state) Department of Public Health to study the relationship between income and cost of living among working people in Recife. This research was conducted by surveying patients at public health clinics about their family size, income, and expenses for necessities (housing, clothing, and food). The resulting, widely cited publication, "Living Conditions of the Working Class in Recife: An Economic Study of Their Diet," was the first of its kind in Brazil and a model for subsequent investigations by de Castro and others.[35] The report concluded that Recife's workers lived in a constant state of debt, with average salaries lower than what was required to sustain a family of five or more. Food absorbed three-quarters of workers' budgets and was nonetheless nutritionally inadequate; the typical diet amounted to 1,650 calories each day with no milk, fruits, or vegetables. De Castro cited malnutrition as the cause of high infant mortality (almost 260/1,000 in Recife – and higher in rural areas) and low life expectancy, which amounted to a tremendous loss of "human capital."[36] What had often been characterized as "*mal de raça*" (racial weakness) was in fact "*mal de fome*," he argued – physical debility caused by poor nutrition.[37] The data from de Castro's study became the empirical basis for passage of a minimum wage law for urban workers, supported by President Getúlio Vargas.

De Castro held a series of academic and political positions over several decades, all centered on problems of nutrition and political economy. From 1937 to 1957, he occupied the Human Geography chair at the new University of Brazil in Rio de Janeiro, where he was appointed director of the Institute of Nutrition. Along with his academic posts, in 1942 he became director of a Brazilian agency intended to address food supply problems during the world war. During this period, de Castro shifted his research focus from the physiological study of malnutrition (its direct causes and impact on the human body) to a political-economic analysis of hunger as a widespread social scourge. He deliberately rejected the term "malnutrition" in favor of the more evocative "hunger" (*fome*), rather than adopting a more objective and distanced approach to questions of food scarcity.[38] His goal was to emphasize the human

---

[35] Josue de Castro, *As Condições de Vida das Classes Operárias no Recife: Estudo Econômico de sua Alimentação* (Rio: Dept. de Estatística e Publicidade, Min. do Trabalho, Indústria e Commercio, 1935).

[36] Ibid., 19–20.

[37] Ibid., 7.

[38] De Castro, *Fome*, 14.

suffering and injustice at the core of this phenomenon. This focus is evident in a fictional sketch de Castro published in the Rio de Janeiro newspaper *Diário de Notícias* during a drought in 1951, in which a farmer is forced to witness his young child's death by dehydration after years of backbreaking labor in the semiarid interior of northeast Brazil. In this and similar (but nonfictional) accounts de Castro worked to elicit sympathy from his countrymen for the chronic tragedy that plagued the most marginal citizens of his home region.[39]

In 1948, de Castro organized a Latin American conference on nutrition in Montevideo, sponsored by the UN FAO and World Health Organization. Two meetings followed, in Rio de Janeiro (1950) and Caracas (1953), all addressing similar themes to those highlighted in his publications.[40] In 1950, conservative landowners blocked de Castro's appointment to become Brazil's Minister of Agriculture, due to his support for controversial land redistribution measures. He then became president of the UN FAO's Executive Council, a post that he held for four years. While at the FAO in Rome, de Castro founded a campaign against world hunger with which he remained involved through 1965. President Vargas appointed de Castro to a National Commission for Agrarian Reform (Comissão Nacional de Política Agraria), which aimed to extend rights to rural workers that had been obtained by urban workers during the 1930s. Vargas also made him vice-president of a national social welfare commission (Comissão Nacional de Bem Estar Social) in 1953.

De Castro was not the only Brazilian nutritionist to criticize the political and economic foundations of malnutrition during the mid twentieth century. Several others conducted similar studies and published books and articles that highlighted undernourishment as a crisis as important as poor sanitation, a public health issue that had received considerable attention and federal funding during the first decades of the twentieth century.[41] Like de Castro, these authors emphasized that what had often been interpreted as signs of racial inferiority among Brazil's poor were instead indicators of malnutrition, stemming from a range of social inequities.[42] Scientists like Pernambuco's

---

[39] Josue de Castro, "A Seca," *Diário de Notícias* (Rio de Janeiro), 1951. Newspaper clipping held in JdC archive, undated.

[40] See *O Estado de S. Paulo*, 27 maio 1953: "Instalado em Campinas o Primeiro Seminario Latino-Americano Sobre os Problemas da Terra," clipping obtained from Centro Josué de Castro, Recife.

[41] Eric D. Carter, "Social Medicine and International Expert Networks in Latin America, 1930–1945," *Global Pubic Health* 14, no. 6–7 (2019): 791–802.

[42] Ruy Coutinho, *O Valor Social da Alimentação* 2nd ed. (Rio de Janeiro: Agir Ed., 1947); Nelson Chaves, *O problema alimentar do Nordeste brasileiro: Introdução ao seu estudo economico social* (Recife: Ed. Medico Cientifica, 1946); Nelson Chaves, *A sub-alimentação no Nordeste brasileiro* (Recife: Imprensa oficial, 1948); and Orlando Parahym, *O Problema alimentar no sertão* (Recife: Imprensa Industrial, 1940). On medicine and racial discourses in northeast Brazil, see Stanley E. Blake, *The Vigorous Core of Our*

secretary of public health Nelson Chaves, a physiology professor at Recife's medical school, mobilized quantifiable data and their own professional status to generate sympathy for the chronically malnourished, in hope of promoting more just access to essential resources.

De Castro's Response to Vogt

De Castro reached international readers with his 1951 book *A Geopolítica da Fome*, which outlined the problem of global hunger as a product of economic relationships within and among nations. Published in English as *The Geography of Hunger* to avoid use of a term tainted by association with Nazi ideology, the book was translated into over twenty languages during the 1950s.[43] Neither Vogt's *Road to Survival* nor Fairfield Osborn's *Our Plundered Planet* was as broadly translated.[44] *The Geography of Hunger* expanded on de Castro's 1946 study of hunger in Brazil, which itself stemmed from his mid-1930s analysis of the diets of Recife's working poor. The book vigorously opposed the view, ascendant within American development organizations in particular, that global hunger was rooted in excessive population growth. Its specific target was *Road to Survival*, which de Castro satirically termed *The Road to Perdition*.

In *A Geopolítica da Fome*, de Castro accused Vogt and other "neo-Malthusians" of laying blame for social unrest and looming environmental crises on the poor for their undisciplined reproductive choices, when they should instead be seen as victims of global economic and political forces that conspired to deprive their families of sufficient food. De Castro insisted that hunger is not a biological or ecological phenomenon; it is the product of human economies. He cited the dedication of vast tracts to monocrop exports as the primary culprit, both because of the soil erosion that ensues and because those crops displace food cultivation. "Hunger has been chiefly created by the inhuman exploitation of colonial riches, by the *latifundia* and one-crop culture which lay waste the colony, so that the exploiting country can take too cheaply the raw materials its prosperous industrial economy requires," de Castro wrote.[45] Those guilty of creating this crisis were not the famished "as Vogt asserts," but rather "those who go in for neo-Malthusian theories while they defend and benefit from the imperialist type of

---

Nationality: Race and Regional Identity in Northeastern Brazil (Pittsburgh: University of Pittsburgh Press, 2011).

[43] These included English, Spanish, French, Italian, German, Swiss, Russian, Polish, Czech, Hungarian, Bulgarian, Yugoslavian, Lithuanian, Norwegian, Swedish, Hebrew, Japanese, Chinese, and Hindi.

[44] *Road to Survival* was translated into at least eleven languages.

[45] Josue de Castro, *The Geography of Hunger* (Boston: Little, Brown, 1952), 7.

economy."[46] The neo-Malthusian recommendations of Vogt and his colleagues "reflect the mean and egotistical sentiments of people living well, terrified by the disquieting presence of those who are living badly," the Brazilian asserted bluntly.[47]

In de Castro's reading of modern history, the dominant global model of "economic colonialism," driven by raw materials export from the tropics, generated hunger among laborers in those peripheral regions. He proposed an alternative model that he termed a "geography of abundance," mutually beneficial to the world's wealthy and poor. Instead of trying to limit the growth of populations in the Global South, de Castro argued, those countries should expand the land and sea areas used for food cultivation and intensify agricultural production with the aid of fertilizers, pesticides, and better soil conservation. By raising the real wages and consumption of the poor, economic exchange would be stimulated worldwide – and population reduction would no longer be necessary or desirable. De Castro disputed Vogt's central claim that all of Earth's fertile soils were already under cultivation. "A great many areas of very good soil are to be found in South America" and elsewhere, he asserted, and not yet farmed.[48] Within Latin America, only Puerto Rico could be said to be overpopulated, and "chronic starvation" there was the result of neocolonial policies by the United States, which displaced food production in favor of sugar, tobacco, and coffee exports.[49] "The fundamental truth can no longer be concealed," De Castro intoned in his conclusion. "The world has at its disposal enough resources to provide an adequate diet for everybody, everywhere. If many of the guests on this earth have not yet been called to the table, it is because all known civilizations, including our own, have been organized on a basis of economic inequality."[50]

De Castro had great faith in human creativity, capacity for technological innovation, and instinct for survival. There is no "impassable limit to human population" fixed by nature, he insisted – in contrast to Vogt's postulation of firm limits to Earth's carrying capacity. Rather, people "transform natural limitations into social opportunities" through science and ingenuity.[51] De Castro deemed chemical fertilizers and other technologies for intensive farming to be critical for increasing food production, along with fish farming

---

[46] Ibid., 17.

[47] Ibid., 312. Historian Piers Hale observes that English socialists also rejected the application of Malthusianism to human societies, believing that "Humanity could raise itself out of the struggle for existence through labor, cooperation, the application of technology to the natural resources that lay all around them, and the fair and equitable distribution of the goods that they produced." Hale, *Political Descent*, 258.

[48] De Castro, *Geography*, 97.

[49] Ibid., 123.

[50] Ibid., 281.

[51] Ibid., 25.

and hydroponic agriculture. Like many other Cold War technocrats he was enthusiastic about the US Tennessee Valley Authority model, through which "rational control of land, water, and all the various resources of the region" raised incomes and improved living conditions for millions of people within the agency's jurisdiction.[52] "When deserts of ice and impenetrable tropical jungles are being turned to gardens and orchards, when the lands we farm and the plants we grow are being made to multiply their yield, and while we are barely learning how to tap the great food reservoirs of the waters, the wild flora, and of artificial synthesis, the Malthusians go on setting up their sinister scarecrows," he chided. "It is nothing to us, since we have no reason to fear them."[53]

Paralleling the organization of Vogt's book, de Castro's substantive chapters analyze deficiencies in the typical diet of poor people from many global regions, based on prevailing understanding among nutritionists in the late 1940s of the importance of minerals, vitamins, and amino acids. The Brazilian viewed hunger as correctable through effective use of modern technologies, but also as a matter of equitable resource distribution (as economist Amartya Sen would argue several decades later).[54] Both of these elements – the techno-logical and social limits on food supply – must be addressed simultaneously to provide adequate food for expanding human communities. De Castro cited a recent UN FAO proposal as one potentially helpful intervention in the global food economy: "What we propose is an international instrument of consult-ation and co-operative action in the commodity field, so that nations may join in concerted efforts to attack the common enemies of mankind – poverty, disease and hunger – instead of each attacking the other's prosperity in a futile effort to defend its own."[55] Ensuring food sufficiency was fundamental to global security, de Castro insisted, deploying one of Vogt's motivating con-cerns in service of an opposing position.

Central to de Castro's argument was the belief that hunger itself contributes to population growth, so reducing hunger would correspondingly slow popu-lation increase: "The psychological effect of chronic hunger is to make sex important enough to compensate emotionally for the shrunken nutritional appetite," he speculated.[56] In support of this theory de Castro cited research on the suppression of fertility in rats that consume a high-protein diet, as

---

[52] Ibid., 137. On the TVA as a regional development model, see Amy C. Offner, *Sorting Out the Mixed Economy: The Rise and Fall of Welfare and Developmental States in the Americas* (Princeton: Princeton University Press, 2019).

[53] De Castro, *Geography*, 299.

[54] Amartya Sen, *Poverty and Famines: An Essay on Entitlement and Deprivation* (Oxford: Clarendon Press, 1981).

[55] De Castro, *Geography*, 306.

[56] Ibid., 69.

wealthier people do.[57] He provided data on the inverse relationship between national birth rates and protein consumption and suggested a mechanism to account for this, based on research by physiologists at the University of Chicago:[58] protein deficiency reduces liver function, which reduces "the liver's ability to inactivate estrogens," thereby *increasing* women's fertility.[59] Higher fertility among less well-fed populations might be an evolutionary protection, de Castro hypothesized, as prospects for the survival of offspring diminished due to food scarcity.[60]

Another cornerstone of de Castro's opposition to overpopulation discourse was his conviction that a better-fed population would be more industrious and could therefore produce more food. "In diet [lie] the origins of Chinese submissiveness, of the fatalism of the lower castes in India, of the alarming improvidence of certain populations in Latin America," he insisted, uncritically referencing widely held stereotypes.[61] "The 500,000,000 Chinese could have a life absolutely free of hunger if they were physically capable of work, if their nutritional and hygienic conditions allowed them to make use of the geographic potentialities of their country."[62] Sons were often deemed essential to farmers without livestock, such as Chinese working tiny plots to feed their families (who could not waste meager food on animals). This produced a vicious cycle of population growth, de Castro argued. Thus, "[t]o wipe out hunger ... it is necessary to raise the productive levels of marginal peoples and groups, and through economic progress to integrate them into the world economic community."[63] In de Castro's developmental imagining, resource constraints would be overcome by the same people whom Vogt targeted for fertility control. Adequately fed human communities would become an asset to greater productivity, rather than a drain on ecological resources or a threat to political stability. Extrapolating from his extensive experience with marginalized communities in Recife, de Castro attributed similar potential for resourcefulness and ingenuity to impoverished people worldwide. Vogt, with little personal experience of the world's poor, projected a pessimistic vision of the future on continents overrun by desperate, starving hordes. De Castro's contrasting vision stemmed from frequent encounters with the chronically hungry and a more sympathetic understanding of their plight.

---

[57] Ibid., 70.
[58] Ibid., 72. The physiologists were Anton Carlson and Fred Hoelzel, with whom de Castro corresponded in the 1950s.
[59] Ibid., 164.
[60] Ibid., 71.
[61] Ibid., 68.
[62] Ibid., 166.
[63] Ibid., 303.

It is notable that de Castro's writing emphasized class location and social marginalization but rarely made explicit mention of race, despite the fact that in his native region African ancestry correlated closely with poverty – and that as a child he knew men who had spent their youth as slaves. De Castro's silence with regard to race is likely traceable to the emphasis in Brazil, particularly from the 1930s through the 1950s, on racial harmony ("racial democracy") as a national characteristic.[64] It is not clear that he embraced northeastern sociologist Gilberto Freyre's depiction (in seminal works like *Casa Grande e Senzala*, about colonial slaveholding) of Brazilian society as racially harmonious.[65] Nevertheless, he may have been disinclined to take direct aim at the widespread assumption that explicitly racial prejudice was not an issue in Brazil. Reference to social class, often in terms of contrasts between elite and marginal groups, sufficed – in de Castro's view – to describe the divisions and injustices that he wished to draw attention to. De Castro himself was sometimes described as a Mulatto man, suggesting that he had both European and African ancestry (and that this mixed lineage was evident in his appearance), but he did not discuss this as a significant aspect of his own identity. In his worldview and experience, class loomed larger than race as a way of conceptualizing the inequities in power and resource access that he wished to draw attention to. De Castro's concerns were with global capitalism and the legacies of imperialism, which he understood in political and economic terms rather than racial ones. He also disregarded gender as an analytical category, which seems astonishing to a contemporary reader, given his subject matter, but was true of virtually all the men who debated population growth as a global issue in the mid twentieth century. Their central focus was the quantitative relationship between human birth rates and food supply, and they debated this with little reference to women – notwithstanding women's essential roles in childbirth and nutrition.[66]

Reception of De Castro's Work

The Geography of Hunger elicited a range of responses from reviewers worldwide. In August of 1952, demographer Kingsley Davis published a blistering condemnation in the *American Sociological Review*, calling de Castro emotional, utopian, and unscrupulous; a dishonest cheat who "makes no fetish of consistency" in his use of facts. Davis accused the Brazilian of masquerading "under the cloak and prestige of science" while discarding "all the canons of

---

[64] George Reid Andrews, "Brazilian Racial Democracy, 1900–90: An American Counterpoint," *Journal of Contemporary History* 31, no. 3 (1996): 483–507.

[65] F. Vasconcelos, "Fome, eugenia e constituição do campo da nutrição em Pernambuco," *História, Ciências, Saúde – Manghinhos* 8, no. 2 (2001): 322–326.

[66] Bashford, *Global Population*, 92–94.

scientific logic and evidence," and he called upon social scientists to issue a robust condemnation of the book. Several critics in the United Kingdom objected to de Castro's indictment of British imperialism despite efforts by British scientists to promote intensified agricultural production of precisely the kind that de Castro was calling for. Others accused him of naive credulity regarding the socialist experiments underway in China, the Soviet Union, and (later) Cuba – and of dishonestly omitting mention of the 1930s famines in Stalinist Russia from his published work. A reviewer in the *New Statesman and Nation* noted that de Castro's idealistic prescriptions for greater global cooperation with regard to food distribution seemed inconsistent with his thesis in the section criticizing European imperialism, namely, that "the heart of man is deceitful above all things and desperately wicked."[67]

Other readers were much more complimentary. The American Political Science Association gave de Castro their FDR Foundation award in 1952. Novelist and Nobel laureate Pearl S. Buck, who contributed a preface to the American edition of *Geography of Hunger*, called it "the most hopeful and generous book I have read in my entire life."[68] Several explicitly anti-Malthusian authors wrote to de Castro commending his important work. "Population pressure is always directly and immediately related to the number of people who are able to make only very meager economic demands ... on the existing soil and other natural resources of any given area or of the whole planet, irregardless [*sic*] of how great or abundant these resources might be," noted Fred W. Smith of Camden, Ohio. The mystery, in his view, was why Malthusianism continued to hold any sway in the twentieth century. However, he concluded:

> When one recalls ... how universally and effectively it supports the status quo and is able to place, with a fair show of science, philosophy, and reason, the responsibility and the blame for vice, misery, and starvation squarely on the shoulders of nature and providence, and finally on the sexual incontinence of the dispossessed and starving people themselves, then the phenomenal success enjoyed by this theory appears to be very much less remarkable.[69]

Other supporters of de Castro's critique highlighted the resource pressures caused by overconsumption (especially of beef) by citizens in the industrialized world and the tremendous waste of food in the United States, suggesting that any global food crisis was not caused by inadequate supply. Daniel Slutzky of the Department of Social Sciences, University of El Salvador, noted that the

---

[67] Walter Elliot in *New Statesman and Nation*, March 8, 1952.

[68] In the *Herald Tribune*, 1952, according to De Castro (*O Drama Universal da Fome* [Rio de Janeiro: ASCOFAM, 1958], 299).

[69] Letter to J. de Castro from Fred W. Smith, JdC archives, física 331 pasta 51.

International Planned Parenthood Federation conducted similar contraceptive promotions throughout Central America regardless of national and local population densities, which differed substantially. This supported his thesis that the central goal of IPPF and related organizations was neither economic development nor improving the welfare of the poor but rather reducing population numbers in places the United States feared as potential incubators of communism. Slutzky compared this to the Nixon administration's contraceptive promotions among poor urban Blacks in the United States.[70]

Establishing Global Networks for Development Action

Following his return from the FAO, de Castro served two terms as a Pernambucan representative in the national legislature and member of the Brazilian Workers' Party (PTB). He was appointed Brazil's ambassador to the UN in 1962. The right-wing military regime that came to power in 1964 stripped de Castro of his political rights, due to his support for leftist causes, including granting voting rights to illiterate people, labor protections for rural workers, and agrarian reform (land redistribution). He took refuge in Paris, remaining there until his death in 1973.[71]

De Castro continued his campaigns against hunger and poverty well beyond his years with the FAO. He established ASCOFAM (Associação Mundial de Luta Contra a Fome) in Brazil to better organize production and distribution of nutritional foods in his native northeast region, home to many of the country's most impoverished people. From 1960 to 1965 he led the UN FAO's World Campaign Against Hunger. He established an international NGO known as CID (Centro Internacional para o Desenvolvimento) in 1962 to address structural issues underlying global poverty and continued this work from his exile in Paris after 1964.[72]

Upon his departure from the FAO in 1955, de Castro chastised his colleagues there for "lacking courage" to confront global hunger as a political issue, asserting that they preferred to view it as a problem of technical know-how, a less contentious position.[73] His critique of the UN and its agencies intensified over the following decade. By 1965, de Castro was referring to the UN as a reactionary organization established by elites with regressive views, one that impeded the development of solutions to the world's problems and

---

[70] Daniel Slutzky, "Politica Demografica y Subdesarrollo en America Latina," *Out* 1969, p. 32. JdC archives, física 331 pasta 51.

[71] Manoel Correia de Andrade, "Josue de Castro: O homem, o cientista, e o seu tempo," in de Castro, *Fome*, 285–321.

[72] See Archive Davies, *A World without Hunger: Josué de Castro and the History of Geography* (Liverpool: Liverpool University Press, 2023).

[73] A 1955 newspaper clipping in Portuguese; no source. JdC archive, física 14 pasta 29.

served instead to obfuscate with statistics.[74] Nevertheless, his years as president of the FAO's Executive Council had immersed him in an international network that he expanded strategically over subsequent decades. While representing the FAO, de Castro visited President Truman to discuss the establishment of an international food reserve and met with Popes Pius XII and Paul VI. The latter (while still a cardinal) reportedly told the Brazilian that *Geography of Hunger* was "the most Christian text I've read in my life."[75]

De Castro's letters from the 1950s until his death reference trips to China in 1957 (he was impressed by how well-fed the rural population appeared to be) and Russia, and a celebration of the Cuban Revolution with Fidel Castro in 1961. He met or corresponded with an array of influential figures, including Americans Eleanor Roosevelt and Robert Kennedy, Italian Roberto Rossellini (who had plans to make a film based on *Geography of Hunger*), Indian prime minister Pandit Nehru, former Argentine president Juan Perón, French philosopher Jean-Paul Sartre, and UN secretary-general U Thant, from Burma. De Castro's correspondence indicates that he was popular among "anti-Malthusians," including Catholic clergy, from numerous countries across the Americas and Europe. These scholars and activists raised the alarm about the Nixon administration's support for coercive sterilization, with backing from Paul Ehrlich (who reportedly proposed that the United States lace its food aid with sterilizing chemicals[76]) – measures that de Castro referred to in some of his writings as contraceptive genocide. But de Castro also maintained cordial correspondence with representatives of the International Planned Parenthood Foundation, including Lady Rama Rau of India.

In 1962, de Castro was instrumental in establishing an International Center for Development (CID) headquartered in Paris. The goal of this nongovernmental organization was to provide an alternative to existing development efforts that were "contaminated by neocolonialism" and, in de Castro's view, amounted to little less than alms.[77] CID hoped to promote development efforts that would be of genuine assistance to Third World populations and to provide a forum for frank discussion of global problems, followed by concerted action. The founders, including representatives from Greece, Hungary, France, Belgium, Peru, Chile, the United States, India, and Senegal, proposed the creation of an International Development University with a focus on education and human resources as key to global development. In addition to his work with this group, during his exile in Paris de Castro

---

[74] Newspaper clipping from *Folha de São Paulo*, May 5, 1965. JdC archive, fisica 96 pasta 40.
[75] Interview with de Castro published in *Prova*, January 8, 1966. JdC archive, fisica 96 pasta 40.
[76] Comment attributed to Ehrlich during the 13th UNESCO-North Am. Conference as reported in *O Globo*, November 26, 1969. JdC archive, fisica 541 pasta 58.
[77] "Centro Internacional para o Desenvolvimento," JdC archive, fis. 314 pasta 118.

became deeply involved in nuclear nonproliferation and global peace movements spearheaded by Robert Oppenheimer and Bertrand Russell. His meetings and correspondence with political leaders and activists around the world maintained his sense of connection to marginalized people and his commitment to opposing injustice, despite geographic distance from the communities that had forged these sensibilities earlier in his life.

By the mid-1960s, the central argument of de Castro's writing was that since underdevelopment and hunger are closely correlated, while population density and hunger are not, population is not the causal factor leading to hunger. Science could be used to solve problems of food supply and distribution, but only where there was political will to do so (as in Britain during World War II). "Neo-Malthusians" attempted to explain political–economic dynamics and their consequences as natural, when they are fundamentally social phenomena. Highly technocratic approaches to development, such as those espoused in the early 1970s by the Club of Rome (which sponsored a widely read publication entitled The Limits to Growth)[78] did not adequately acknowledge the political variables that affect resource access. De Castro had become increasingly fascinated by the United States as an exemplar of underdevelopment; a draft chapter of one unpublished manuscript written near the end of his life was entitled "Misery in the Midst of Abundance" and focused on impoverished minority communities within that wealthy nation.

Conclusion

From 1948 until his death in 1973, de Castro sought to elevate alternative ways of conceptualizing global development, decentering the interests, priorities, and cultural assumptions of the most powerful governments and their populations. His critique emphasized that many variables can be analyzed in considering the relationship between human populations, natural resources, global security, and ecological health – and that the decision to problematize one of those variables rather than another (e.g., the fertility rates of women in less industrialized nations rather than the consumption habits of Americans) is a political choice, influenced in the mid twentieth century by geopolitical concerns, nascent modernization theories, and the geographic and social positions of the most highly resourced participants in this debate.

There are provocative parallels between the early Cold War debate about overpopulation and more recent discourse about climate change. In both cases, significant regional and social-class differences in contribution to the problem are elided to suggest that human communities face a shared challenge

---

[78] Dennis Meadows et al., Limits to Growth: A Report for the Club of Rome's Project on the Predicament of Mankind (New York: Universe Books, 1972).

for which they are similarly culpable and must sacrifice in the interest of a secure future. Anna Tsing has written about climate change discourse in the following terms, focusing on debates that preceded the 1992 Earth Summit in Rio de Janeiro:

> Spokespeople for the global south argue that global climate models are an articulation of northern interests. Global climate models show everyone invested in the same reductions of greenhouse gases; they cover up the fact that most of these gases are emitted in northern countries. In blaming southern countries for a share of the greenhouse gas problem, the models also obscure differences between northern and southern emissions. Many greenhouse gases emitted in southern countries are "subsistence emissions," in contrast to the "luxury emissions" of northern countries. Global modeling, they imply, is not neutral.[79]

This assertion by Tsing's "spokespeople" is somewhat analogous to de Castro's arguments decades before. There are contrasts, notably that de Castro thought overpopulation was not nearly as significant a concern as Vogt believed it to be, whereas Tsing's protagonists did not aim to diminish the significance of the climate crisis. But in both cases, voices from postcolonial regions highlight the culpability of industrialized nations in precipitating catastrophe. And in both instances, the objectivity of scientific paradigms emanating from centers of global power is questioned by advocates for marginal regions who highlight the political agendas embedded in those first-world discourses. Like Tsing's southern voices, de Castro questioned what countries such as the United States stood to gain through an overpopulationist framing of resource scarcity, and he drew attention to dynamics that were elided by emphasizing Third World fecundity as the root cause of rising hunger. In a world of vastly unequal wealth and power, any implication of shared responsibility for accelerating resource scarcity should reasonably be viewed with skepticism; de Castro's life's work stands as a reminder of this.[80]

In a recent critique of the economic ideologies underlying twentieth-century population policies, historian Michelle Murphy describes Notestein's advice to Pakistan's government in the late 1950s as "symptomatic of an *economized* reformulation of Foucault's description of the violent purifications of state racism as some must die so that others might live into *some must not be born so that future others might live more abundantly (consumptively)*."[81] She emphasizes that American economists in the 1960s valued "averted births"

---

[79] Anna Tsing, *Friction: An Ethnography of Global Connection* (Princeton: Princeton University Press, 2005), 105–106.

[80] Ethnographer Jade Sasser offers a similar critique in *On Infertile Ground: Population Control and Women's Rights in the Era of Climate Change* (New York: New York University Press, 2018).

[81] Murphy, *Economization of Life*, 41. Italics in the original.

more highly than lives lived in poverty, which were often evaluated negatively in relation to GDP. In a section reflecting on overpopulation concerns in relation to climate change Murphy asks, "What kind of population control practices and racisms are reactivated by pointing the finger at human density in a moment when wealthy human-capital assemblages with often low levels of fertility are responsible for the vast bulk of [carbon] emissions?"[82] Murphy's analysis highlights the demographic abstractions wielded to justify a range of population control measures in the interest of economic growth, limiting births in some communities so that other people (whiter, wealthier, and more connected to centers of global power) can live more lavishly. What de Castro strove to broadcast during the years when theories promulgated by Notestein and others were gaining influence was that "population" is not abstract. To critique public health efforts in a city like Recife, as Vogt did, was to advocate for the painful and avoidable deaths of thousands of infants, deaths witnessed by family members powerless to intervene. This searing reality was obvious to de Castro because he had lived among people who benefited enormously from cleaner water, vaccinations, and other public health interventions in his native city. *The Geography of Hunger* implored readers to reckon with this reality. De Castro hoped to invert Vogt's apocalyptic portrayal of a world overburdened by hungry bodies, emphasizing instead what those people could contribute if reasonably provided for – and pointing out the significant per-capital resource drain of the world's wealthiest people.

[82] Ibid., 47 and 138.

10

Bureaucratic Vulnerability
Possession, Sovereignty, and Relationality in Brazilian Research Regulation

ROSANNA DENT*

Introduction

Geneticist Fabrício Rodrigues dos Santos rushed through his words as he told me about his experience of fieldwork in A'uwẽ territory. From his office at the Universidade Federal de Minas Gerais, he wove an entrancing story of his time in Aldeia Etênhiritipá. His eyes shone as he recounted a hunting trip, stargazing, and a movie night; I was fascinated. Rather than focusing on days filled with collecting genealogical data and genetic samples, Santos's narrative centered on what he called "the most interesting part ... the anthropological experience."[1] Santos's tale did not fit with my preconceptions of genetic sampling for the Genographic Project.

I had been introduced to Santos by one of his colleagues, Maria Cátira Bortolini, a fellow scholar of human genetics who was hosting me for a period of research in her department at the Universidade Federal do Rio Grande do Sul in Porto Alegre. As Bortolini followed my growing interest in the history of genetics research in A'uwẽ (Xavante) communities as well as its ethical oversight, she suggested I interview Santos.

The South American branch of the Genographic Project, a global program sponsored by National Geographic and IBM, had run into a complex and slow

* This work was supported by Fulbright IIE, the Social Science Research Council, the American Council on Learned Societies, a postdoctoral fellowship at McGill University funded by the Mellon Foundation, and a National Science Foundation grant (#2147284). Any opinions, findings, and conclusions expressed here do not necessarily reflect the views of the National Science Foundation. My deepest gratitude to Aldeia Pimentel Barbosa for hosting and teaching me and for participating in this work, along with leaders from Aldeia Etênhiritipá. Researchers generously recorded oral histories. Coparticipants and especially editors improved this chapter with thoughtful feedback, as did the anonymous reviewers. A'uwẽ is the auto-denomination of a group referred to as Xavante by most in the broader Brazilian and international scholarly publics. Here I use "A'uwẽ," and maintain "Xavante" in quotations.
[1] Fabrício Rodrigues dos Santos, interview with Rosanna Dent, Belo Horizonte, March 6, 2014.

process of regulatory approval in Brazil. As a transnational, corporately sponsored program that focused on genetic sampling of Indigenous groups across the continent, the project was subjected to multiple additional levels of ethical oversight. As the coordinator for the program, Santos navigated the four-year process to obtain official permissions to conduct the embattled research project. However, while he was happy to tell me about the difficulties of navigating regulatory bureaucracies, it was talking about his time in the field that made his eyes shine. As I listened to him explain his fieldwork, it seemed to me that there was another process of research regulation underway as well – though perhaps not explicitly articulated. The A'uwẽ *aldeia* – village or autonomous political community – that hosted him and his team, Etênhiritipá, was also working to instill a framework for their interactions, a relational and affective basis for knowledge making.

At the time of my interview with Santos, I was conducting my own participant-observation through oral history and archival research in labs and academic departments around Brazil. It would be a year before my fieldwork would extend to overlap explicitly with Santos's with my first visit to Terra Indígena (T.I., Indigenous Land) Pimentel Barbosa, the A'uwẽ terri-tory where the Genographic team conducted their research. The aldeias of this territory have been hosting researchers since shortly after they established diplomatic relations with the Brazilian government in 1946. The first anthro-pologist arrived in the aldeia of Wedezé in 1958. Since then, the community – and subsequently communities as the population grew and aldeias divided – of Terras Indígenas Pimentel Barbosa and Wedezé have hosted dozens of researchers. Geneticists and biomedical researchers James V. Neel, Francisco Salzano, and colleagues followed the first anthropologist in 1962. Subsequently, scientists from almost every discipline of the human sciences have visited, from social and cultural anthropology, linguistics, and education, to public health, biomedicine, and human genetics, creating an extensive published literature of *warazú* (non-A'uwẽ) understandings of A'uwẽ life, language, health, biological differentiation, and history.

These communities are a classic example of "overstudied Others," as described by Eve Tuck and K. Wayne Yang.[2] However, T.I. Pimentel Barbosa has become a hub of scholarly attention not only because of the wealth of past studies and data sets, but also due to the interest of aldeia residents in cultivating relationships with researchers. While A'uwẽ in the 1950s and 1960s had little context to understand the actions of the scientists who arrived to study them, with time and experience they developed their own expertise in research. In the context of ongoing Brazilian colonialism in

---

[2] Eve Tuck and K. Wayne Yang, "R-Words: Refusing Research," in *Humanizing Research: Decolonizing Qualitative Inquiry with Youth and Communities*, eds. Django Paris and Maisha T. Winn (Thousand Oaks, CA: SAGE, 2014), 223–248, 223

Indigenous territories, residents and especially certain leaders came to engage scholars for their own reasons. They developed strategies to guide and direct researchers.

This chapter explores the two different systems of research oversight that applied to Santos and the Genographic Project. It does so by situating these systems within the broader relational nature of field sites and fieldwork that has shaped the experience of warazú researchers and A'uwẽ participants. As the other chapters in this book have shown, affective relationality shapes knowledge in the human sciences across the scales of the transnational and national. Both the assertion and recognition of expertise are bound up with appeals to affect: about the racial construction of national character, or the politics of consumption within the international order. Here I combine attention to the bureaucratic interface of the transnational and national with the very personal level at which Indigenous actors modulate affect to attend to pressures of Brazilian state administration. Attention to the "complex moral sensibilities and structures of feeling" of research participants does much to illuminate the limitations of abstracted ethics and formalized medical research regulation, as Warwick Anderson has shown in his explorations of the relationships between Fore communities afflicted by the neural disease kuru and the scientists who sought to study and sample them.[3]

In this case, sets of normative considerations stretch and vary across fields of scholarship including genetics, public health, and anthropology. Members of the Genographic Project encountered both the formalized bureaucracy of ethics regulation of the Brazilian state, and the systems of relational ethics that A'uwẽ leaders and community members in T.I. Pimentel Barbosa have developed to try to instill or compel responsible research from warazú like myself who seek knowledge in A'uwẽ territory. By relational ethics, I refer to principles that some A'uwẽ articulate, both explicitly in conversation and through their demands and actions. These principles hold that researchers should center relationships and the responsibilities that accompany these relationships when engaging or seeking to engage with members of the aldeias.

In exploring these two systems, I focus primarily on genetics-based research that began in 1962 and continues to the present. I consider other forms of scholarship and research methodology in my discussion of what I see as A'uwẽ regulation because these interactions have profoundly influenced aldeia residents' experiences of knowledge production. This is particularly true of the work of anthropologists who lived in A'uwẽ territory for extended periods, as well as public health scholarship that has evolved into repeated, ongoing programs of study.

---

[3] Warwick Anderson, *The Collectors of Lost Souls: Turning Kuru Scientists into Whitemen* (Baltimore: Johns Hopkins University Press, 2008), 7.

My focus here is on genetics research because it brings into relief two facets of what I call bureaucratic vulnerability. First, this history demonstrates the role of the state and the regulatory system in adjudicating research, as well as researchers' attempts to continue their inquiries even when that oversight could constrain current research. I argue that the way some geneticists have interpreted state regulatory systems regarding biosamples creates additional risks for Indigenous people under study. At the same time, Indigenous groups are placed in a bureaucratic double bind, where non-Indigenous experts are called on to justify and validate their claims in the eyes of the state. For all the flaws inherent in its conceptualization, the Genographic Project allows us to see ways A'uwẽ have responded to the dual and interrelated challenges of recognition under a colonial state and the management of outside researchers. This is the second axis of bureaucratic vulnerability: The implicit requirement to be documented in certain ways pushes community members to engage and cultivate relationality with researchers.[4]

Here I explore the possessive logics of both the systems of the state and the actions of researchers as Indigenous heritage, genes, lands, or knowledge come to be the focus of study and documentation. In studying and writing about these systems, which also shape my own research engagements both with the Brazilian state and with A'uwẽ community members, I do not aim to place blame or exonerate – I too am implicated and embroiled in Brazilian state regulation and relations-building in Pimentel Barbosa.[5] Rather, I hope history and the work of the historian have a role to play to make sense of the contexts, unintended consequences, and possible alternate futures that emerge from seriously considering how A'uwẽ actors build affective and political connections with the scholars who visit them.

I begin by exploring the concept of bureaucratic vulnerability and how regulatory structures and their avoidance are conditioned by the possessive logics of Brazilian colonialism. From the protectionist regulation of the Brazilian state, I turn to examine a set of relationship-based practices that A'uwẽ interlocutors have developed over repeated interaction and years of collaboration with a group of anthropologists and public health researchers. Finally, I turn to how and why

---

[4] Margaret Bruchac highlights parallel logics in *Savage Kin: Indigenous Informants and American Anthropologists* (Tucson: University of Arizona Press, 2018), 180.

[5] My own relations are primarily constituted through Aldeia Pimentel Barbosa, where I have spent approximately four months over five trips since 2015. As Warren describes in Chapter 3, this project was not initially conceptualized through Indigenous Studies methodologies, and I was initially hesitant to engage A'uwẽ for fear of replicating research harms. However, since working directly with Aldeia Pimentel Barbosa, my work has changed and I have also begun collaborations with Aldeias Etênhiritipá and Paraíso. While I have not explicitly discussed this paper in the aldeias, I have talked with them about many episodes herein, including ongoing use of old biological samples, and community experiences of Genographic research.

A'uwẽ aldeia members embraced the Genographic Project, even as other similarly well-informed groups declined to participate. Contextualized in prior experience with scholars, and the mandates of bureaucratic thinking that constrain Indigenous rights to land, health, and education in Brazil, the relational work that aldeias of T.I. Pimentel Barbosa performed looks different.

Bureaucratic Vulnerability

In discussions and formal interviews between 2012 and 2014, I quickly came to see that the regulatory hurdles that the Genographic Project faced loomed large in the imagination of other geneticists with interest in studying Indigenous genes. The layers of bureaucracy and the approach of the regulatory body, the Comissão Nacional de Ética em Pesquisa (CONEP, the National Commission of Ethics in Research) proved such a perceived barrier that various labs stopped proposing new sampling. Instead, they used workarounds to continue their research, whether on samples from collaborators in other countries where official approvals are easier to attain or by using stored samples collected under prior ethical and regulatory regimes.

This dynamic drew my attention to questions of vulnerability: How are Indigenous people positioned as uniquely vulnerable research subjects within Brazilian legislative frameworks? How do regulatory bureaucracies and the people that interact with them simultaneously construct and respond to perceived vulnerability, while also creating new kinds of risk for Indigenous groups? And how are broader bureaucracies of recognition related to and dependent on expert knowledge production? In this section, I explore the concepts of vulnerability and bureaucracy as they relate to what Aileen Moreton-Robinson and Maile Arvin have each referred to as a *logic of possession* over Indigenous peoples.[6] This logic is practised by both the Brazilian state in its oversight of researchers and the extension of federal recognition of Indigenous lands, and by non-Indigenous scholars in human genetics research that aims to tell a universal history of humankind. Those who, in Ailton Krenak's words, are not full members of the *clube da humanidade* (the club of humanity) can be the subjects of research to illuminate a history of humanity that scientists hold to be universal.[7]

The logic of possession is enacted through discourses of both biology and national history. It provides a counterpart to more common discussions of

---

[6] Aileen Moreton-Robinson, *The White Possessive: Property, Power, and Indigenous Sovereignty* (Minneapolis: University of Minnesota Press, 2015); Maile Arvin, *Possessing Polynesians: The Science of Settler Colonial Whiteness in Hawai'i and Oceania* (Durham, NC: Duke University Press, 2019).

[7] Ailton Krenak, "Tragédia Yanomami mostra que clube da humanidade não é para todos," interview with Eduardo Sombini, Illustradíssima Conversa, January 28, 2023.

the settler colonial logic of elimination, as articulated by Patrick Wolfe.[8] Arvin highlights how scientific classification of Native Hawaiian people as "almost white" has served to naturalize the presence of white settlers in Polynesia. She argues, possession expresses "more precisely the permanent partial state of the Indigenous subject being inhabited (being known and produced) by a settler society."[9] Jenny Reardon and Kim TallBear have extended a similar analysis to the politics and practices of genetic research on Indigenous peoples including the Genographic Project. They highlight how (usually) white scientists make claims on the genes of Native peoples in service of Western creation stories that do not serve and may even undermine Indigenous epistemologies.[10] In the Brazilian context, Tracy Devine Guzmán has discussed how the paternalistic logics of expansionism were articulated through an "anti-imperial imperialism." Framed in terms of their opposition to foreign interests in Amazonia or (other) Indigenous territories, Brazilian state actors constructed Native people as "our Indians in our America," justifying and extending their own ongoing internal colonialism.[11]

The regulation and production of expert knowledge is central to enacting logics of possession. As Joanne Barker has so convincingly written, mandates for documentation to "prove" Indigeneity create conflicted and conflictual relationships between Native peoples and scholarship produced by outsiders.[12] When the object of study is an Indigenous group, Aileen Moreton-Robinson writes, the product is cultural difference, which serves nation-states by producing "manageable forms of difference."[13] Indigenous peoples must demonstrate their claims in specific ways to be recognizable and to access even the limited rights afforded to them.[14]

In twentieth- and twenty-first-century Brazil, difference has most often been managed through bureaucracy. At the highest level, state-led arbitration of knowledge provides the foundation for the twin processes of defining

---

[8] Patrik Wolfe, "Settler Colonialism and the Elimination of the Native," *Journal of Genocide Research* 8, no. 4 (2006): 387–409, https://doi.org/10.1080/14623520601056240.

[9] Arvin, *Possessing Polynesians*, 16.

[10] Jenny Reardon and Kim TallBear, "'Your DNA Is Our History': Genomics, Anthropology, and the Construction of Whiteness as Property," *Current Anthropology* 53, no. S5 (2012): S233–S245.

[11] Tracy Devine Guzmán, *Native and National in Brazil: Indigeneity after Independence* (Chapel Hill: University of North Carolina Press, 2013), 105–130.

[12] Joanne Barker, "The Specters of Recognition," in *Formations of United States Colonialism*, ed. Alyosha Goldstein (Durham, NC: Duke University Press, 2014), 33–56.

[13] Moreton-Robinson, *White Possessive*, xvii.

[14] Elizabeth A. Povinelli, *The Cunning of Recognition: Indigenous Alterities and the Making of Australian Multiculturalism* (Durham, NC: Duke University Press, 2002), 12 and 173–185.

and racializing *indígenas* and justifying the expropriation of Indigenous land.[15] The introduction of a system of oversight for explorers, scientists, and artists wishing to visit Indigenous lands in the 1930s served as a mechanism by which the Brazilian state could claim Indigenous peoples and cultures as national patrimony, while also asserting the role of protector. At the same time, the state reasserts its sovereignty through the regulation of researchers. I describe this process in more detail in the next section. However, for the purposes of understanding the second axis of bureaucratic vulnerability, I highlight here that some geneticists respond to the seemingly ever-increasing bureaucracy of regulation by using less-regulated strategies to continue to research Indigenous groups, but without consultation or ongoing consent. Citing bureaucratic barriers, they have continued to use historical biosamples, or tissues or DNA from Indigenous groups beyond Brazil's borders to research without engaging in fieldwork or the direct accountability of relationship-building.

In framing these dynamics as "bureaucratic vulnerability," I am interested in the creation of risk and harm through state administration.[16] The concept of vulnerability in bioethical discourse has been problematized for its insinuations of weakness, a focus on participants at the expense of attention to structures of research inquiry, and its limitation in addressing the specificity of particular groups' experiences.[17] Here I seek to move beyond classifying Indigenous people as "vulnerable," rejecting damage narratives,[18] to focus instead on how the concept of vulnerability and the bureaucracies built around it create risks for A'uwẽ communities. The term bureaucratic vulnerability intends to draw readers' attention to the structures of research inquiry and state knowledge production that marginalize A'uwẽ knowers and knowledge. This focus on bureaucratic structures also helps me grapple with what I see as A'uwẽ desire to engage with researchers as well as community members' refusal to talk about or dwell on negative perceptions of researchers or damage enacted by researchers.[19]

---

[15] On the co-construction of "*índigena*," and "Spaniard" in the sixteenth century and the corresponding radical change in governmentality through bureaucratization, see Irene Marsha Silverblatt, *Modern Inquisitions: Peru and the Colonial Origins of the Civilized World* (Durham, NC: Duke University Press, 2004).

[16] Anthropologist Rosana Castro highlights an interesting parallel in her analysis of Brazilian state prioritization of pharmaceutical trails that renders racialized Black and Brown Brazilians *biodisponible* (bioavailable) for clinical research in a context of scarcity of care. See Castro, *Economias políticas da doença e da saúde: uma etnografia da experimentação farmacêutica* (São Paulo: Hucitec Editora, 2020).

[17] Alexis K. Walker and Elizabeth L. Fox, "Bioethics, 'Vulnerability' and Marginalization," *AMA Journal of Ethics* 20, no. 10 (2018): E941–E947.

[18] Tuck and Yang, "R-words," 223.

[19] Tuck and Yang draw on Audra Simpson's work to emphasize that "refusals are not subtractive, but theoretically generative." "R-words," 223. See Audra Simpson, *Mohawk*

Settler Knowledge and Bureaucracies of Possession and Recognition

The history of Brazilian state regulation of research in Indigenous territory is tightly bound up with possessive logics, as well as a nationalist concern about the presence of foreign researchers. More recent classifications of Native peoples as vulnerable cannot be divorced from this and a broader history of Brazilian *tutela* or tutelage. Tutela and its proponents – identifying Indigenous people as child-like and in need of protection, education, and moral uplift – justified post-Independence colonization and provided terms in which they claimed moral authority.[20] As Antonio Carlos de Souza Lima has argued, the development of the administration of Indigenous peoples and lands in the early twentieth century was part of a "massive siege of peace," which also helped to form the Brazilian state.[21]

Official legislation and the institutionalization of research oversight was not implemented until 1933 under the dictatorship of Getúlio Vargas with the formation of the Conselho de Fiscalização das Expedições Artísticas e Científicas (Conselho, Council for Control of Artistic and Scientific Expeditions).[22] The proposal's architects promoted the project as one aimed at those foreign expeditions that did not follow "established norms and ethics," by which they meant those that failed to support technical cooperation and knowledge sharing.[23]

The new *Conselho* was fundamentally concerned with protecting what its members and legislators saw as patrimony, which included mineral, botanical, and ethnological specimens. Once established, it worked with the Ministry of Foreign Relations and the Serviço de Proteção aos Índios (SPI, Indian Protective Service) to ensure, among other things, that an equal half of all specimens and materials collected were deposited with the Brazilian government before export permits would be granted.[24] The government also required copies of all resulting reports and publications, contributing – at least in theory – to the legibility of Indigenous groups to the state. As Luís Grupioni explains, ". . . the Indians interested the Council as a testimony, as an

Interruptus: Political Life across the Borders of Settler States (Durham, NC: Duke University Press, 2014), 95–114.

[20] Antonio Carlos de Souza Lima, *Um grande cerco de paz: Poder tutelar, indianidade e formaçao do estado no Brasil* (Petrópolis: Vozes, 1995); Seth Garfield, *Indigenous Struggle at the Heart of Brazil: State Policy, Frontier Expansion, and the Xavante Indians, 1937–1988* (Durham, NC: Duke University Press, 2001).

[21] Lima, *Um grande cerco de paz.*

[22] Luís Donisete Benzi Grupioni, *Coleções e expedições vigiadas: Os etnólogos no Conselho de Fiscalização das Expedições Artísticas e Científicas no Brasil* (São Paulo: Hucitec Editora, 1998), 50–53.

[23] Ibid., 51–52.

[24] Ibid.

inheritance, transformed into patrimony that needed to be preserved; it is in the action of collecting artifacts and depositing them in museums that this organ occupies a place in the indigenist field."[25] In the case of the Conselho, nationalism and a paternalistic and assimilationist mandate served as moral justification for regulating researcher access to Indigenous groups in Brazil.

During the military dictatorship (1964–1985), a variety of institutions were charged with collectively controlling scientific expeditions into Indigenous territories.[26] Bureaucratic demands grew, but the framing was similar: preventing scientists, particularly foreign ones, from visiting militarily strategic areas or absconding with valuable patrimony was enmeshed with protecting Indigenous groups from the same visitors.

The exercise of Brazilian sovereignty over Indigenous peoples was also intimately dependent on the knowledge created by researchers, particularly in relation to the identification and demarcation of Indigenous lands. Although there were strong fluctuations over the course of the dictatorship, FUNAI administrators were sometimes supportive of research, conceptualizing it as a resource for the *indigenista* organization and government more broadly to understand – and by implication govern – Indigenous groups. For example, as president of FUNAI in 1975, the unusually progressive General Ismarth de Araújo Oliveira supported the view that anthropologists should be required to provide information in the form of field reports and final publications and that, "the organization of this documentation will be one of the greatest weapons that FUNAI has for the defense of Indigenous land."[27] FUNAI even prioritized research in certain areas where the government lacked knowledge about Indigenous inhabitants.

Federal recognition of Indigenous lands – demarcation – has been an essential factor in the protection of Indigenous lives and lands.[28] However, it also partially remediated the problem created by the state's programs of westward expansionism. The military dictatorship promoted demarcation because it facilitated the regularization of legal claims by ranchers and new

---

[25] Ibid., 269.
[26] They included the Conselho and the SPI until both institutions were dissolved, in 1969 and 1967, respectively. The Fundação Nacional do Índio (FUNAI, National Indian Foundation), the Conselho Nacional de Pesquisa (CNPq, National Research Council), and the Conselho Nacional de Segurança (National Security Council) took over oversight of scientific expeditions.
[27] Conselho Indigenista, "Sessão 1 do Conselho Indigenista," AVESON 222 F lado B, compact disk, SEDOC-MI/FUNAI.
[28] As Garfield points out, it is important to recognize that government actors have played essential roles in preventing murder and dispossession, even as broader policies created conditions for these confrontations. See Garfield, *Indigenous Struggle*, 51.

agribusiness entrepreneurs to surrounding land they had occupied at the invitation of the government.[29]

The bureaucracy changed again in 1988 with the ratification of the current Constitution, and in 1996 with the National Health Council's Resolution 196/96. This resolution added to the existing system a mandatory review by an institutional ethics committee and CONEP that would oversee certain kinds of research considered to be of higher risk.[30] This legislation detailed a series of "special thematic areas" designated for higher scrutiny, most of which focused on biological concerns such as research involving human genetics, or dealing with populations seen as biologically vulnerable, such as pregnant women or children. However, the legislation also identified any research in Indigenous territory as a special thematic area. The law defines Indigenous peoples – the only category determined by sociological parameters – as inherently vulnerable.[31] The implementation of the added layers of review meant that a single protocol dealing with research in Indigenous territory had to pass through at least four processes of review before FUNAI forwarded it to the community in question for consultation. The result was an extensive, slow, and complex approval process.

State reliance on knowledge from the human sciences thrusts Indigenous groups into a bureaucratic double bind. This is an administrative version of the dilemma highlighted by Guzmán in her examination of Indigenous leaders' and activists' political action in Brazil. Guzmán describes this double bind as "knowing that any intervention they might undertake in that system – already against great odds and at great personal cost – unavoidably reinscribes, to some degree, the erasure, exclusion, and delegitimization that has characterized the indigenous–state relationship since its inception."[32] Faced with demands to be legible in certain ways, many Indigenous groups cultivate relationships with anthropologists or other scholars who they hope will become allies in moments of bureaucratic need.

Adopted Warazú and Regulatory Affect

Aldeia Etênhiritipá's reception of Fabrício Rodrigues dos Santos and the Genographic Project was predicated on years of experience of interacting with

---

[29] Ibid., 140–142.
[30] Conselho Nacional de Saúde, "Resolução nº 196, de 10 de outubro de 1996," February 10, 2017, http://bvsms.saude.gov.br/bvs/saudelegis/cns/1996/res0196_10_10_1996.html.
[31] Ricardo Ventura Santos, "Indigenous Peoples, Bioanthropological Research, and Ethics in Brazil: Issues in Participation and Consent," in *The Nature of Difference: Science, Society and Human Biology*, eds. George Ellison and Alan H. Goodman (London: Taylor & Francis Books, 2006), 191.
[32] Guzmán, *Native and National in Brazil*, 166.

researchers. Here, before examining how the Genographic researchers recounted their enrollment in the next section, I trace another case of scholars who developed deep connections in Terra Indígena Pimentel Barbosa. Their acceptance and training by the community has taken place over more than thirty years. As anthropologists and public health researchers Carlos Coimbra and Ricardo Ventura Santos have visited, researched, and collaborated in T.I. Pimentel Barbosa, over time A'uwẽ community members shaped their approaches through the construction of an affective field.

I describe this as an "affective field," to emphasize the dynamic play of experience (being affected) and action (affecting change).[33] Rather than referring to the cultivation of specific emotions such as happiness, nostalgia, or pity, the affective field emphasizes the ever-evolving qualities of human relationships and relationships with place. This space of being affected and affecting others overlaps and works to constitute the field. It compels researchers to center relations; research becomes, fundamentally, a question of "self-in-relation."[34] Whether this involves cultivating field experiences such as stargazing, or a sense of inclusion in ceremony and community events, or a deeper sense of obligation through the articulation of new kin relations and their invocation in the face of community challenges, a complex emotional landscape permeates the researchers' "fields." Following the story of Santos and Coimbra's research allows us to see how aldeia residents developed their own regulatory system that is affectively based.[35] Their request that Coimbra and Ventura Santos conduct a delimitation study for the demarcation of an adjacent A'uwẽ territory in the mid-2000s also underscores how taking a relational approach to researchers is a precarious but important technique in the face of bureaucratic imperatives of documentation.

In 1990, Coimbra and Santos were introduced in Aldeia Pimentel Barbosa by Nancy Flowers, an anthropologist who had spent fourteen months there during a critical period of the 1970s when residents of T.I. Pimentel Barbosa faced particularly acute challenges to their land rights. Community members indicated that the first step for the researchers upon arriving was to present themselves and their plans at the *warã*, the men's council meeting, a twice daily gathering of men who have completed spiritual initiation that serves as a

---

[33] Michael Hardt, "Forward: What Affects Are Good For," in *The Affective Turn: Theorizing the Social*, eds. Jean Halley and Patricia Ticineto Clough (Durham, NC: Duke University Press, 2007), ix.

[34] Fyre Jean Graveline as quoted in Margaret Kovach, *Indigenous Methodologies: Characteristics, Conversations and Contexts* (Toronto: University of Toronto Press, 2009), 14.

[35] Of particular relevance is Anderson's excavation of how Fore cultivated connections, kinship, and networks of gift exchange and indebtedness with biomedical and social scientists that arrived to study kuru: Anderson, *Collectors of Lost Souls*.

space to discuss political happenings of the community. Santos explained this saying, "If we arrived one day, the next day at five-thirty in the morning we were there in the warã, introducing ourselves, recounting our news, with them wanting to know what we wanted to do there, what our plans were."[36] Having already hosted researchers dozens of times, the warã had come to function as part of the A'uwẽ system of oversight. Scholars were expected to publicly present their ideas and plans for their work, and then be present for what sometimes were long, formal discussions in the A'uwẽ language. The warã became a space of accountability, where at the end of a period of investigation, scholars are called back to update the community on their activities and expectations moving forward.

As they undertook their first years of research in Pimentel Barbosa, recent PhDs Coimbra and Santos were busy settling into new positions at the Escola Nacional de Saúde Pública (ENSP, National School of Public Health) in Rio de Janeiro. There they built a research program in Indigenous health with particular attention to the social determinants of health, and T.I. Pimentel Barbosa came to play a central role in their work.[37] Part of what made ongoing visits to the aldeia possible was the open-ended quality of their relationships.

Coimbra and Santos were welcomed and supervised by a variety of Elders, leaders, and others in the aldeia. Over time, "our perspective really changed," Coimbra explained to me. "[At the beginning,] we went into the A'uwẽ community without knowing anything, just with the Neel and Salzano references [from a 1962 study] in hand to repeat that research ... In contrast, today the exchange is really intense."[38] Coimbra and Santos' approach was also evolving in concert with changing notions about the practice of public health in Brazil. The early 1990s was a time of widespread changes in the Brazilian public health system in the wake of re-democratization, with discourses about health rights and equity centered by many public health researchers and practitioners.[39]

In 2014, Tsuptó, a young leader who officially represents Aldeia Pimentel Barbosa to FUNAI, described a shift in how researchers and community members have engaged over time. Thinking back to the earliest ethnographic publications, which documented his community and other A'uwẽ aldeias, he said, "There was a time when there was disrespect. Without the knowledge of the Indigenous population, some works were published, which I see today as a

[36] Ricardo Ventura Santos, interview with Rosanna Dent, Rio de Janeiro, April 15, 2014.
[37] Carlos E. A. Coimbra Jr., interview with Rosanna Dent, Rio de Janeiro, March 19, 2014.
[38] Ibid.
[39] Ana Lúcia de Moura Pontes, Felipe Rangel de Souza Machado, and Ricardo Ventura Santos, *Políticas antes da política de saúde indígena* (Rio de Janeiro: Editora Fiocruz, 2021).

lack of respect. But now, today we are talking. This is respect. It is through dialogue that the work is done, and that is important."[40] This dialogue increasingly shapes the research projects that the Pimentel Barbosa Elders accept. In the case of Santos and Coimbra, as diabetes and other metabolic issues became increasingly prevalent, key interlocutors within the aldeia started insisting that the public health researchers turn their attention to chronic health problems. "Really, to be honest," Coimbra recounted,

> At first, when they started to talk to me about diabetes, it took me about two years to come to terms with the fact that I could not escape, because I had always strongly focused on the ecology of infectious disease. I knew something about metabolic disorders, but I did not know the field intimately . . . It took me a while to get up the nerve, but then we did it.[41]

By the 2010s, research on metabolic issues was a major aspect of the work of the ENSP research group. It was in the context of these sustained interactions that leaders in Pimentel Barbosa were able to advocate for a new direction in research that would address issues they prioritized.

Tsuptó emphasized that, yes, while he thought the research topics were important, the process of research also created enduring connections: "The work got deeper and won our confidence. And the relationship of friendship with Carlos and Ricardo . . . it's not just through the work or research, but through our relationship of friendship as people."[42]

Another key strategy that community members have applied to warazú researchers is the incorporation into the A'uwẽ affective field through adoption. Increasingly, over time, Elders in Pimentel Barbosa have adopted scholars who come to research. "Adoption" here is a process of claiming a researcher, by which an Elder (or sometimes more than one) asserts a relationship of kinship by publicly announcing their chosen relationship to the researcher. Whether or not a researcher is formally adopted depends on a variety of factors, from age to whether we stay in the aldeia or at the nearby government post, to the length of our visits. There is no formula.

What is clear is that by asserting kinship, A'uwẽ Elders call on researchers to behave according to social norms of family; they invoke these terms to emphasize their moral authority.[43] As researchers, we understand and attempt

---

[40] Tsuptó Buprewên Wa'iri Xavante, Barbosa Sidówi Wai'azase Xavante, Luiz Hipru Xavante, and Agostino Seseru Xavante, interview with Rosanna Dent, Água Boa, MT, June 4, 2014. Tsuptó referred – I think – not only to the abstracted relationship between aldeia and researcher, but also as an invitation to me to take a relational approach.

[41] Coimbra, interview.

[42] Tsuptó Buprewên Wa'iri Xavante, interview with Rosanna Dent, Canarana, MT, July, 10, 2019.

[43] Jeffrey Kaufmann and Annie Philippe Rabodoarimiadana make this point also for the case of Malagasy reception of fieldworkers in "Making Kin of Historians and

to meet these expectations to varying levels. As others have discussed, these forms of chosen kinship function in multiple ways. They help to bridge boundaries across power differentials, and in Cristian Alvarado Leyton's words, institutionalize "a moral imperative of loyalty and solidarity ... [through] affective relatedness, promising in turn an enduring relationship."[44] While some scholars express skepticism about the authenticity of anthropologists' or other researchers claims to adoption,[45] I understand A'uwẽ use of kin terms as a pragmatic invocation of relationality and expression of affective connection that helps clarify understandings of what constitutes ethical or moral work according to A'uwẽ expectations.

Santos and Coimbra were not initially incorporated into specific families, but after many years of collaboration, there was consensus among community members that they belonged to one moiety, poreza'õno, rather than öwawe, the other.[46] Their place in the social system of the aldeia was based not on the facts of their research, but on the facts of their persons. "Researcher is what you do. Poreza'õno is who you are," Tsuptó explained to me as I asked how we are separated into moieties. Likewise speaking of Coimbra, Tsuptó explained, "He would come independent of the research ... I would say, 'we need help. Carlos ... this thing is happening. Can you do this work?' 'No, Tsuptó. I'll do it.' And through that, Barbosa – as an Elder – he spoke: 'Carlos is my brother ... Everyone will respect him the way they respect me' ... So he decided. He spoke this to the warã."[47] Tsuptó's uncle Barbosa Sidówi Wai'azase Xavante claimed Carlos as family and extended his protection to the researcher.

Other researchers, especially those who have arrived at a younger age and lived more time in the aldeia, have been adopted and named, claimed by an A'uwẽ Elder as their child. Since the year 2000 at least seven researchers associated with the ENSP team – including myself – have been publicly adopted. In discussion, Elders Marilda, Solange, Angélica and Agostinho explained how this adoption works through the translation of Goiano, saying, "It is the person who arrives first, to encounter, greet, and love the person. So it is the one who arrived, greeted, and loved, and there they become family, they

---

Anthropologists: Fictive Kinship in Fieldwork Methodology," *History in Africa* 3 (2003): 179–194. On anthropologists as kin, see also Bruchac, *Savage Kin*.

[44] Cristian Alvarado Leyton, "Fictive Kinship," in *Encyclopedia of Human Relationships*, eds. Harry Reis and Susan Sprecher (Thousand Oaks, CA: SAGE, 2009), https://doi.org/10.4135/9781412958479.

[45] Kaufmann and Rabodoarimiadana detail the skepticism of Clifford Geertz, for example, but assess that taking seriously fictive kinship helps show the contingency, movement, and pragmatic-dynamic quality of research relationships. See "Making Kin," 192–193.

[46] A'uwẽ belong to one or the other moiety, passed to them from their father, and may only marry someone from the other.

[47] Tsuptó, interview.

begin to transform into family ... For example, Serebură arrived, greeted, and liked you, and already established [*colocou*] to call you daughter. So there you already are transformed."[48] As Elders extend this formal familial belonging, they inculcate obligation and esteem. Adoption implies a great deal of work, as adoptive family members often take an outsized role in caring for us warazú researchers, from educating us about how to behave and how to understand things that are unfamiliar to preparing materials for our participation in community ceremonial life.

This adoption draws researchers into an affective field, where a researcher's decisions and actions take place within a profoundly relational system. These actions to adopt, incorporate, teach, and oblige researchers are an A'uwẽ praxis that in my interpretation works to destabilize the researcher–subject binary. I see researcher and A'uwẽ creation of the affective field as working toward Kim TallBear's eloquent urging that "we must soften that boundary erected long ago between those who know versus those from whom the raw materials of knowledge production are extracted."[49]

Interest in incorporating researchers, however, is also a response to the failings and mandates of the state. Tsuptó recounted the work of one of Ventura Santos and Coimbra's students, Rui Arrantes, on oral hygiene: "now children, school-aged children, they use toothbrushes ... with their [ENSP's] work it awoke [us]. We had to do something, we had to take action."[50] Tsuptó emphasized that the researchers arrived with such a high level of training and technical competence that they were able to implement good programs that government employees failed to realize for their lack of relationships, dedication, and training.[51]

The ENSP researchers also responded to calls from the communities of T.I. Pimentel Barbosa to support their efforts to reclaim a large portion of their territory by conducting a delimitation study for submission to FUNAI and the judicial system. "As one of the leaders," Tsuptó recounted, "I needed help. For this delimitation study, a field study, I needed an anthropologist, I needed an environmentalist, I think a biologist ... I did not want FUNAI to choose someone. I did not trust them."[52] After consulting with the other aldeias of T.I. Pimentel Barbosa, Tsuptó invited Coimbra, Ventura Santos, and another member of their team, anthropologist James Welch, to conduct the study:

---

[48] Angélica Wautomouptabio Xavante, Agostinho Seresu Xavante, Marilda Peuzano Xavante and Solange Penepe Xavante, interview with Rosanna Dent with translation by Goiano Tserema'a Xavante, Canarana, MT, July 3, 2019.
[49] Kim TallBear, "Standing with and Speaking as Faith: A Feminist-Indigenous Approach to Inquiry," *Journal of Research Practice* 10, no. 2 (2014): 2.
[50] Tsuptó, interview.
[51] Ibid.
[52] Ibid.

"We went to Brasília, but at the time FUNAI had no funding to pay them . . . Beyond research, they were doing it because they are honest . . . A [different] anthropologist would not do this for free."[53]

"My first inclination was not to do it!" James Welch told me in a good-natured tone, "I thought it would be a huge amount of time . . ." he paused, "which it was."[54] Coimbra told me, "We could not say no, of course, so we did it."[55] Their collective expertise made them ideal for the project, Welch explained: "We decided together that it was important and it was the right thing and we probably had the best data to do it. We were probably the people that could produce a high-quality report."[56] In 2008, Santos, Coimbra, and Welch joined a group of FUNAI employees to complete the Wedezé delimitation study in collaboration with the eleven aldeias of T.I. Pimentel Barbosa.

The formal process for legal demarcation of Indigenous Territory in Brazil begins with a multidisciplinary delimitation study, which combines ecological, archeological, and anthropological expertise. First, individuals with state-recognized epistemic authority, usually framed in terms of training in the relevant academic fields, are appointed to a working group. They produce a delimitation study for FUNAI. Once approved, this study then passes to the courts, where it faces often-extensive legal challenges from affected landowners. If the courts accept the study, the land becomes officially demarcated, with timelines for non-Indigenous occupants to vacate the area.[57] Indigenous groups do not have complete legal sovereignty over their lands, which remain under federal control.

For the Wedezé delimitation study, the anthropologists were able to draw on their extensive experience as well as historical source material: Field notes from researchers who had witnessed critical land struggles of the late 1970s helped corroborate A'uwẽ explanations of how they had lost Wedezé, and how important it continued to be for them. The earliest anthropological study was central in supporting community claims to the longevity of their connection to Wedezé. To complement the historical data, ENSP researchers drew on publications and data sets that they and their students had produced over the preceding two decades. Months of collaborative work produced further evidence including ethnobotanical surveys, oral histories from Elders, and technical surveying of cemeteries and ritual spaces. The study was comprehensive, and thoroughly backed up by years of respected research.[58]

---

[53] Ibid.
[54] Welch, interview, March 27, 2014.
[55] Coimbra, interview.
[56] Welch, interview, March 27, 2014.
[57] FUNAI, "Entenda o processo de demarcação," www.funai.gov.br/index.php/2014-02-07-13-24-53.
[58] Ricardo Ventura Santos et al., "Relatório circunstanciado de identificação e delimitação: Wedezé, população indígena Xavante" (Brasília: FUNAI, 2011).

The report led to the delimitation of 150,000 hectares of A'uwẽ land in 2011, at a time when few new Indigenous lands were being recognized. With FUNAI's acceptance of the report, the process moved to the courts. As of 2024 the proposal still faces legal challenges and a long and precarious road to official demarcation. However, the strong case that the researchers were able to build in cooperation with the aldeias was a major step toward demarcation. Interlocutors in the aldeias of Pimentel Barbosa recognized Coimbra, Santos, and Welch not only as friends of the aldeia, but as scholars whose authority would be recognized by the state. They differentiated the academics from other warazú as those most well prepared to present data on territorial claims to Wedezé. At the same time, Coimbra's sense that "we could not say no," underlines the obligation the researchers felt.[59] In part, this dedication was cultivated by the investment of A'uwẽ leaders and community members in the researchers, and by the aldeia's ongoing work to find common ground.

The bureaucratic processes of recognizing land – albeit small portions of prior territories – served and serve to inscribe Brazilian sovereignty over Native lands and reinscribe the power of the state to grant recognition. It is an area where administrators call on what they hold to be apolitical expert knowledge: Scientific empiricism is called upon "to manage the existence and claims" of groups like A'uwẽ communities. This mandate to be documented means that Indigenous groups have incentives to not only allow, but even to actively seek out relationships with scholars.

A'uwẽ community members shape knowledge production by engaging with the hopes, desires, and fears of the scholars who come to study them. This is not to say that there are never moments of refusal – questions are avoided, researchers are turned away from certain topics, projects are allowed to perish in inaction.[60] However, A'uwẽ actively work with scholars and so exercise agency, even (and perhaps especially) within a context of unequal access to power, resources, and knowledge. As Sherry Ortner points out, citing Laura Ahearn, the point "is not that domination and resistance are irrelevant, but that human emotions, and hence questions of agency, within relations of power and inequality are always complex and contradictory."[61] In a system where academics are considered among the most reliable experts to consult on land demarcations or lobby for better education, health, or environmental

---

[59] Anthropologists in Brazil have frequently been called upon to provide different kinds of technical reports, for land delimitation or in criminal cases. For a critical overview of the conflicts this positioning produces, see João Pacheco de Oliveira, "The Anthropologist as Expert: Brazilian Ethnology between Indianism and Indigenism," in *Empires, Nations, and Natives: Anthropology and State-Making*, eds. Benoît de L'Estoile, Federico Neiburg, and Lygia Maria Sigaud (Durham NC: Duke University Press, 2005), 223–247.

[60] Simpson, *Mohawk Interruptus*.

[61] Sherry B. Ortner, *Anthropology and Social Theory: Culture, Power, and the Acting Subject* (Durham NC: Duke University Press, 2006), 138.

management policies, A'uwẽ community members have chosen to draw researchers into an affective field in order to compel them to "stand with" rather than "give back" to the community.[62]

The Xavante Genographic

As mentioned in the opening to this chapter, Genographic researchers Santos and Vieira spoke of their fieldwork in T.I. Pimentel Barbosa with relish. The researchers brought enthusiasm to discussing all their fieldwork, but in interviews they repeatedly set their time in Pimentel Barbosa apart. This suggests, I think, that their A'uwẽ hosts have been particularly adept at modulating the affective field of engagement. The two scientists articulated a sense of connection and belonging. This section examines researchers' personal reports of fieldwork in T.I. Pimentel Barbosa to explain the affective experience of research and explore the investment of time and effort that aldeia residents dedicated to the visiting researchers.

It is important to contextualize my interviews with the researchers within the fierce debates about genetic sampling of Indigenous groups generally and the Genographic Project specifically. Native activists and social scientists have objected to the premises of the project, citing the fraught Human Genome Diversity Project (HGDP) of the early 1990s, and a long history of scientific abuses of Indigenous subjects.[63] Some who were invited to participate articulated their own objections, including the Q'eros communities in Peru discussed by Adam Warren in Chapter 3 of this book.[64] Social scientists and historians have situated the Genographic Project within a longer trajectory of human biology, highlighting continuities with previous research agendas from the 1960s and 1990s.[65] In Brazil, journalists also picked up on the contested

---

[62] TallBear, "Standing with and Speaking as Faith," 2. On relationality in research, see also Kovach, *Indigenous Methodologies*; Shawn Wilson, *Research Is Ceremony: Indigenous Research Methods* (Winnipeg: Fernwood Publishing, 2009).

[63] See Indigenous Peoples Council on Biocolonialism, "Human Genetics Issues," *Indigenous Peoples Council on Biocolonialism*, 2017, www.ipcb.org/issues/human_genetics/index .html; Jenny Reardon, *Race to the Finish* (Princeton: Princeton University Press, 2004), 2 and 205; TallBear, *Native American DNA: Tribal Belonging and the False Promise of Genetic Science* (Minneapolis: University of Minnesota Press, 2013), especially 149–176.

[64] Groups that chose not to participate cited concerns about control over the use of the samples. In Brazilian press coverage, the most cited group to reject participation in South America was the Hatun Q'eros community. See Warren, Chapter 3, and TallBear, *Native American DNA*, 189–196.

[65] On the genealogy of the Genographic in the context of the 1960s Human Adaptability Arm of the International Biological Program, the Human Genome Project and the Human Genome Diversity Project, and broader attempts to study, characterize, and so construct "Native American DNA," see Joanna Radin, "Latent Life: Concepts and Practices of Human Tissue Preservation in the International Biological Program,"

nature of the project. They cast the initiative as a second *Projeto Vampiro*, citing the nickname of the HGDP. They drew comparisons to other controversial scientific endeavors, including the collection and storage of Yanomami blood, and the use of biosamples from Karitiana and Paiter (Suruí) to create immortalized cell lines for research.[66] It was in large part due to these critiques that the regulatory process for the project was so belabored.

At different moments in their interactions with me, the Genographic researchers sought to rearticulate their defenses of the project through our conversations.[67] Santos's version of the A'uwẽ fieldwork also included exaggerations and reasons to read the interviews critically. However, Santos and Vieira's sentiments of excitement and longing were genuine, and many portions of their accounts match up with reports from other researchers, including (in some ways) my own, about how they were received and treated by aldeias in T.I. Pimentel Barbosa. While some Indigenous groups were wary of participating, or chose not to, the A'uwẽ aldeias that the Genographic Project visited embraced the project.

The scientists' initial connection with T.I. Pimentel Barbosa was through Jurandir Siridiwẽ Xavante, a leader from Aldeia Etênhiritipá. Jurandir had participated in an advisory board made up of leaders from five different Indigenous groups that consulted with the Genographic during the regulatory process.[68] The scientific team was composed of four men: geneticist Fabrício Rodrigues dos Santos as the principle investigator; biophysicist and postdoctoral researcher Pedro Paulo Vieira; Francisco Araújo, a graduate student in social anthropology at the Universidade Federal do Rio de Janeiro; and Peruvian Aymara graduate student José Sandoval.[69] As their primary interlocutor, Jurandir played a central role in the researchers' experience of their work in the *Terra Indígena*. He coordinated their stay, officially presenting them to the warã on their first night in Etênhiritipá so that aldeia Elders could consider the researchers' proposal.[70] Jurandir exercised his political influence

---

Social Studies of Science 43, no. 4 (2013): 484–508; Reardon, *Race to the Finish*; and TallBear, *Native American DNA* respectively.

[66] Marcelo Leite, "Projeto Genográfico e 'Projeto Vampiro'," Folha de São Paulo, April 17, 2005; María Amparo Lasso, "Indígenas em guarda ante o projeto Genográfico," www .adital.com.br/site/noticia2.asp?lang=PT&cod=16334.

[67] For example, Santos emphasized the slow process of introducing the project to the A'uwẽ leadership, and with their approval, presenting the project to the community. Santos responded, implicitly, to critics who considered the time allotted for community consent processes in the project inadequate. See TallBear, *Native American DNA*, 190–191.

[68] Santos, interview. In 2008, the Genographic team held an initial meeting with the advisory group while the rounds of review continued at CONEP; Jurandir joined representatives from Kaingang, Tariana, Wapixana, and Pareci communities.

[69] Ibid.

[70] Vieira, interview.

in favor of the research project: He rallied aldeia residents to show up for the scientists' explanations of the project; he helped coordinate the support that the project would need at each turn.

Other aldeia residents also provided support. Two assumed the roles of guides and guards, helping the researchers with daily tasks, and protecting them and their equipment from overly curious children.[71] These men took the researchers out to explore the *cerrado*, and taught them about A'uwẽ fire hunting practices. The researchers also worked with two leaders from each of the nine aldeias that participated in the study.

Santos' narrative of his research experience in Etênhiritipá included a wide variety of interactions that had little to do with the project's goals of collecting genealogical data and genetic samples:

> It was really good because we interacted a lot. I brought a movie, I brought my computer. I have a film that tells the story of first contact of an uncontacted Indian group over in Rondônia ... And they loved it ... It was an all-night movie session, with that incredible starry sky, everyone sitting. The whole aldeia, you know? A lot of people.[72]

Later in our interview, Santos continued to describe stargazing with a laser star pointer and an iPhone constellation application. "So it was a moment fully lived in every minute," Santos sighed.

But Santos and Pedro Paulo Vieira felt most deeply drawn in by what they understood as their inclusion in the aldeia. "And not only that," Santos told me, following up on his account of the movie night, "We participated in rituals with them. Not the rituals they put on for tourists, ones that they were really doing." Santos explained: "There were two rituals going on at the same time. One was [like] a baptism ... In that ceremony I was baptized too. I'm öwawe," he said, referring to his incorporation into one of the A'uwẽ moieties.[73] Santos's story of his time in the aldeia betrayed the joy, excitement, and sense of engagement that set the A'uwẽ experience apart for the Genographic researcher. What Santos conveyed to me was not a series of emotional responses, but the researchers' movement through an affective field.

"The Genographic was adopted by the Xavante of South America in Brazil," Pedro Paulo Vieira told me, framing the A'uwẽ fieldwork as the pinnacle of the Genographic in Brazil:

> They are a people with an extremely strong culture – extremely ancestral, extremely rich – who, instead of wanting to understand what we were doing, simply absorbed the Genographic into their own culture. Fabrício, myself, and some other members of our team were even assigned to clans

[71] Santos, interview.
[72] Ibid.
[73] Ibid.

within the aldeia. I was given a name. We participated in Xavante rituals. That is to say, we became part of the Xavante community because of the project.[74]

Vieira emphasized Etênhiritipá's adoption of both the researchers and the project. This inclusion was compelling to the biophysicist because it was both personal and intellectual; it incorporated an invitation to remarry his wife in the aldeia alongside a perceived interest and investment in the scientific work itself. The researchers were struck by what they perceived as the authenticity of their hosts. At the same time, they felt embraced, included in this authenticity.

Focusing on the application, uses, and cultivation of affect highlights the extensive care work involved for A'uwẽ communities to host outsiders. The narratives that the Genographic researchers offered suggest that A'uwẽ subjects went to substantial trouble to inculcate certain affective states in the researchers who visited them. Rituals needed explanation. Equipment had to be protected. Even spontaneous decisions, such as a hunting or fishing trip, involved extensive work.

This experience of belonging was essential to the researchers' understandings of their fieldwork, and they mobilized this perceived acceptance and belonging to make claims about the legitimacy of their work through our conversations. But this sense of belonging and acceptance was also – and continues to be – marshalled by A'uwẽ with the expectation of mutuality. The investment to bring researchers into the affective field is about building enduring relationships to shape research through dialogue, and form experts who can be called upon to address community needs. However, when community members invest in Santos or me, they face the possibility that we will disappoint or betray. It is a precarious strategy within the double bind of state recognition.

Conclusion

The use and reuse of blood samples is a pressing moral question of late twentieth- and early twenty-first-century genetics. In the words of Emma Kowal, Joanna Radin, and Jenny Reardon, "Within biomedicine, indigenous biospecimens are increasingly the crucibles in which ethical practice is determined."[75] An extensive and thoughtful literature in STS (science and technology studies), history of science, and Native Studies has explored the stakes of genetic research for defining Indigeneity,[76] grappling with questions of

---

[74] Vieira, interview.
[75] Emma Kowal, Jenny Reardon, and Joanna Radin, "Indigenous Body Parts, Mutating Temporalities, and the Half-Lives of Postcolonial Technoscience," *Social Studies of Science* 43, no. 4 (2013): 477.
[76] Kim TallBear, "Genomic Articulations of Indigeneity," *Social Studies of Science* 43, no. 4 (2013): 509–533, https://doi.org/10.1177/0306312713483893; Arvin, *Possessing Polynesians*.

scientific responsibilities toward communities whose DNA is under study and the value (often conflicting) that geneticists and communities place on stored samples.[77] There are many important critiques that have been raised regarding the Genographic Project, and it seems Santos, Vieira, and colleagues had little latitude to adjust their scientific practices in conversation with residents of T.I. Pimentel Barbosa. However, their navigation of both the regulatory systems of the Brazilian state and the oversight of their A'uwẽ hosts – even if limited by structural constraints – exists in contrast to the work of some other human geneticists.

In the wake of the Genographic's hurdles with the ethics council, other geneticists expressed intense concerns and pessimism regarding the regulatory system and the ability to continue their work. In 2012, I asked senior geneticist Francisco Salzano, widely considered a founder of Brazilian human genetics, about timelines for getting new regulatory permission for sampling in Indigenous communities. He lamented, "Now there is what I call *geneticophobia* ... When you speak of DNA, everyone is horrified, and thinks that the genome of a person will be patented. This is also reflected in the regulations of the National Commission for Ethics in Research, CONEP." The regulatory experience of the Genographic Project was the first example he gave of how research regulations apply to genetics work with Indigenous populations: "For approval of these studies in Brazil, it took at least three years, with requests for information going back and forth."[78] Maria Cátira Bortolini described this saying, "it's an incredibly complicated thing to get approval from the Ethics Council to study Indigenous communities. So, like I said, we do not do new collections. We have used samples called historical samples with an approval to use these historical samples."[79] Among the samples still in use by Salzano and Bortolini's team as of 2017 were those first collected by Salzano and colleagues in T.I. Wedezé in 1962.

Salzano received ethics committee approval through his institution shortly after the National Council of Health instituted the new oversight system in 1996. The collections, which he had made over four decades were one of the foundations of his career until his death in 2018, and also provided material for analysis by dozens of doctoral students, some of whom went on to become

On broader issues of genetic articulations of race in Latin America, see Peter Wade et al., eds., *Mestizo Genomics: Race Mixture, Nation, and Science in Latin America* (Durham, NC: Duke University Press, 2014).

[77] Warwick Anderson, "Objectivity and Its Discontents," *Social Studies of Science* 43, no. 4 (2013): 557–576, https://doi.org/10.1177/0306312712455732; Emma Kowal, "Orphan DNA: Indigenous Samples, Ethical Biovalue and Postcolonial Science," *Social Studies of Science* 43, no. 4 (2013): 577–597; Joanna Radin, *Life on Ice: A History of New Uses for Cold Blood* (Chicago: University of Chicago Press, 2017).

[78] Salzano, interview.

[79] Maria Cátira Bortolini, interview with Rosanna Dent, Porto Alegre, January 28, 2014.

close collaborators.[80] In most cases, the researchers did not return to the Indigenous communities where the original sampling took place to conduct ongoing consent processes. Because the intellectual questions of the research group have remained focused exclusively on the genetic history of the populating of the Americas and questions of human evolution and differentiation, the university ethics committee approved ongoing use.[81] This is despite the fact that most of the technologies and techniques now used did not exist when the scientists collected the original samples. When geneticists claim, "what genome scientists are trying to obtain is a history of humankind in general, not of only one ethnic group,"[82] and position historic collections as *patrimônio da humanidade* or human patrimony, they make claims through a logic of possession.

In my discussions with Bortolini and other members and former members of her lab, I emphasize what I see as the minimum imperative for ongoing consent from communities whose members and ancestors are being studied. While she and others have used the regulatory difficulties of the Genographic Project as evidence for the bureaucratic impossibility of obtaining such consent, I have used Santos's engagement by Etênhiritipá as an example of why such ongoing discussions are an ethical necessity. Neither of our positions is innocent, and the conversation is ongoing. Likewise, before the pandemic, I began to discuss the stored samples with community members in the aldeias of Etênhiritipá and Pimentel Barbosa in what I imagine will be a much longer-term conversation. Already community members have taken delight in certain aspects of the history of the genetics research, including work underway to return anthropometric photographs from the 1960s in digital format.[83] They have also begun to draw parallels between the presence of their samples in laboratories and the histories of collection, use, and sale of samples from Paiter (Suruí) and Karitiana aldeias.[84]

For Santos or Bortolini or for me, even if we embrace meaningful long-term engagement in Pimentel Barbosa, we may not be able to disavow the possessive logics of the state. However, by working through the affective field of A'uwẽ regulations of research, we may open ourselves up to being changed.

---

[80] Francisco Mauro Salzano and Sidia M. Callegari-Jacques, *South American Indians: A Case Study in Evolution* (Oxford: Oxford University Press, 1988); Francisco M. Salzano and Maria C. Bortolini, *The Evolution and Genetics of Latin American Populations* (Cambridge: Cambridge University Press, 2005).

[81] José Roberto Goldim, interview with Rosanna Dent, Porto Alegre, June 16, 2014.

[82] Francisco M. Salzano, "Bioethics, Population Studies, and Geneticophobia," *Journal of Community Genetics* 6, no. 3 (2015): 199, doi:10.1007/s12687-014-0211-3.

[83] This NSF-funded project, "Digital Archives and Indigenous Afterlives of Scientific Objects" is a collaboration with six aldeias and James R. Welch (ENSP-FIOCRUZ), Lori Jahnke (Emory University), and Laura R. Graham (University of Iowa).

[84] Fieldnotes, July 2019.

As TallBear writes, "A researcher who is willing to learn how to 'stand with' a community of subjects is willing to be altered, to revise her stakes in the knowledge to be produced."[85] I am unsure what will be asked of me as I collaborate on a digital archive project in six aldeias, but while I may not be able to shift the possessive logics of recognition, at least with A'uwẽ community members we can work to shift the logics of the research itself.

[85] TallBear, "Standing with and Speaking as Faith," 2.

PART IV

Conclusions and Epilogues

11

Unsettling Encounters

STEPHEN T. CASPER

The *Journal of the History of the Behavioral Sciences* was founded in 1965 and *History of the Human Sciences* was first published in 1988. Much can be learned by surveying both across their histories. The *Journal of the History of the Behavioral Sciences*, established in 1965, has a mission rooted in the idea of disciplines and disciplinary history. During its early years, notable figures in psychology, sociology, and anthropology contributed articles that offered reminiscences on their fields or revisited landmark debates. Despite its wide chronological breadth, the journal recognized that concepts from antiquity or the early modern world did not always fit neatly into disciplinary chronologies. *History of the Human Sciences*, first published in 1988, in contrast emphasized the intersectional study of the way scientists constructed the human world. It focused on the years after 1850 and employed social science methodologies to frame historical investigations. In its first decade, it took an interdisciplinary approach, drawing inspiration from Foucauldian engagement, science studies, and postmodernism.

Both journals drew on traditions in intellectual history and the history of biology that foregrounded internalism, the history of the inner workings of the sciences.[1] Many articles in both contributed to internalist debates within the human sciences, treating epistemological and methodological trends in past science as important in their own right. Anthropology, psychology, and sociology, as enterprises seeking ontological insight into notions like "nature/nurture" or "human nature," were explored in depth, as were developments and debates in geography, economics, population studies, neuroscience, and even medicine. Each emphasized scientists who specifically encountered people as subjects or who drew on human sciences knowledge to elaborate on technoscientific societies and cultures that characterized the period of late industrialization or after. Both took up the challenge of the

---

[1] Steven Shapin, "Discipline and Bounding: The History and Sociology of Science as Seen through the Externalism-Internalism Debate," *History of Science* 30, no. 4 (1992): 333–369.

construction of scientific facts as that trend became magnified in the 1980s and after.[2]

To me, looking back on the history of both journals, both show a surprising lack of engagement with ethics, gender, and sexuality, even as books published from the 1980s onwards made these subjects salient.[3] Equally, I observe in both slight influence from settler colonialism, alterity, decolonialism, and postcolonity – despite significant literature on these ideas since the 1970s, and which this book shows to be crucial to understanding the context of human science work.[4] The irony of this latter observation is poignant given that the human sciences' past is closely intertwined with imperial conquest, racial supremacy, colonial governance, Indigenous affairs, integration, and apartheid.[5] Moreover, by 2010 it was clear that the methods and questions of postcolonial, subaltern, and Indigenous Studies are intellectual projects that qualify for – even demand – inclusion in the narrative of the methodologies of the human sciences.[6]

The fact that this irony exists leads me rather inevitably into a contemplation of my own position in the academy. As a white male, a gray-haired scholar, and an editor of this book, I am in the necessary position to reflect upon the way I have personally contributed to this unsettling shaping of the canonical history of the human sciences. Through such reflexivity, I hope to shed light on inherent biases within the history of the human sciences that this anthology reveals to be untenable. Looking back on all of those matters that slowly brought me into a career in the history of medicine and science, I see that for much of my career, I surrounded myself with historical texts authored by white men. I am often also teaching the history of the human sciences primarily to science students, many of whom are white, male, and hail (as I did) from rural areas of the United States. I see in them, in other words, a

---

[2] Ian Hacking, *The Social Construction of What?* (Cambridge, MA: Harvard University Press, 2001).

[3] This comment is made after reading through the titles, abstracts, and often first pages of the articles in both journals from their origins to the present. Changes began to appear only in the 2010s, and appear to have followed the global turn that took place in the first decade of the twenty-first century. Please understand that I am not saying there were no contributions on these topics in other areas of scholarship. I am merely observing that the trend in both has been directed toward preserving a particular kind of Westernized ideal about the disciplinary autonomy of the human sciences and their universalist claims.

[4] Michel-Rolph Trouillot, *Silencing the Past: Power and the Production of History* (Boston, MA: Beacon Press, 2015).

[5] Patrick Wolfe, *Settler Colonialism and the Transformation of Anthropology: The Politics and Poetics of an Ethnographic Event* (London: Cassell, 1999).

[6] Margaret M. Bruchac, *Savage Kin: Indigenous Informants and American Anthropologists* (Tucson: University of Arizona Press, 2018); J. Kēhaulani Kauanui, "'A Structure, Not an Event': Settler Colonialism and Enduring Indigeneity," *Lateral* 5, no. 1 (2016), https://doi .org/10.25158/L5.1.7.

reminder of myself. As I once was, I find many of my students are drawn to the historical secularism of European and American techno-science. As I once did, I suspect they view technocracy as a path toward economic opportunity and security.

Obviously, any attempt to generalize from my personal experience would be of limited import. Still I suppose many of the deceased white males who contributed to the human sciences, and their successors who documented the history of the human sciences, found solace in the idea that science and its methods, when applied to human nature, could provide universal tools for plumbing human existence and even improve the human condition. In essence, my personal encounter with the human sciences continues to be an encounter with ideas, abstractions, and ideologies primarily.

Having never knowingly been a subject of human science research myself, the subjects of those many past encounters, like those discussed across the book's chapters, proved incidental to my own fascination with the human sciences, fields supposedly on a quest to use scientific knowledge to bestow dignity on humans and humanity. For me, the intellectual allure of the human sciences lay in the way they revealed the intrinsic value of cultures or psychological processes through their variability, making the difference and otherness of the human form the source of dignity and value. This alternative dignity provided a last defense against the commodification of being in the face of artificial intelligence, synthetic life, transhuman studies, or capitalism. Because I never felt myself the object or subject of study, I could extrapolate on future risks to "the human" rather than face any immediate structural violence caused by human scientists who use me as an incidental presence for their studies of our nature.

My hope is that the readers of this anthology have realized by now (I assume most readers knew this before they opened this book) that almost every aspect of my privileged frame is questionable – morally, intellectually, methodologically, and ideologically. For those who share my origin story, I hope this self-reflection generates productive discomfort in the face of the question: How then can we frame the history of the human sciences moving forward, knowing or at least suspecting that we must? To think through an answer to that question, I will focus my argument on the larger organization of this book. This is no impartial review of each author's individual case. Instead, I am hoping to take what I see as the larger persuasive argument of this whole book and state it succinctly to you. I think this book's authors in total are collectively calling for a revisionist stance against an epistemological conceit within the human sciences: namely, the narrative that a particular scientist's situational context allows them to retain their own moral reasoning and imperatives even as they seek universal knowledge from others who are incidentally available. This book insists that knowledge made through encounters simply does not work like that.

The Savage Expedition to Civilization

The opening of our anthology centers around the theme of expedition science. While this phrase may seem innocuous, it holds significant meaning in its relational context, as demonstrated in the chapters by Rodriguez, Warren, and Gil-Riaño. The romanticized notion of Victorian and Edwardian anthropologists, naturalists, and geographers as adventurers on expeditions has persisted in our cultural consciousness, perhaps stemming from a confluence of cultural representations in such examples as Muscular Christianity, Rudyard Kipling, H. G. Wells, *National Geographic Magazine*, Tom Swift, Indiana Jones, colonial clothing chic, and David Attenborough. However, a critical examination of the term "expedition science" shows its position within a particular Eurocentric power dynamic. Those who were colonized, captured, and then traveled to Europe (think Jemmy Button the Yámana boy taken hostage by Captain Robert Fitzroy of the HMS *Beagle*) were not considered to be on expeditions, nor were individuals like C. L. R. James, Gandhi, or Aché children who found themselves in settings imagined civilized.[7]

These observations highlight a range of euphemisms within the human sciences that similarly obscure the relational nature of research. Expressions such as "going native," for example, are overtly racist, while other terms like "ethnoscience," "ethnomethodology," "cultural competency," or "participant-observer" may appear more benign but still rely on the premise that those being studied are not fully aware of or able to participate in the research being conducted on them. Such language reinforces the power dynamics inherent in research that studies marginalized communities or exploits colonized subjects.

Three potential responses to the romantic ideal of scientific adventurers are described by Rodriguez, Warren, and Gil-Riaño. The first response is that the adventurers participated in projects larger than science, such as imperialism, colonialism, or genocide. Rodriguez clarifies that the pursuit of Indigenous skulls in the late nineteenth century cannot be regarded merely as a typical event in the history of craniometry when, during the same period, the Argentinian military (by then a nationalist and settler force) exterminated and displaced thousands of Indigenous people. Gil-Riaño further illustrates this point by emphasizing that the relationship between scientists and children in twentieth-century Paraguay concealed the relationship between scientists and their military contacts, as well as the relationship between settler societies and groups compelled to assimilate to settler logic and governance. The fact that the children became loot from expeditionary violence, and subsequently

---

[7] Ruth Mayer, "The Things of Civilization, the Matters of Empire: Representing Jemmy Button," *New Literary History* 39, no. 2 (2008): 193–215.

became evidence against hereditarian racial determinism, demonstrates that even antiracist human science drew strength and substance from colonial violence.

The second response is that visible patterns of resistance and refusal to cooperate with the logic of expeditionary science demonstrate the subjects/objects' shrewd recognition of, and thus subversive autonomy within, these episodes of normal science. The subjects shaped human science knowledge. Warren's case highlights that frustrated scientists interpreted their subjects' autonomy and resistance as evidence of irrationality. Their label of irrationality fits well with other similar terms, such as "savage," "child-like," "innocent," "happy carefree," "insane," and so on.

There is no scholarship so far as I am aware that has considered such labels through the postcolonial insight that such alterity could be a type of doublespeak by the subject that generated evidence that contradicted the objectives and reasoning of human science investigation. However, to even begin to comprehend how such actions may have influenced European or settler scientific knowledge, it is necessary to have a nuanced understanding of how subjugated peoples created and practiced resistance, recognize that it occurred, and accept that this autonomy shaped scientific knowledge. Such a project of reclamation of the human sciences would have profound consequences.

Finally, historians of expeditionary science have perhaps asymmetrically accepted that the scientists possessed knowledge while the objects/subjects did not. With the political and economic upheavals occurring from the late 1870s (as depicted by Rodriguez) to the 1930s (as illustrated by Riaño), it is plausible that relationality increasingly implied a shared desire for scientific knowledge by all parties involved – albeit for different purposes. The violence of colonial structures, as described by Rodriguez and Warren, shaped the work of scientists while supporting these structures' ulterior logics and goals. It is not unreasonable to think individuals encountered by these so-called adventurers may well have had analogous objectives in mind for their own human sciences.

One of the peculiarities of colonial and settler science is its certainty that science possessed universal characteristics while simultaneously assuming that those characteristics were not understood, valued, or desired by those who encountered it as objects and subjects. As noted by Warren, many individuals photographed for the purposes of racial science were actively involved, even holding rulers in specific ways. Despite the deplorable experiences of expeditionary science, it is possible that some people who encountered it became interested in creating useful knowledge to understand why human scientists were prone to violence and supportive of oppression. While such speculation may appear unfounded or absurd, Riaño's conclusion, though centered on a child of assimilation, testifies to the fact that such desires fueled future resolve.

Externalism in the History of the Human Sciences Is Internalism

So far, my analysis of positionality has focused on "expedition science" as an example to illustrate how postcolonial, Latin American, and Indigenous Studies have redefined internal frameworks within the history of human sciences, critiquing internalist claims. My inquiries into the ownership of scientific knowledge, methodological limitations in historical arguments, and the impact of violent contexts on scientific knowledge may appear to stem from externalist criticism when it comes to the natural sciences. Externalism in the history of science argues that historians must consider social contexts to comprehend the conditions for scientific progress. However, matters become more complex in the human sciences where both the context and the human are the subject and object of study. Studying context is simply part of the internal logic of the human sciences. This makes it difficult to recognize that the intellectual history of the human sciences is never inherently externalist.

Traditionally, historians of the human sciences accepted institutions, disciplines, and their archives as useful ontological constructs to shape their historiography. Intellectual schools, institutes, and disciplinary origins played significant roles for storytelling the history of the human sciences. While this may suffice for historiography, it is important to recognize the normative oddity that the concept of disciplines was itself a human science analytic. Studies of discipline formation originated in sociology, and what appears to be contextual is now in historical writing, in fact, an historical acceptance of an abstract construct construed as possessing ontological recognizability. Critiques, if any, have primarily drawn from Foucauldian analysis, which is often applied to total institutions such as asylums, museums, prisons, and schools, but less frequently to central ideas or entire projects, as Foucault explored in works such as *The Order of Things* and *The Archaeology of Knowledge*.

This tendency has resulted in an unusual framework for externalism. Historians of science often use an *alternative human science category* as a framework against which to study the *history of its alternatives*. Thus, one finds political science, economics, or biomedical frameworks and theories analyze sociological, anthropological, or psychological claims to declare past scholarship limited. Anthropology is often employed to evaluate the veracity of sociology or economics; medical knowledge is brought into dialogue with economics. To simplify, the history of the human sciences often amounts to little more than an argument about a different form of internalism as getting at the "truer context." Occasionally, this trend becomes known as a "turn," a kind of cultural vogue – such as the "cultural turn," "linguistic turn," or "neuroscience turn."

Much of this work ends up in a self-referential loop, often only made evident through postcolonial criticism. Yet the leap from there should not be to conclude that the territorial location of these sciences matters and thus that they were somehow different or purer within European nations or the United States. Instead, it appears that the circularity of these sciences contributed to their power, utility, productivity, and violence everywhere.

Part II of this book addresses the self-referential ambiguities present in the human sciences. While the organization of this section may suggest a traditional exposition of encounters within institutions, each of the three authors of the cases in this section asks why power was necessary if the sciences worked. Equally they ask why the sciences seemed to have had mixed successes. Each author finds a distinct obliviousness within the sciences themselves that raises the unsettling question about whether that oversight was contextually generated or whether it is baked into the nature of the objectivity human science researchers have historically claimed.

Arvin exposes the inadequacies of the human sciences through Indigenous and Disabilities Studies frames. She recounts the history of the human sciences in Hawai'i as a backdrop for territorial violence and assimilation, while also directly focusing on the desires of Native Hawaiians and the fact that new structural violence contended with extant (if vanishing) institutions and structures as well. What stands out clearly in Arvin's analysis is the poverty of imagination the human scientists brought to their observations and applications. The inadequacies of their methods led them increasingly backward toward the racial determinism and normative Calvinism that characterized the origins of human sciences. Psychology and social science, which strove (admittedly failing) by the mid twentieth century for nuanced pluralism, perception, cognition, and understanding of interpersonal and community interactions and dynamics, pathologized differences rather than seeking to understand them when confronted by an encounter with alternative worldviews. In other words, the scientists denied the agency their sciences sought to recognize in generalizable ways as characteristic of humans.

Ortiz Díaz explores the territorial violence and assimilation process in twentieth-century Puerto Rico, examining how social scientists from imperial and nationalist backgrounds tried to create a more nuanced human science that matched their societal experience, expectations, and identities. Their goal was to apply it toward prisoner rehabilitation in carceral settings and deliver humane ends, restorative justice, and repair. However, the self-referential logic of looking for a science modeled upon itself proved to be problematic. Ortiz Díaz demonstrates that the available tools for such a reimagined science had to be reforged ideologically, but even then, the tools were unfit for purpose. Unfortunately, the scientists seeking reparative methods ultimately also fell into pathologizing tropes.

Stark's chapter provides an additional frame for understanding how human sciences are rooted in violence. She argues that bioethics in the United States is an extension of American settler colonialism. Her case study follows the biography of Carolyn Mathews, who began as an undergraduate at NIH and eventually participated in studies of the Akimel O'odham people in Arizona. Mathews in her later life became skeptical about the use of human subjects in experimentation, but her earlier position as a settler was not visible to her. Stark infers from this example that as bioethics emerged in the 1970s and after, it sanctioned an understanding of ethics that aligned with settler colonial precepts in the US Empire. The creation of a bioethics discourse led to self-referential ambiguity, as human subject research could be deemed ethical simply because it had declared it so.

Awareness of these legacies of empire, coloniality, and nationalism adds a critical lens to the study of the human sciences. Postcolonial and decolonial theory and methods run the risk of simply replacing old concepts with new ones in the human sciences. Silva, for instance, envisions neocolonial versions of the human sciences, powered by a correct linguistic currency, which create a global pastiche of elite academic discourse that recognizes sensibilities while perpetuating extractive practices in service to global capital.[8] These concepts, theories, and methods may not be sufficient on their own to articulate a new relationality within the knowledge/power dynamic woven into the human sciences. In other words, these languages may merely place historians of the human sciences in a position not dissimilar from Stark's interlocutor, Mathews.

The archive of the human sciences comes with no warning, but perhaps there should be one. Focusing solely on the scientists results in a history that may critique their logic but still duplicates their story. Concentrating only on the scientists' subjects ends up accepting the scientists' ventriloquism. What sets these sciences apart is that the archives and the theories they create are often a product of unsettling encounters. Ortiz Díaz's case makes this clear, as the scientists in his study fail to elide the problematics of their work. Similarly, Arvin's story ends with Stanley Porteous, the scientist she highlights, having his name stripped from a building named after him in 1974 at the University of Hawai'i and renamed after Allan Saunders, a noted faculty activist associated with politically left-wing causes who nonetheless regarded slaveholder Thomas Jefferson as "one of my heroes."[9]

The Human in the Mirror

In the history of the human sciences, there is often a duality of nature. In studying "the Other" (whomever "the Other is supposed to be"), human

---

[8] Guilherme C. Silva, "The South as a Laboratory (Again)? Dealing with Calls for 'alternatives' in the North," *4S Reflections* (2022).

[9] "Pau Hana Years: Dr. Allan Saunders," PBS HAWAI'I, December 16, 2020, http://uluulu .hawaii.edu/titles/24652.

scientists hoped to reveal larger truths about themselves-*cum*-the human. Indeed, this book has sought to make clear that when human scientists used nature as a mirror for the purposes of studying human nature, they often ended up studying their own reflections. Their mirror of nature reflected their sciences' circular relationality; almost everything seemingly returned to the observer. This hermeneutics of these past encounters now produce the historian's text, and historical explanations shed light on why the historical claims made within these sciences adopted forms of ownership and governance. In Part III of this book, Karin Rosenblatt, Eve Buckley, and Rosanna Dent explore what happens when both sides claim possession of the mirror.

In Rosemblatt's case, the discovery of the last Mexica emperor's bones, Cuauhtémoc, in Ixcateopan, Mexico, resulted in conflicting claims of ownership and self-referential truth claims by opponents. The Indigenous community and Eulalia Guzmán, Rosemblatt's biographical subject, celebrated the discovery as an anticolonial assertion of identity through ownership of the bones. However, masculine and cosmopolitan science accused Guzmán and the community of perpetuating a fraud. The history of bone gathering in the human sciences takes on a new significance in this case. It highlights the longevity of meaning that bones can acquire, a phenomenon familiar in the conventional history of the human sciences. The bones acquired a double rationality, invested with political meaning through conflicting scientific claims and methods. Craniology, now widely regarded as a pseudoscientific relic of racist, racial science, is a prime example of the larger pattern behind Rosenblatt's argument. In her story, the pursuit of hard knowledge to legitimate claims of authenticity came from the marginal and vulnerable. The resulting collision occurred through scientific claims, methods, disciplinary differences, and innuendo, with gender playing a significant role in attacks against Guzmán. Ultimately, the claim that the bones were a fraud prevailed, but for the community of Ixcateopan the bones retained their meaning. Rosenblatt's case might encourage some reckoning with the way bones continue to contain meanings for settlers as well, with even their rightful repatriation extracting new symbolic meanings while mediating novel new forms of erasure within often hallowed cultural institutions.

Eve Buckley's work similarly showcases dual rationality. Her case focuses on population and development studies during the Cold War era and exposes the longevity and reach of neo-Malthusian tenets through dialogues on fertility and scarcity. She examines the writings and advocacy of Josué de Castro, a Brazilian physician and intellectual who challenged the theories of a population crisis in poor nations, arguing that the crisis was due to agricultural fertility rather than human fertility. He called for structural reforms of the global food system to balance out the observable overabundance in wealthy nations. Intriguing in de Castro's encounter with population studies is the reciprocal forms of abstraction it shows, an infinity mirror of nature reflecting a relationality of object and subject purely determined by the holder. While Buckley rightfully highlights how population control emerged as neocolonial benevolence, denying its human costs, de Castro's technique for holding the mirror reversed the direction of power by subjecting his wealthy interlocutors in the Global North to their own form of armchair theorizing, albeit about their nature rather than his own. De Castro, in so doing, employed one of human science's cherished rhetorical practices – creating a "big picture" account of humanity's condition. De Castro's inversion of center and periphery articulated global cosmopolitanism against European cosmopolitanism, and thereby submitted Northern and Western cosmopolitan intellectuals in the 1950s to an unusual examination of themselves as subjects in a technocratic, world historical analysis. Like Rosenblatt's Guzmán, de Castro's reasoning and writing did not emerge as a winning position, but it did demonstrate the way supposedly factual social science theory assumed ideologically the naturalness of imperial and colonial relations.

Rosanna Dent concludes this anthology's third section on an optimistic note that extends Rosenblatt and Buckley's narratives toward particularized alliances, as she analyzes the relational conditions that emerge from the Genographic Project, a recent study proposing a general evolutionary history of humanity based on blood samples. Dent explores the logic of possession and bureaucratic vulnerability, which creates opportunities for abuse for A'uwẽ people, as they navigate their sovereignty with the Brazilian state and transnational researchers, while insisting on a relational and affective ethics of their own. Despite the risks, both the Brazilian state and the A'uwẽ recognize the potential benefits of bioprospecting, as it establishes a foundation for the community's recognition with the state. However, the scientists navigate Brazil's regulatory state with a logic of possession of their own, and they historicize samples taken before ethical guidance and modern technologies existed to continue their research. Dent describes how the researchers seek to embrace the affective ethics of the A'uwẽ through long engagement and acceptance of their desires, in a convincing relational shift. While this solidarity may seem puzzling at first, it makes sense as a liberatory project for A'uwẽ sovereignty. Dent hopes that this solidarity will shift the logic of research toward a more ethical approach.

Together, Rosenblatt, Buckley, and Dent point to a growing recognition that the phenomenological characteristics of encounters can sometimes escape or transcend the physical and structural violence of the human sciences. Although neither Rosenblatt nor Buckley can do more than recognize the double rationality of the human sciences, which can create epistemic discomfort, both authors show that the human sciences can be used as decolonizing tools too. In Dent's case, the rejection of the violence of the human sciences may lead to the hermeneutics of the encounter, producing identification, solidarity, and knowledge – although the entire story is becoming rather than told.

Concluding Clearly

As Warren, Rodriguez, and I were working on Chapter 1, I was reminded to read finally an English-language essay by Silvia Rivera Cusicanqui criticizing Aníbal Quijano. My initial engagement with Quijano had been revelatory, and thus Cusicanqui's critique equally gave me pause. It stimulated in me a reconsideration of what historians usually mean when they demand clarity from each other. Cusicanqui highlighted that the jargons and languages of critical theory and human sciences are more than just alienating for many; they represent a form of re-colonialism by language on colonial subjects. According to her, colonial subjects in these jargons became caricatures of the West, and "multicultural adornments for neoliberalism," unrecognizable to people in the upside-down worlds of colonialism.[10] The authors of the essays in this book may experience discomfort as they read Cusicanqui's words. I understand their perspective. We all are in a position with jargon not dissimilar to Aníbal Quijano's, the subject of her critique.

For historians, too, the complaint hurts. How many times have historians been told to avoid jargon for the benefit of clarity. The request for clarity may be viewed as an attempt at aesthetics. But clarity in the way that historians demand it of each other is obviously a source of power too. It has its own double meaning. Who is deemed unclear? Who must strive to write with acceptable clarity? Who possesses the definition of clarity? Who gets to demand it? Clearly, clarity constitutes a form of dominance over meaning, and thus Cusicanqui's demand for it feels stubbornly like a shoe that fits uncomfortably. However, this book intends to elicit a different type of discomfort regarding how historians of the human sciences present a history that is profoundly tied to varieties of Enlightenment liberalism. Our book, as a whole, unsettles encounters by insisting that it is ethical to listen in the way Cusicanqui suggests we listen.

It is important to note that this book is not anti-science or illiberal. However, it argues that the human sciences, including their history, must either adopt inclusive methods and nonnormative means of quantification and measurement, or acknowledge that they are not sciences but rather technologies created to perpetuate specific forms of domination.[11] We have argued collectively that while these technologies of domination may have limitations, they are also useful for facilitating capricious uses – and by implication that the people who use them now do so at their own risk. This interpretation suggests that future human sciences, fueled by bio-recognition technologies,

---

[10] Silvia Rivera Cusicanqui, "Ch'ixinakax utxiwa: A Reflection on the Practices and Discourses of Decolonization," *South Atlantic Quarterly* 111, no. 1 (2012): 99.

[11] Paul Forman, "(Re)cognizing Postmodernity: Help for Historians – Of Science Especially," *Berichte zur Wissenschaftsgeschichte* 33, no. 2 (2010): 157–175.

sensor data, neuropharmacology, large-scale data analyses, keyboard and mouse patterns, geo-tracking, and artificial intelligence, are likely to become harbingers of new forms of domination. The danger is that these technologies may frame normal and pathological behaviors, with the goal of predicting patterns with a degree of accuracy that may overlook novel forms of resistance, double rationality, and survivance.

How then can we frame the history of the human sciences moving forward, knowing or at least suspecting that we must? At the risk of being too clear, I offer then in answer to this question a final summary of the argument of the book you have now read: The history of the human sciences is primarily a history of physical and structural violence in which historians of the human sciences are also implicated. Writing about this history requires recognizing that accounts of these sciences may make the violence invisible or cast it as incidental or the result of a few bad actors, while hoping that the larger program will prove viable and emancipatory. The self-referential nature of the human sciences reveals that structural violence is fugitive within the linguistic and epistemic frameworks of the human sciences regardless of intent. Applying the hermeneutics of encounters modeled in these essays to other populations who became objects of study, and according frameworks that draw on affect, desires, agency, needs for reparative justice, and a place in the narrative, is thus crucial and ethical. The object of the encounter is not incidental or abstract, nor is the encounter apolitical, and thus what can be said about it, should be said accordingly.

12

Feel It in Your Bones

The Difference Indigenous Studies Makes

MARÍA ELENA GARCÍA

Empire, Colonialism, and the Human Sciences is a timely, impressive, and generative contribution for thinking about the politics of historical scientific engagement and especially its intimate entanglement with coloniality, power, indigeneity, race, and gender. Reading these chapters, and especially my affective engagement with them, brought up a lot. In this short epilogue, however, I want to focus on three main themes: (1) the encounter with Indigenous Studies; (2) the importance of engaging with Native ideas of affect – what Dian Million calls "felt theory"; and (3) the significance of thinking with haunting and ghosts as central to reimagining the history of science in the Americas.

Some of the central questions that emerged for me as I read echo the questions and concerns many of the authors directly address in their contributions to this book. To begin where the book ends, that is, in the spirit of "productive discomfort," I offer some thoughts about what an encounter with Indigenous Studies might, or maybe even should, produce. To phrase it perhaps a little provocatively, if the "human sciences" are to engage meaningfully with Indigenous Studies – with theorists from the Native North and Native South – then I would suggest that that engagement must be transformative, not just additive. To be clear, I think many of the contributions here do reflect this move toward transformation. To give an example of what I mean, we can consider the work of K'iche' Maya scholar Emil' Keme who is cited in the introduction. In his influential essay, "For Abiayala to Live, the Americas Must Die," he calls for a reconsideration of the geographies of knowledge we continue to work with.[1] He asks us to question the very category of Latin America and proposes a shift toward the concept of Abiayala for thinking otherwise and toward what he calls a transhemispheric Indigenous bridge. This raised questions for me about terminology, naming, and language. While contributors to this book do not use this terminology, I do think in many ways the book gestures toward the transhemispheric

---

[1] Emil Keme, "For Abiayala to Live, the Americas Must Die: Toward a Transhemispheric Indigeneity," *Native American and Indigenous Studies* 5, no. 1 (2018): 42–65.

Indigenous bridge Keme is calling for in placing discussions of Indigenous Brazil, Paraguay, Peru, and Hawai'i (for example) into conversation. But perhaps there is a way to more explicitly engage with or address this call.

In other words, it is not enough to simply cite Native theorists. How does thinking with Native theorists, Native epistemologies and ontologies, radically transform the work we are doing? How does it shift the why, the audience, the approach or method used? How does it transform the way we think about knowledge production? About what counts as knowledge? Who is this for? Some chapters answer these questions more directly than others, but as a whole they encourage us to think along these lines.

Another question the book raises is about scholarly representation. Who is at the table? Who is inviting whom? What are the networks and processes that have already shaped who participates in this conversation? Let me be clear that I am not questioning the editorial decisions that led to inviting this group of unquestionably talented scholars. My point is about the broader workings of disciplinarity and academic boundaries that makes specific projects legible in specific ways.

I wondered too about the tensions inherent in placing Indigenous Studies and decolonial scholarship in the same frame without more fully unpacking both the possibilities and tensions that exist. In his conclusion, Stephen T. Casper mentions Silvia Rivera Cusicanqui's critique of Aníbal Quijano's use of "decolonial" jargon. But Rivera Cusicanqui's critique went beyond that, and focused in particular on Walter Mignolo and other scholars of decoloniality for what she understands as extractive/imperial knowledge production. Here is one memorable quote: "Walter Mignolo and company have built a small empire within an empire, strategically appropriating the contributions of the subaltern studies school of India and the various Latin American variants of critical reflection on colonization and decolonization."[2] Moreover, she was concerned not only with neologisms but with structures of power. Let me quote Rivera Cusicanqui once more:

> Equipped with cultural and symbolic capital, thanks to the recognition and certification from the academic centers of the United States, this new structure of academic power is realized in practice through a network of guest lectureships and visiting professorships between universities and also through the flow – from the South to the North – of students of indigenous and African descent from Bolivia, Peru, and Ecuador, who are responsible for providing theoretical support for racialized and exoticized multicultural-ism in the academies. Therefore, instead of a "geopolitics of knowledge," I propose the task of undertaking a "political economy" of knowledge.[3]

---

[2] Silvia Rivera Cusicanqui, *Ch'ixinakax utxiwa: On Practices and Discourses of Decolonization* (Boston: Polity Press, 2020), 51.
[3] Ibid., 59–60.

I think it is in these tensions and in explorations of the limits of this project that we might find some interesting possibilities. I really appreciated Casper's conclusion and the importance of productive discomfort. He writes: "I hope this self-reflection generates productive discomfort in the face of the question: how can we frame the history of the human sciences moving forward, knowing that we must?"

One of the opportunities this language offers – the language of discomfort as an embodied, affective response – is to engage with what Athabascan literary scholar and poet Dian Million famously called "felt theory." Million's work came forcefully to mind, for example, when I read Eve Buckley's chapter discussing the significance of emotion and affect in debates about overpopulation, hunger, and poverty between Brazilian geographer Josué de Castro and American conservationist William Vogt. But the discomfort that Casper describes is a kind of understanding that is fueled by a *decolonizing* desire, and I mean this very much following Million who wrote about a key part of decolonization. "To 'decolonize'," she writes, "means to understand as fully as possible the forms colonialism takes in our own times."[4] And I would add in our own lives. Naming the embodied responses our work provokes is not just self-flagellation but an indicator that signals the need for new forms of relationality and repair. Holly Barker, cultural anthropologist and curator for Oceanic & Asian Culture at the University of Washington's Burke Museum, once told some of our students during a visit to the Burke that she feels a "knot in her stomach" every time she sets foot in the building that houses Native artifacts and ancestors without Native permission. Yet, Barker used that discomfort to create a form of knowledge production she calls "research families," a form that has mentored an incredible number of Pacific Islander students and shown them how to use museum collections to reconnect with their own peoples, waters, and lands.

Like Barker's work, many of the contributions in this book model responses to this productive discomfort and some possible paths forward. In his chapter, "Subverting the Anthropometric Gaze," Adam Warren offers an explicit engagement with Indigenous methods, even or especially when they are *not* part of the work. I found in Warren's direct and honest discussion of his methodological choices and decisions, of the practical issues raised by considering how his research could or should shift through engagement with Indigenous methods, a model for seriously considering the possibilities and the limits of this kind of work, for thinking through the implications of designing research that is not situated within decolonial or Indigenous frameworks from the beginning, but also for what can change moving forward. For

---

[4] Dian Million, "Felt Theory: An Indigenous Feminist Approach to Affect and History," *Wicazo Sa Review* 24, no. 2 (2009): 55.

example, his determination to translate his work into Quechua to share with communities for very specific political uses is inspiring and offers a concrete and important path forward.

Maile Arvin's chapter, "Replacing Native Hawaiian Kinship with Social Scientific Care," is also a beautiful gift. The care she begins with is a striking gesture of relationality and accountability, and a model for others working in this field. "Throughout this essay," she writes, "I do not use the real names of children who are named in archival records given I have not (at least not yet) been able to contact descendants or other kin who may have particular desires about sharing these stories." Her work reflects an elegant move away from damage-centered research, without ignoring the harm and brutality of what she terms the "settler colonial transinstitutionalization" of children in the territory of Hawai'i. Her chapter was particularly inspiring for me as I begin working with the archived testimonies of Indigenous survivors of the recent war in Peru.

Similarly, Rosanna Dent's engaging discussion of the history of genetics research in A'uwẽ communities in Brazil, underscores the importance of affect and relations in discussions about ethics and bureaucracy. She explores the various affective dynamics that have shaped the experience of researchers and the communities and peoples impacted by this research, as well as the bureaucratic regulation of research itself. And she insists on centering the agency of A'uwẽ *aldeias* as they work toward developing their own frameworks for regulating and overseeing research conducted in their communities. Dent's discussion of adoption and kinship in relation to research, and her emphasis on affective fields were particularly compelling. She describes the adoption of researchers by community members as an Indigenous strategy for "claiming" researchers, and thus asserting "a relationship of kinship by publicly announcing their chosen relationship to the researcher." This claim compels researchers to behave according to particular familial and community social norms, thus offering the A'uwẽ moral authority and a dimension of control over researchers' actions. And, she writes, in taking these relations seriously, in "working through the affective field of A'uwẽ regulations of research, we may open ourselves up to being changed." And quoting Kim TallBear, she continues: "A researcher who is willing to learn how to 'stand with' a community of subjects is willing to be altered, to revise her stakes in the knowledge to be produced." This is a central concern in Indigenous Studies, a concern that Dent takes seriously and tackles thoughtfully and with great care.

These and other contributors in this book invite serious, careful, detailed reflection on approaches that can transform the field in significant ways. But let me move to the next theme, which is the significance of thinking through and with haunting. As I read the chapters in this book, I kept returning to what thinking with ghosts might offer. In "A Glossary of Haunting," Eve Tuck and C. Ree write that haunting does not "hope for reconciliation. Haunting lies precisely in its refusal to stop . . . For ghosts, the haunting is the resolving, it is not what needs to be resolved."[5] In my own work in Peru on the afterlives of war, and more specifically on the impact of political violence on more-than-human life, I have found this an incredibly generative conceptual tool. In Peru – and in so many other places – we need ghosts to continue their work, to continue to haunt so that we may never forget, so that we may continue to work not toward any kind of false reconciliation, but rather toward altogether reimagining possible worlds. Perhaps, we can think about haunting as a kind of healing, or better, as a disruption, a call to wake up. We may even begin to think about haunting as a kind of anticolonial practice, especially if we think with and from Indigenous standpoints. As Tuck and Ree put it, haunting is a "relentless remembering and reminding" that "with some crimes of humanity – [such as] the violence of colonization – there is no putting to rest."[6] This, to me, also includes thinking about the disruption of relations, and the repairing of those relations, not only among humans, but also in relation to nonhuman kin. In my own work, I want to think about the many nonhuman ghosts that may also wander through Andean valleys and rivers. I wonder, do they too demand justice? Do they too ask to be remembered? How do they figure in these histories and politics? What are the nonhuman relations disrupted? And what are the ghastly memories embedded in lands, in rivers; what do glaciers remember and how do they respond?

I read and feel this book as an invitation to sit with and think with ghosts, to take them seriously. From haunted institutions in Hawai'i and Puerto Rico (Arvin and Ortiz Díaz), to the potential haunting of/by the ghosts of kidnapped Aché girls (Gil-Riaño) or of/by the Akimel O'odham people from Arizona – Carolyn Matthews' "human subjects of research" (Stark), to the afterlives of Cuauhtémoc's bones (Rosemblatt), and skulls as "uncanny objects," (Rodriguez), many of the contributors gesture to this and in various ways ask what it means to think of how the human sciences are haunted. However, it is worth slowing down to think with the many entities that haunt the entire book: bones, skulls, DNA samples, spirits. I hesitated to name these as human, nonhuman, or once-human, since the very category of the human (and "the living") seems to be one that is being interrogated by this project, and also by radical Black and Indigenous scholarly traditions.[7] Indeed, it might be worth

---

[5] Eve Tuck and C. Ree, "A Glossary of Haunting," in *Handbook of Autoethnography*, eds. Stacey Holman Jones, Tony E. Adams, and Carolyn Ellis (Oakland, CA: Left Coast Press, 2013), 642.

[6] Ibid., 642, 648.

[7] Bénédicte Boisseron, *Afro-Dog: Blackness and the Animal Question* (New York: Columbia University Press, 2018); Zakiyyah Iman Jackson, *Becoming Human: Matter and Meaning in an Anti-Black World* (New York: New York University Press, 2020); Patty Krawec, *Becoming Kin: An Indigenous Call to Unforgetting the Past and Reimagining Our Future* (Pine Bush, NY: Broadleaf Books, 2022); Elizabeth Povinelli, *Geontologies: A Requiem to*

complicating the human subject/nonhuman object distinction made a few times throughout the book. Taking Indigenous Studies, epistemologies, and ontologies seriously, means reframing and rethinking who or what is animate and inanimate, alive or dead. Bones are not things, they are ancestors who can be dangerous, restless, or at peace. If thinking with "the materiality of human remains expands the historian of science's toolkit" (Rodriguez, this book), that toolkit expands even further when we attend to what Rodriguez calls evocatively the "spiritual materiality" of bone, which I take to mean the ontological, epistemological, and metaphorical possibilities that come with thinking of bones as more-than-material. It is notable that in several Polynesian languages the word for bone is also the same word for people, tribe, or nation (e.g., Kanaka 'Ōiwi in Hawai'i; iwi for Maori in Aotearoa/New Zealand). In the supernatural and historical Peruvian novella *Adios Ayacucho* by Julio Ortega, a tortured, murdered, and disappeared Andean campesino searches for his own remains.[8] He fails to find them and resorts to stealing the bones of the conquistador Francisco Pizarro and laying in his tomb to reassemble himself, metaphorically linking the violence of the late twentieth century with that of the sixteenth. Rosemblatt's discussion of the afterlives of Cuauhtémoc's bones could be placed in interesting conversation with Ortega's novella and its broader implications. This is not the place to add more flesh to these bones, but possibilities are many, and the essays here push us to think expansively.

Rodríguez's description of the "haunting effects" of bones also inspired me to think with photographs as haunted. In particular, the work of anthropologist Lisa Stevenson came to mind. In an essay titled "Looking Away," she draws on Roland Barthes and John Berger to describe the care that can be part of anthropological encounters and ethnographic writing. This kind of approach, she writes, "addresses how images – whether, photographic, painted, or written – may come to be seen as 'just'."[9] And she considers the possibility that "it might be necessary to *look away* from our interlocutors, or

Late Liberalism (Durham, NC: Duke University Press, 2016); Daniel Ruiz-Serna, *When Forests Run Amok: War and Its Afterlives in Indigenous and Afro-Colombian Territories* (Durham, NC: Duke University Press, 2023); Kim TallBear, "Why Interspecies Thinking Needs Indigenous Standpoints," *Cultural Anthropology* (2011), https://culanth.org/field sights/why-interspecies-thinking-needs-indigenous-standpoints; Zoe Todd, "Fish, Kin, and Hope: Tending to Water Violations in Amiskwaciwaskahikan and Treaty Six Territory," *Afterall* 43, no. 1 (2017): 102–107; Alexander Weheliye, *Habeas Viscus: Racializing Assemblages, Biopolitics, and Black Feminist Theories of the Human* (Durham, NC: Duke University Press, 2014); Sylvia Wynter, "Unsettling the Coloniality of Being/Power/Truth/Freedom: Towards the Human, after Man, Its Overrepresentation – An Argument," *CR: The New Centennial Review* 3, no. 3 (2003): 257–337.

[8] Julio Ortega, *Adiós Ayacucho* (Lima: Grupo Cultural Yuyachkani and Fondo Editorial de la UNMSM, 2008).

[9] Lisa Stevenson, "Looking Away," *Cultural Anthropology* 35, no. 1 (2020): 12.

the images we have of them, in order to be able to sense, and then communicate to others, their singularity. The traces they leave behind in our memories can allow us to register an aliveness that exceeds our existing labels, categories, and styles of thinking."[10] In other words, Stevenson calls for "looking away" from those we are trying to understand, in order to more fully "see." For her, "looking away" – from a photograph, a person, someone we are trying to represent – can be a form of "seeing with our eyes closed" that gestures to "the singularity of another being."[11] As she writes, this form of looking – or not looking – might "allow us to go beyond seeing someone as a specimen from a social category."[12]

Stevenson is writing about photographs she encounters in the archives of McMaster University's Health Sciences Library, photographs from the mid-1900s of Inuit patients in Canadian sanatoriums. She explores the idea of "looking away" as one form of refusing the "look" of the colonial gaze, refusing "categorical ways of looking" that reduce lively beings to specimens, ethnic categories, anthropological types. She is searching for an "un-stately, unseemly, un-fixative" way of looking; for a way to move "beyond the clinical label or social category [that] involves a play between seeing with our eyes and seeing with our soul."[13]

I find inspiration in Stevenson's invitation to look away to more fully "see" individual beings. Gil-Riaño's work in particular brought Stevenson to mind as I wondered what the photographs of some of the Aché girls he writes about would reveal if approaching them through and with Stevenson's framing. And yet, the privileged focus on sight raises important concerns about this approach. We must find ways to push beyond ableist language and framings that are so often part of scholarly discourses. Perhaps, then, we might read Stevenson in multi-sensorial conversation with Tina Campt's powerful work on "listening" to images. For Campt, "'listening to images' … designates a method of recalibrating vernacular photographs as quiet, quotidian practices that give us access to the affective registers through which these images enunciate alternate accounts of their subjects."[14] "To listen to them," she writes, "is to be attuned to their unsayable truths, to perceive their quiet frequencies of possibility …"[15] This last point, "to be attuned to their unsayable truths, to perceive their quiet frequencies of possibility," is what I think Stevenson is trying to do through her play with the language and practice of looking and seeing. It invites a move toward a multilayered affective

---

[10] Ibid.
[11] Ibid., 8.
[12] Ibid., 11.
[13] Ibid.
[14] Tina Campt, *Listening to Images* (Durham, NC: Duke University Press, 2017), 5.
[15] Ibid., 45.

attunement to others, one that perhaps allows for a more complex relation to and with others, one that recognizes the emotional richness of their lives, as well as their multifaceted experiences of and in life. What would it mean for historians of science to think more explicitly with such an approach?

And to return once again to Dian Million, this kind of looking, not looking, and hearing is also a kind of feeling, a kind of "felt analysis." This work helps us take seriously the "structures of feeling" that were both part of the extractive and colonial mode of the human sciences that all the contributors describe so well, and also the new kind of structures of feeling that emerge once we center Indigenous Studies values like radical relationality, reciprocity, and accountability in the writing, teaching, and mentoring we do. I think that the conversations modeled in this book can help reveal how Indigenous Studies have transformed our work and can signal alternative ways forward.

13

The Pole Is Back Home

GABRIELA SOTO LAVEAGA

On May 10, 2023, a delegation from the Gitxaała Nation entered Harvard's Peabody Museum. Wearing ceremonial red and black, the seven members had traveled from their territory in the Pacific Northwest to the ground floor of the museum where a small group of museum staff, invited guests, and chance visitors observed a palpably momentous event. Even unsuspecting museum-goers who wandered out of the Lakota images exhibit to the galleries on the first floor seemed to sense the importance of the occasion. Surrounded by Penabscot canoe exhibits and the Kaats' and Brown Bear Totem Pole, the Gitxaała Nation and university representatives acknowledged in a poignant ceremony the return of a Gitxaała house frontal pole and associated fragments. Repatriation of the totem pole had taken 126 years.[1]

This moving ceremony was not to hand over the totem pole; its actual return had taken place months earlier. Rather, this observance served as an acknow-ledgment of closure, and, potentially, a blueprint for possible future encounters. In fact, the Gitxaała Nation had come to leave a marker, a *gan niidza*, "to explain what happened." In selecting an object to represent them as a living people, the Gitxaała Nation was purposefully disrupting the official narrative that others had constructed about them. They are still *present*, a delegate asserted; not simple objects to be collected or studied. The Gitxaała Nation representatives were there "closing a circle that should never have been opened," when sacred objects were stolen from their territory in the nineteenth century and transformed through new claims of ownership into scientific "artifacts."

What can repatriation of a sacred object like the Gitxaała house pole tell us about troubling encounters and their possible afterlives? Rather than delve into the rich themes of data collection, knowledge making, research ethics in science, and other topics covered so well in this book, I wish to center the ceremony at the Peabody as one of meaning making and repair that situates respect and reciprocity as an ethical frame for possible futures. Many of the

---

[1] The exact date is not known as some news outlets said 126 years. www.ominecaexpress.com/news/gitxaala-house-post-returning-after-138-years-and-decades-in-harvard-storage.

chapters in this book foreground disquieting interactions whose future endings are yet to be written. Who can collect and who gets collected, studied, and exhibited are certainly stories about domination, but they are also tales about the manifold constructions – evidentiary "slivers" of both material objects and human remains (blood, skulls, or other genetic material) – needed to cement the colonizing idea of difference. How might historians of the human sciences think about these stories as we move forward?

The Totem Pole

To the Gitxaała and other Indigenous peoples, totem poles or house posts (or pts'aan) possessed history and power. Not only were they territorial markers but they also contained stories of the clans they represented. Purchased by a skipper of a New England Fish Company in 1885, the totem pole, an integral part of Gitxaała culture, was sold under duress to be "displayed as a historic oddity" in the offices of the Boston harbor fishing company.[2] As the Gitxaała museum webpage explained, "It was cut down, saved, sold, abused, given away, and displayed." Though the returned pole stood at an already imposing 12 feet, it is thought to have been as tall as 50 feet when it originally stood outside a clan home.

On the day of the ceremony at the Peabody, I learned more about the history of the pole from the director of the museum and a member of the Gitxaała Nation. I am not a scholar of the Gitxaała Nation but I was there as a mere member of the faculty museum committee. In other words, I, like the many scientists in this book, am rendering my interpretation of the events wholly aware of my power in being able to write an account. Having said that, though I was in the first row, my view was partially blocked by someone who, despite arriving late, planted themselves in front of occupied chairs and stood there for the duration of the ceremony. To see, those who were seated needed to strain, leaning to either side of the man, who turned out to be a *Boston Globe* reporter. From my uncomfortably slanted viewing position, I could not help but reflect what it means to be able to see/to witness/to rewrite histories with an unobstructed view. As I moved my body to the left and right, trying to capture as much of the experience as possible, there were inevitable moments when I missed key details and was left to infer what had happened. Even my advantaged seating precluded me from fully understanding subtle movements, words spoken, and the meaning of songs and rituals. In short, I lacked a broader historical context and training to give a truer account, made more difficult by the reporter who, now part of my account, continued to obstruct my view.

Though these blocked views were mere moments, afterwards I found myself thinking how they rather neatly captured the partial views we have been given

---

[2] https://gitxaalamuseum.com/journeyhome.

from archives, oral histories, ethnographies, and erstwhile collecting expeditions of those who have long been the living subjects of science. Indeed, what I was witnessing was a mere instant in a much longer and complex history – described in most of the chapters of this book – of colonizers, sailors, businessmen, intermediaries, and social scientists, among others, who trafficked in objects, facts, and conjecture.

The pole or pts'aan was the first of seventy-three culturally significant objects that are currently held in more than twelve national and international museums and which are at different stages of being returned to the Gitxaała Nation. Those who had come to acknowledge the return were also there to reframe the history, to retell it from their perspective. Hearing the spokesperson explain the theft of the pole was a powerful reminder of what was lost when objects – and people – were captured during, as a Nation member explained, "one of our darkest moments." The loss was more profound for it was more than the material object displayed first by the fishing company and later the museum; it is what the objects represented, the history and the spirits of the ancestors. For the Gitxaała Nation the pole was living, holding the spirits of those who came before.

Indigenous newspaper coverage of the return of the totem pole eschewed words like "artifact" and "repatriate," replacing them instead with Sm'algyax words to describe the return of the pole as putting the "value and sacredness back in" and most importantly reasserting "we are still alive."[3] Historian Vera Candiani reminds us that words, especially in extractive economies, have deeper meanings. She writes, "Using such descriptors not only treats objects, processes, and people as products of random genetic combinations, but also, by so swiftly categorizing it thus, also prematurely forecloses on deeper understandings of the things in themselves and their relationships to processes around them."[4] Candiani uses this argument to urge historians to think more broadly about environments, but deliberate word choice, as used here, is a powerful assertion against imposed scientific terminology and categorization.

Explaining why the use of "artifact" erased multiple meanings, a member told a reporter:

> The word "artifact" is such a loaded colonial word that implies that our culture died off. It's not an artifact, it's part of our living culture. "We're

---

[3] https://coastalfirstnations.ca/gitxaala-bring-treasure-home-a-historic-turning-point-for-the-nation/.

[4] Vera Candiani here was explaining how certain "artifacts" used for transforming colonized landscapes were described as being either "hybrid" or "Mestizo" while knowledge was always described as being "European or creole." Though she uses the word artifact to reference technologies, Candiani acknowledges that this forecloses other meanings. "Reframing Knowledge in Colonization: Plebians and Municipalities in the Environmental Expertise of the Spanish Atlantic," *History of Science* 55, no. 2 (2017): 234–252.

still alive," Wilg'oosk says. "We came up with phrases in our language . . . we had to differentiate between what's already come home, what's on its journey home, and what we still need to bring home. We were really challenging our committee, with our Elders, to try and find the right phrase, the right way to explain it in our culture."[5]

The afterlives of troubling encounters hence necessitate new words for ruptures created in communities but also for, as in this case, the jigsaw-like pieces needed to bring them together. Words, like the sacred objects now being returned to US Tribal Nations through the 1990 Native American Graves Protection and Repatriation Act (NAGPRA), have power.

In her introductory comments at the Peabody ceremony, the director of the museum, Jane Pickering, raised the question of what ethical stewardship of Indigenous objects might mean in the twenty-first century. She spoke not just about acknowledging that many of the objects the museum obtained and withheld came to it under duress and violence, but also argued for an ethics of repair. To establish open communication with communities whose objects had long been on display, repair, the director clarified, "sometimes meant *return* to the community."[6]

But the return of sacred objects opens the possibilities for other portrayals, other interpretations in museum spaces. For instance, on that day as the ceremony neared its ending, the representatives unfurled the marker they had brought to leave at the museum: a flag of their nation. With the gifting of the flag, the Gitxaała Nation reclaimed power over those who had taken their culture to display as an exotic talisman, both the shipping company and the museum. By requesting that it be the Gitxaała Nation flag, a representation of a sovereign nation, that is displayed where the totem pole once stood is laden with meaning. It creates another encounter, one infused this time with a deeper understanding, a fuller picture of both the historical context and cultural meaning. It affirms sovereignty and the dignity and power of telling one's own story. It was also a reminder for future museumgoers that the Gitxaała people are not relics of the past but, as the delegates remarked several times, "we survived."[7]

---

[5] https://coastalfirstnations.ca/gitxaala-bring-treasure-home-a-historic-turning-point-for-the-nation/.

[6] On museums and collecting, see Amy Lonetree, *Decolonizing Museums* (North Carolina: University of North Carolina Press, 2012).

[7] For a notion of survivance, see Gerald Vizenor, ed., *Survivance: Narratives of Native Presence* (Lincoln: University of Nebraska Press, 2008). For pushing back against damage narratives see Eve Tuck, "Suspending Damage: A Letter to Communities," *Harvard Educational Review* 79, no. 3 (2009): 415 and Eve Tuck and K. Wayne Yang, "Decolonization Is Not a Metaphor," *Decolonization: Indigeneity, Education & Society* 1, no. 1 (2012): 1–40.

SELECT BIBLIOGRAPHY

Adams, David Wallace. 1995. *Education for Extinction: American Indians and the Boarding School Experience, 1875–1928.* Lawrence: University Press of Kansas.

Adams, William Y. 2016. *The Boasians: Founding Fathers and Mothers of American Anthropology.* New York: Hamilton Books.

Aguirre, Carlos. 2005. *The Criminals of Lima and Their Worlds: The Prison Experience, 1850–1935.* Durham, NC: Duke University Press.

Aguirre, Carlos, and Ricardo D. Salvatore. 1996. *The Birth of the Penitentiary in Latin America: Essays on Criminology, Prison Reform, and Social Control, 1830–1940.* Austin: University of Texas Press.

Akien, Lewis R. 2003. *Tests psicológicos y evaluación,* 11th ed. Mexico City: Pearson.

Alain Liogier, Henri. 1996. "Botany and Botanists in Puerto Rico." *Annals of the New York Academy of Sciences* 776 (1): 42–45.

Alberto, Paula L., and Eduardo Elena. 2016. *Rethinking Race in Modern Argentina.* New York: Cambridge University Press.

Alden Mason, John, and Aurelio Macedonio Espinosa. 1921–1929. *Porto Rican Folklore: Folktales.* New York: American Folklore Society.

Alegría, Ricardo E. 1975. *Los dibujos puertorriqueños del naturalista francés: Augusto Plée, 1821–1823.* S.I.: s.n.

Alemán, Jesse. 2013. "The Other Country, Mexico, the United States, and the Gothic History of Conquest." In *The Spectralities Reader: Ghosts and Haunting in Contemporary Cultural Theory,* edited by María Pilar Del Blanco and Esther Peereen, 507–526. London: Bloomsbury.

Alonso Bejarano, Carolina, Lucia López Juárez, Mirian A. Mijangos García, and Daniel M. Goldstein. 2019. *Decolonizing Ethnography: Undocumented Immigrants and New Directions in Social Science.* Durham, NC: Duke University Press.

Álvarez, Ana Isabel. 1993. "La enseñanza de la psicología en la Universidad de Puerto Rico, Recinto de Río Piedras: 1903–1950." *Revista Puertorriqueña de Psicología* 9 (1): 13–29.

Álvarez-Curbelo, Silvia, and María Elena Rodríguez-Castro. 1993. *Del nacionalismo al populismo: cultura y política en Puerto Rico.* Río Piedras: Ediciones Huracán.

Amador, José. 2015. *Medicine and Nation Building in the Americas, 1890–1940.* Nashville, TN: Vanderbilt University Press.

Amparo Lasso, María. 2016. "Indígenas em guarda ante o projeto Genográfico." October 8. www.adital.com.br/site/noticia2.asp?lang=PT&cod=16334.

Andermann, Jens. 2007. *The Optic of the State: Visuality and Power in Argentina and Brazil.* Pittsburgh: University of Pittsburgh Press.

Anderson, Warwick. 2006. *Colonial Pathologies: American Tropical Medicine, Race, and Hygiene in the Philippines.* Durham, NC: Duke University Press.

_____. 2006. *The Cultivation of Whiteness: Science, Health, and Racial Destiny in Australia.* Durham, NC: Duke University Press.

_____. 2012. "Objectivity and Its Discontents." *Social Studies of Science* 43 (4): 557–576.

_____. 2019. *The Collectors of Lost Souls: Turning Kuru Scientists into Whitemen.* Baltimore: John Hopkins University Press.

_____. 2020. "Finding Decolonial Metaphors in Postcolonial Histories." *History and Theory* 59 (3): 430–438.

Andree, Christian. 2002. "Vorwort." *Vol. 52, in Rudolf Virchow, Sämtliche Werke.* Berlin: Blackwell Wissenschafts.

Andrews, George Reid. 1996. "Brazilian Racial Democracy, 1900–90: An American Counterpoint." *Journal of Contemporary History* 31 (3): 483–507.

Anon. 1963. "Arthritis among the Blackfeet." *Modern Medicine* 45–46.

_____. 2014. "'William M. O'Brien' Obituaries." *Yale Alumni Magazine,* March 27. https://yalealumnimagazine.com/obituaries/1335-william-m-o-brien-56md.

Appelbaum, Nancy P., Anne S. Macpherson, and Karin Alejandra Rosemblatt. 2003. *Race and Nation in Modern Latin America.* Chapel Hill: University of North Carolina Press.

Archibald, Jo-Ann, Jenny Bol Jun Lee-Morgan, and Jason De Santolo. 2019. *Decolonizing Research: Indigenous Storywork as Methodology.* London: Zed Books.

Argeri, María E. 2005. *De Guerreros a Delincuentes. La desarticulación de las jefaturas indígenas y el poder judicial. Norpatagonia, 1880–1930.* Madrid: CSIC.

Arias, Arturo, ed. 2001. *The Rigoberta Menchú Controversy.* Minneapolis: University of Minnesota Press.

Arnold, David. 1993. *Colonizing the Body: State Medicine and Epidemic Disease in Nineteenth-Century India.* Berkeley: University of California Press.

Arvin, Maile. 2019. *Possessing Polynesians: The Science of Settler Colonial Whiteness in Hawai'i and Oceania.* Durham, NC: Duke University Press.

Asad, Talal. 2011. *Anthropology and the Colonial Encounter.* New York: Humanity Books.

Ashford, Bailey K. 1903. *La uncinariasis en Puerto Rico.* Puerto Rico: Tip. "El Pais."

_____. 1934. *A Soldier in Science: The Autobiography of Bailey K. Ashford.* London: G. Routledge.

Ayala, César J., and Rafael Bernabe. 2007. *Puerto Rico in the American Century.* Chapel Hill: University of North Carolina Press.

Baatz, Simon. 1996. "Imperial Science and Metropolitan Ambition: The Scientific Survey of Puerto Rico, 1913–1934." *Annals of the New York Academy of Sciences* 776 (1): 1–16.

Baker, David, and Ludy Benjamin. 2013. *From Seánce to Science: A History of the Profession of Psychology in America.* Akron: University of Akron Press.

Baretta, Silvio R. Duncan, and John Markoff. 2005. "Civilization and Barbarism: Cattle Frontiers in Latin America." In *States of Violence,* edited by Fernando Coronil and Julie Skurski, 587–620. Ann Arbor: University of Michigan Press.

Barkan, Elazar. 1996. *The Retreat of Scientific Racism: Changing Concepts of Race in Britain and the United States between the World Wars.* Cambridge: Cambridge University Press.

Barker, Joanne. 2014. "The Specters of Recognition." In *Formations of United States Colonialism,* edited by Alyosha Goldstein, 33–56. Durham, NC: Duke University Press.

Barnard, Alan J., and Jonathan Spencer, editors. 2010. *Routledge Encyclopedia of Social and Cultural Anthropology,* 2nd ed. London: Routledge.

Bashford, Alison. 2014. *Global Population: History, Geopolitics, and Life on Earth.* New York: Columbia.

Bateman, Fiona, and Lionel Pilkington. 2011. *Studies in Settler Colonialism: Politics, Identity and Culture.* Basingstoke: Palgrave Macmillan.

Becker, David V., Lewis E. Braverman, John T. Dunn, et al. 1984. "The Use of Iodine as a Thyroidal Blocking Agent in the Event of a Reactor Accident: Report of the Environmental Hazards Committee of the American Thyroid Association." *JAMA* 252 (5): 659–661.

Beinart, William. 1984. "Soil Erosion, Conservationism and Ideas about Development: A Southern African Exploration, 1900–1960." *Journal of Southern African Studies* 11: 52–83.

Beisel, Nicola Kay. 1998. *Imperiled Innocents.* Princeton: Princeton University Press.

Bell, Hugh M. 1928. "An Experimental Analysis of the Ability of University Students to Study." MA thesis, Stanford University.

   1939. *The Theory and Practice of Personal Counseling, with Special Reference to the Adjustment Inventory,* 2nd ed. Stanford: Stanford University Press.

Ben-Zvi, Yael. 2007. "Where Did Red Go?: Lewis Henry Morgan's Evolutionary Inheritance and U.S. Racial Imagination." *CR: The New Centennial Review* 7 (2): 201–229.

Bennett, Peter H., and Thomas A. Burch. 1968. "The Distribution of Rheumatoid Factor and Rheumatoid Arthritis in the Families of Blackfeet and Pima Indians." *Arthritis and Rheumatism* 11 (4): 546–553.

Benzi Grupioni, Luís Donisete. 1998. *Coleções e expedições vigiadas: Os etnólogos no Conselho de Fiscalização das Expedições Artísticas e Científicas no Brasil.* São Paulo: Editora Hucitec ANPOCS.

Bergland, Renee. 2000. *The National Uncanny: Indian Ghosts and American Subjects*. Dartmouth, NH: UPNE.

Bergman, Abraham B., David C. Grossman, Angela M. Erdrich, John G. Todd, and Ralph Forquera. 1999. "A Political History of the Indian Health Service." *The Milbank Quarterly* 77 (4): 571–604.

Bernal, Guillermo. 2006. "La psicología clínica en Puerto Rico." *Revista Puertorriqueña de Psicología* 17 (1): 342–388.

2013. "60 Years of Clinical Psychology in Puerto Rico." *Interamerican Journal of Psychology* 47 (2): 212.

Bieder, Robert E. 1986. *Science Encounters the Indian, 1820–1880*. Norman: University of Oklahoma Press.

Bingham, Walter V. 1935. "MacQuarrie Test for Mechanical Ability." *Occupations: The Vocational Guidance Journal* 14 (3): 202.

Birn, Anne-Emanuelle. 2007. "Child Health in Latin America: Historiographic Perspectives and Challenges." *História, Ciências, Saúde-Manguinhos* 14 (3): 677–708.

Black, Edwin. 2012 [2003]. *War against the Weak: Eugenics and America's Campaign to Create a Master Race*. Washington, DC: Dialog Press.

Blackhawk, Ned, and Wilner Isaiah Lorado, editors. 2018. *Indigenous Visions: Rediscovering the World of Franz Boas*. New Haven, CT: Yale University Press.

Blake, Stanley E. 2011. *The Vigorous Core of Our Nationality: Race and Regional Identity in Northeastern Brazil*. Pittsburgh: University of Pittsburgh Press.

Blanchard, Pascal, Nicolas Bancel, Eric Deroo, and Sandrine Lemaire. 2008. *Human Zoos: Science and Spectacle in the Age of Empire*. Liverpool: University of Liverpool Press.

Boake, Corwin. 2002. "From the Binet-Simon to the Wechsler-Bellevue: Tracing the History of Intelligence Testing." *Journal of Clinical and Experimental Neuropsychology* 24 (3): 383–405.

Boas, Franz. 1920. *The Anthropometry of Porto Rico*. Philadelphia: Wistar Institute of Anatomy and Biology.

Boisseron, Bénédicte. 2018. *Afro-Dog: Blackness and the Animal Question*. New York: Columbia University Press.

Bonham, Valerie H., and Jonathan D. Moreno. 2008. "Research with Captive Populations: Prisoners, Students, and Soldiers." In *The Oxford Textbook of Clinical Research Ethics*, edited by Ezekiel J. Emanuel, Christine Grady, Robert A. Crouch, Lie Reidar, Franklin Miller, and David Wendler, 461–474. New York: Oxford University Press.

Boulon-Díaz, Frances. 2006. "Hacia una propuesta para el fortalecimiento escolar." *Revista de la Asociación de Psicólogos Escolares de Puerto Rico* 8: 3–4.

2015. "A Brief History of Psychological Testing in Puerto Rico: Highlights, Achievements, Challenges, and the Future." In *Psychological Testing of*

Hispanics: Clinical, Cultural, and Intellectual Issues, edited by Kurt F. Geisinger, 51–66. Washington, DC: American Psychological Association.

Boulon-Díaz, Frances, and Irma Roca de Torres. 2007. "School Psychology in Puerto Rico." In *The Handbook of International School Psychology*, edited by Shane R. Jimerson, Thomas Oakland, and Peter Thomas Farrell, 309–322. Thousand Oaks, CA: Sage.

———. 2016. "Formación en psicología en Puerto Rico: historia, logros y retos." *Revista Puertorriqueña de Psicología* 27 (2): 230–241.

Boyd, Byron A. 1991. *Rudolf Virchow: The Scientist as Citizen*. New York: Garland.

Brau, Salvador. 1972. *Ensayos: disquisiciones sociológicas*. Río Piedras: Editorial Edil.

Briggs, Laura. 2002. *Reproducing Empire: Race, Sex, Science, and U.S. Imperialism in Puerto Rico*. Berkeley: University of California Press.

———. 2020. *Taking Children: A History of American Terror*. Oakland: University of California Press.

Briones, Claudia, and José Luis Lanata. 2002. "Living on the Edge." In *Archaeological and Anthropological Perspectives on the Native Peoples of Pampa, Patagonia, and Tierra del Fuego to the Nineteenth Century* , edited by Claudia Briones and José Luis Lanata, 1–12. Westport, CT: Bergin and Garvey.

Brody, David. 2010. "Building Empire: Architecture and American Imperialism in the Philippines." In *Visualizing American Empire: Orientalism and Imperialism in the Philippines*, 140–163. Chicago: University of Chicago Press.

Bronfman, Alejandra. 2004. *Measure of Equality: Social Science, Citizenship, and Race in Cuba, 1902–1940*. Chapel Hill: University of North Carolina Press.

Brown, Stephen M., D. Carleton Gajdusek, Webster C. Leyshon, Arthur G. Steinberg, Kenneth S. Brown, and Cyril C. Curtain. 1974. "Genetic Studies in Paraguay: Blood Group, Red Cell, and Serum Genetic Patterns of the Guayaki and Ayore Indians, Mennonite Settlers, and Seven Other Indian Tribes of the Paraguayan Chaco." *American Journal of Physical Anthropology* 41 (2): 317–343.

Bruchac, Margaret. 2014. "My Sisters Will Not Speak: Boas, Hunt, and the Ethnographic Silencing of First Nations Women." *Curator* 57 (2): 153–171.

———. 2018. *Savage Kin: Indigenous Informants and American Anthropologists*. Tucson: The University of Arizona Press.

Brugge, Doug, and Rob Goble. 2002. "The History of Uranium Mining and the Navajo People." *American Journal of Public Health* 92 (9): 1410–1419.

Brunsma, David L., David Overfelt, and J. Steven Picou. 2007. *The Sociology of Katrina: Perspectives on a Modern Catastrophe*. Lanham, MD: Rowman & Littlefield.

Buchanan, Joseph R. 1885. *Manual of Psychometry: The Dawn of a New Civilization*. Boston: Holman Brothers.

Bulhan, Hussein Abdilahi. 1985. *Frantz Fanon and the Psychology of Oppression*. New York: Plenum.

Bunzl, Matti, and H. Glenn Penny. 2003. "Rethinking German Anthropology, Colonialism, and Race." In *Worldly Provincialism: German Anthropology in the Age of Empire*, edited by Matti Bunzl and H. Glenn Penny. Ann Arbor: University of Michigan Press.

Burch, Susan. 2021. *Committed: Remembering Native Kinship in and beyond Institutions*. Chapel Hill: University of North Carolina Press.

Burkhead, Michael Dow. 2007. *The Treatment of Criminal Offenders: A History*. Jefferson, NC: McFarland.

Burrows, Geoff C. 2014. "The New Deal in Puerto Rico: Public Works, Public Health, and the Puerto Rico Reconstruction Administration, 1935–1955." PhD dissertation, Graduate Centre, City University of New York, New York.

———. 2017. "Rural Hydro-Electrification and the Colonial New Deal: Modernization, Experts, and Rural Life in Puerto Rico, 1935–1942." *Agricultural History* 91 (3): 293–319.

Burton, Antoinette, ed. 2005. *Archive Stories: Facts, Fictions, and the Writing of History*. Durham, NC: Duke University Press.

Buss, Allan R. 1980. *Psychology in Social Context*. New York: Irvington.

Butte, George C. 1927. *Report of the Attorney General of Porto Rico for the Fiscal Year Ending June 30, 1927*. San Juan: Bureau of Supplies, Printing, and Transportation.

Cabán, Pedro A. 1999. *Constructing a Colonial People: Puerto Rico and the United States, 1898–1932*. Boulder: Westview Press.

Calderón Aponte, José. 1903. *El crímen de Bairoa*. Puerto Rico: Imprenta de Heraldo Español.

Campbell, Nancy D., and Laura Stark. 2015. "Making up 'Vulnerable' People: Human Subjects and the Subjective Experience of Medical Experiment." *Social History of Medicine* 28 (4): 825–848.

Campos del Toro, Enrique. 1947. *Report of the Attorney General of Puerto Rico for the Fiscal Year Ending June 30, 1946*, vol. 44. San Juan: División de Imprenta.

Campt, Tina. 2017. *Listening to Images*. Durham, NC: Duke University Press.

Cañizares-Esguerra, Jorge. 2004. "Iberian Science: Ignored How Much Longer?" *Perspectives on Science* 12 (1): 86–125.

Carlson, Elizabeth. 2017. "Anti-Colonial Methodologies and Practices for Settler Colonial Studies." *Settler Colonial Studies* 7 (4): 496–517.

Carpenter, Daniel P. 2010. *Reputation and Power: Organizational Image and Pharmaceutical Regulation at the FDA*. Princeton: Princeton University Press.

Carson, John. 1993. "Army Alpha, Army Brass, and the Search for Army Intelligence." *Isis* 84 (2): 278–309.

2014. "Mental Testing in the Early Twentieth Century: Internationalizing the Mental Testing Story." *History of Psychology* 17 (3): 249–255.

Carter, Eric D. 2017. "Population, Neo-Malthusianism, and Public Health: Latin American Perspectives." Meeting of the Conference of Latin Americanist Geographers (CLAG), New Orleans.

2019. "Social Medicine and International Expert Networks in Latin America, 1930–1945." *Global Public Health* 14 (6–7): 791–802.

Castellanos, M. B. 2017. "Introduction: Settler Colonialism in Latin America." *American Quarterly* 69 (4): 777–781.

Castro, Rodríguez. 1889. *La embriaguez y la locura, ó, consecuencias del alcoholismo.* San Juan: Imp. del "Boletín Mercantil."

Chakrabarty, Dipesh. 2008. *Provincializing Europe: Postcolonial Thought and Historical Difference.* Princeton: Princeton University Press.

Chan, Beatriz, and Barba de Piña. 1988. "Eulalia Guzmán Barrón." In *La antropología en México, v. 10, Panorama histórico: Los protagonistas*, edited by García Mora and Lina Odena Güemes, 255–272. Mexico City: Instituto Nacional de Antropología e Historia.

Chandrasekhar, Sripati. 1954. *Hungry People and Empty Lands: Population Problems and International Tensions.* London: George Allen & Unwin.

Chaves, Nelson. 1946. *O problema alimentar do Nordeste brasileiro: Intro ao seu estudo economico social.* Recife: Ed. Medico Científica.

1948. *A sub-alimentação no Nordeste brasileiro.* Recife: Imprensa oficial.

Chazkel, Amy. 2009. "Social Life and Civic Education in the Rio de Janeiro City Jail." *Journal of Social History* 42 (3): 697–731.

Chenoweth, William L. 2011. "Navajo Indians Were Hired to Assist the U.S. Atomic Energy Commission in Locating Uranium Deposits." Arizona Geological Survey Contributed Report Series: US Department of Energy, September. http://repository.azgs.az.gov/sites/default/files/dlio/files/nid1290/cr-11-n_navajousaec_u_deposits_final.doc.pdf.

Chesterton, Bridget María, and Anatoly V. Isaenko. 2014. "A White Russian in the Green Hell: Military Science, Ethnography, and Nation Building." *Hispanic American Historical Review* 94 (4): 615–648.

Child, Brenda J. 1998. *Boarding School Seasons: American Indian Families, 1900–1940.* Lincoln: University of Nebraska Press.

Cisney, Vernon W., and Nicolae Morar. 2016. *Biopower: Foucault and Beyond.* Chicago: University of Chicago Press.

Clark, Burton R. 1970. *The Distinctive College: Antioch, Reed & Swarthmore.* Chicago: Aldine.

Clastres, Pierre. 1987. *Society against the State: Essays In Political Anthropology.* New York: Zone Books.

1998. *Chronicle of the Guayaki Indians.* New York: Zone Books.

Coello y Mesa, José. 1913. "Crania peruana por José Coello y Mesa." *Revista Universitaria* 2 (5): 2–30.

Coleman, Kevin. 2016. *A Camera in the Garden of Eden: The Self-Forging of a Banana Republic*. Austin: University of Texas Press.

Commoner, Barry. 1971. *The Closing Circle: Nature, Man and Technology*. New York: Random House.

Conferencia Latinoamericana sobre Agricultura y Alimentación. 1965. "Conferencia Latinoamericana sobre Agricultura y Alimentación: Proyecto de Informe – Preámbulo." *Cuadernos Médico Sociales (Chile)* 6 (2): 11–16.

Conklin, Alice L. 2013. *In the Museum of Man: Race, Anthropology, and Empire in France, 1850–1950*. Ithaca, NY: Cornell University Press.

Connell, Raewyn. 2009. *Southern Theory: The Global Dynamics of Knowledge in Social Science*. Cambridge: Polity.

Connelly, Matthew. 2008. *Fatal Misconceptions: The Struggle to Control World Population*. Cambridge, MA: Harvard University Press.

Cope, Douglas R. 1994. *The Limits of Racial Domination: Plebeian Society in Colonial Mexico City, 1660–1720*. Madison: University of Wisconsin Press.

Córdova, Isabel M. 2017. *Pushing in Silence: Modernizing Puerto Rico and the Medicalization of Childbirth*. Austin: University of Texas Press.

Coronado, Jorge. 2009. *The Andes Imagined: Indigenismo, Society, And Modernity*. Pittsburgh: University of Pittsburgh Press.

———. 2018. *Portraits in the Andes: Photography and Agency: 1900–1950*. Pittsburgh: University of Pittsburgh Press.

Correia de Andrade, Manoel. 1996. "Josue de Castro: O homem, o cientista, e o seu tempo." In *Fome: Um tema proibido*, edited by Ana Maria de Castro, 285–321. Recife: CONDPE/CEPE.

Cosio, José Gabriel. 1913. "Delegado del Supremo Gobierno y de la Sociedad Geográfica de Lima, ante la Comisión Científica de 1912 enviada por la Universidad de Yale, acerca de los trabajos realizados por ella an el Cuzco y Apurimac." *Revista Universitaria* 2 (5): 22.

Costa Mandry, Oscar G. 1971. *Apuntes para la historia de la medicina en Puerto Rico: breve reseña histórica de las ciencias de la salud*. San Juan: Departamento de Salud.

Coutinho, Ruy. 1947. *O Valor Social da Alimentação*, 2nd ed. Rio de Janeiro: Agir Ed.

Cox Hall, Amy. 2017. *Framing a Lost City: Science, Photography, and the Making of Machu Picchu*. Texas: University of Texas Press.

Creager, Angela N. H. 2013. *Life Atomic: A History of Radioisotopes in Science and Medicine, Synthesis: A Series in the History of Chemistry, Broadly Construed*. Chicago: The University of Chicago Press.

Cueto, Marcos. 1989. *Excelencia científica en la periferia: Actividades científicas e investigación biomédica en el Perú, 1890–1950*. Lima: GRADE.

Cullather, Nick. 2010. *The Hungry World: America's Cold War Battle against Poverty in Asia*. Cambridge, MA: Havard University Press.

Dahm, Clair. 2017. "Community Corrections: The Rise and Fall of the Rehabilitative Ideal in Postwar St. Louis." PhD dissertation, Brandeis University.

Danzinger, Kurt. 1994. *Constructing the Subject: Historical Origins of Psychological Research.* New York: Cambridge University Press.

Daws, Gavan. 1974. *Shoal of Time: A History of the Hawaiian Islands.* Honolulu: University Press of Hawai'i.

Dawson, Alexander S. 2004. *Indian and Nation in Revolutionary Mexico.* Tucson: University of Arizona Press.

2012. "Histories and Memories of the Indian Boarding Schools in Mexico, Canada, and the United States." *Latin American Perspectives* 39 (5): 80–99.

de Barros, Jaunita, Steven Palmer, and David Wright. 2008. *Health and Medicine in the Circum-Caribbean, 1800–1968.* New York: Routledge.

de Castro, Ana Maria, ed. 1996. *Fome: Um tema proibido,* 3rd ed. Recife: CONDPE/CEPE.

de Castro, Josue. 1935. *As condições de vida das classes operárias no Recife: Estudo econômico de sua alimentação.* Rio de Janeiro: Dept. de Estatística e Publicidade, Min. do Trabalho, Indústria e Commercio.

1952. *The Geography of Hunger.* Boston: Little Brown.

1970. *Of Men and Crabs.* Translated by Susan Hertelendy. New York: Vanguard Press.

de Grosourdy, René. 1864. *El médico botánico criollo.* Paris: F. Brachet.

de la Cadena, Marisol. 2000. *Indigenous Mestizos: The Politics of Race and Culture in Cuzco, Peru, 1919–1991.* Durham, NC: Duke University Press.

2010. "Indigenous Cosmopolitics in the Andes: Conceptual Reflections beyond 'Politics'." *Cultural Anthropology* 25 (2): 334–370.

2015. *Earth Beings: Ecologies of Practice across Andean Worlds.* Durham, NC: Duke University Press.

de León, Pablo Roca. 1951. *Manual escala de inteligencia Wechsler para niños.* San Juan: Departmento de Instrucción Pública.

1953. *Escala de inteligencia Stanford-Binet para niños.* San Juan: Departmento de Instrucción Pública.

de Souza Lima, Antonio Carlos. 1995. *Um grande cerco de paz: Poder tutelar, indianidade e formação do estado no Brasil.* Petrópolis: Vozes.

de Torres, Irma Roca. 2006. "Algunos precursores/as de la psicología en Puerto Rico: reseñas biográficas." *Revista Puertorriqueña de Psicología* 17 (1): 63–88.

2008. "Perspectiva histórica sobre la medición psicológica en Puerto Rico." *Revista Puertorriqueña de Psicología* 19 (1): 11–48.

de Torres, Ramón A., and Rodríguez Castro. 1893. *El crimen de 'Las Lomas' (Juana Díaz): informe pericial sobre las facultades mentales de Francisco Corchado.* Ponce: Tip. "El Vapor."

Debaene, Vincent. 2014. *Far Afield: French Anthropology between Science and Literature,* translated by Justin Izzo. Chicago: University of Chicago Press.

DeJong, David H. 2009. *Stealing the Gila: The Pima Agricultural Economy and Water Deprivation, 1848–1921.* Tucson: University of Arizona Press.

2011. *Forced to Abandon Our Fields.* Salt Lake City: University of Utah Press.

Del Valle Atiles, Francisco. 1887. *El campesino puertorriqueño: sus condiciones físicas, intelectuales y morales, causas que las determinan y medios para mejorarlas*. San Juan: Tipografía de José González Font.

Deloria, Philip Joseph. 1998. *Playing Indian*. New Haven, CT: Yale University Press.

——— 2004. *Indians in Unexpected Places*. Lawrence: University Press of Kansas.

Departamento de Justicia, Oficina del Procurador General. 1933. *Reglamento para el régimen y gobierno de la Penitenciaría de Puerto Rico en Río Piedras*. San Juan: Negociado de Materiales, Imprenta y Transporte.

Diacon, Todd A. 2004. *Stringing Together a Nation: Cândido Mariano Da Silva Rondon and the Construction of a Modern Brazil, 1906–1930*. Durham, NC: Duke University Press.

Dubow, Saul. 1995. *Scientific Racism in Modern South Africa*. Cambridge: Cambridge University Press.

Dumont, Enrique. 1875–1876. *Ensayo de una historia médico-quirúrgica de la isla de Puerto Rico*, 2 vols. La Habana: Habana, Imp. La Antilla.

Dungy, Kathryn R. 2015. *The Conceptualization of Race in Colonial Puerto Rico, 1800–1850*. New York: Peter Lang.

Dussel, Enrique. 1995. *The Invention of the Americas: Eclipse of "the Other" and the Myth of Modernity*. New York: Continuum.

Dussel, Enrique, Javier Krauel, and Virginia Tůma. 2000. "Europe, Modernity, and Modernism." *Nepantla* 1 (3): 465–478.

Earle, Rebecca. 2007. *The Return of the Native: Indians and Myth-Making in Spanish America, 1810–1930*. Durham, NC: Duke University Press.

Edmunds, R. David. 1995. "Native Americans, New Voices: American Indian History, 1895–1995." *The American Historical Review* 100 (3): 717–740.

Eghigian, Greg. 2015. *The Corrigible and the Incorrigible: Science, Medicine, and the Convict in Twentieth-Century Germany*. Ann Arbor: University of Michigan Press.

Ehrlich, Paul R. 1968. *The Population Bomb*. San Francisco: Sierra Club/Ballantine Book.

Elkins, Caroline, and Susan Pedersen. 2005. *Settler Colonialism in the Twentieth Century: Projects, Practices, Legacies*. New York: Routledge.

Ellenbogen, Josh. 2012. *Reasoned and Unreasoned Images: The Photography of Bertillon, Galton, and Marey*. Pennsylvania: Penn State University Press.

Emre, Merve. 2018. *The Personality Brokers: The Strange History of Myers-Briggs and the Birth of Personality Testing*. New York: Doubleday.

Epstein, Steven. 2007. *Inclusion: The Politics of Difference in Medical Research*. Chicago: University of Chicago Press.

Espinosa, Mariola. 2013. "Globalizing the History of Disease, Medicine, and Public Health in Latin America." *Isis* 104 (4): 798–806.

——— 2015. "The Caribbean Origins of the National Public Health System in the USA: A Global Approach to the History of Medicine and Public Health in Latin America." *História, Ciências, Saúde-Manguinhos* 22 (1): 241–253.

Evans, John Hyde. 2002. *Playing God?: Human Genetic Engineering and the Rationalization of Public Bioethical Debate,* 1st ed. Chicago: University of Chicago Press.

2011. *The History and Future of Bioethics: A Sociological View,* 1st ed. Oxford: Oxford University Press.

Fabian, Ann. 2016. *The Skull Collectors: Race, Science, and America's Unburied Dead.* Cambridge, MA: Harvard University Press.

Fabian, Johannes. 1983. *Time and the Other: How Anthropology Makes Its Object.* New York: Columbia University Press.

Fanon, Frantz. 1965. *The Wretched of the Earth.* New York: Grove Press/Atlantic Inc.

Farro, Máximo. 2009. *La formación del Museo de La Plata: Coleccionistas, comerciantes, estudiosos y naturalistas viajeros a fines del siglo XIX.* Rosario, Argentina: Prohistoria Rosario.

Fassin, Didier. 2008. "Beyond Good and Evil?: Questioning the Anthropological Discomfort with Morals." *Anthropological Theory* 8 (4): 333–344.

Felitti, Karina. 2008. "La 'Explosión Demográfica' y la Planificación Familiar a Debate: Instituciones, Discusiones y Propuestas del Centro y la Periferia." *Revista Escuela de História* 7 (2): 1–16.

Feliú, Fernando. 2002. "Rendering the Invisible Visible and the Visible Invisible: The Colonizing Function of Bailey K. Ashford's Antianemia Campaigns." In *Foucault and Latin America: Appropriations and Deployments of Discursive Analysis,* edited by Benigno Trigo, 153–166. New York: Routledge.

Fernández Méndez, Eugenio. 1963. *Franz Boas y los estudios antropológicos en Puerto Rico.* Mexico City: Editorial Cultura.

Fernández Prieto, Leida. 2013. "Islands of Knowledge: Science and Agriculture in the History of Latin America and the Caribbean." *Isis* 104 (4): 788–797.

Ferracuti, Franco, and Christina Maria Giannini. 1968. *L'incesto padre-figlia: studio clinico-criminologico.* Río Piedras: University of Puerto Rico.

Ferracuti, Simon Dinitz, and Esperanza Acosta de Brenes. 1975. *Delinquents and Nondelinquents in the Puerto Rican Slum Culture.* Columbus: Ohio State University Press.

Ferris, Harry Burr. 1916. "The Indians of Cuzco and the Apurimac." *Memoirs of the American Anthropological Association* 3 (2): 59–148.

Figueroa Colón, Julio C. 1996. "Introduction." *Annals of the New York Academy of Sciences* 776: vii–viii.

Finkelstein, Gabriel. 2014. *Emil du Bois-Reymond: Neuroscience, Self, and Society in Nineteenth-Century Germany.* Cambridge, MA: MIT Press.

Flores Galindo, Alberto. 2010. *In Search of an Inca: Identity and Utopia in the Andes.* Edited and translated by Carlos Aguirre, Charles F. Walker, and Willie Hiatt. Cambridge: Cambridge University Press.

Ford, Will. 2020. "A Radioactive Legacy Haunts This Navajo Village, Which Fears a Fractured Future." *Washington Post,* January 19. www.washingtonpost.com/national/a-radioactive-legacy-haunts-this-navajo-village-which-fears-

afractured-future/2020/01/18/84c6066e-37e0-11ea-9541-9107303481a4_
story.html.

Forman, Paul. 2010. "(Re)cognizing Postmodernity: Help for Historians –
Of Science Especially." *Berichte zur Wissenschaftsgeschichte* 33 (2): 157–175.

Foucault, Michel. 1990. *The History of Sexuality, Volume 1: An Introduction.* New
York: Vintage.

  1995. *Discipline and Punish: The Birth of the Prison*, 2nd ed. New York:
  Vintage.

  2009. *Security, Territory, Population: Lectures at the Collège de France
  1977–1978.* New York: Picador.

Fowler, Don D. 2000. *A Laboratory for Anthropology: Science and Romanticism in
the American Southwest, 1846–1930*, 1st ed. Albuquerque: University of New
Mexico Press.

Fox, Renee C., and Judith P. Swazey. 2008. *Observing Bioethics.* New York: Oxford
University Press.

Fuentes, Marisa J. 2016. *Dispossessed Lives: Enslaved Women, Violence, and the
Archive.* Philadelphia: University of Pennsylvania Press.

Gajdusek, D. Carleton. 1963. *Paraguayan Indian Expeditions to the Guayaki and
Chako Indians, August 25, 1963 to September 28, 1963.* Bethesda, MD:
National Institute of Neurological Diseases and Blindness, National
Institutes of Health.

Gale, Thomas, ed. 2007. *American Men & Women of Science: A Biographical
Directory of Today's Leaders in Physical, Biological, and Related Sciences*,
vol. 5, 23rd ed. Detroit, MI: Gale.

Garcia, Angela. 2010. *The Pastoral Clinic: Addiction and Dispossession along the
Rio Grande*, 1st ed. Berkeley: University of California Press.

Gard, Raymond L. 2016. *Rehabilitation and Probation in England and Wales,
1876–1962.* London: Bloomsbury.

Garfield, Seth. 2001. *Indigenous Struggle at the Heart of Brazil: State Policy,
Frontier Expansion, and the Xavante Indians, 1937–1988.* Durham, NC:
Duke University Press.

Gerbi, Antonello. 1978. *La naturaleza de las indias nuevas: Cristóbal Colón a
Gonzalo Fernández de Oviedo.* Mexico City: Fondo de Cultura Económica.

Gershengorn, Marvin C. 2012. "History of the Clinical Endocrinology Branch of
the National Institute of Diabetes and Digestive and Kidney Diseases:
Impact on Understanding and Treatment of Diseases of the Thyroid
Gland." *Thyroid* 22 (2): 109–111.

Gibby, Robert E., and Michael E. Zickar. 2008. "A History of the Early Days of
Personality Testing in American Industry: An Obsession with Adjustment."
History of Psychology 11 (3): 164–184.

Gil-Riaño, Sebastián. 2018. "Relocating Anti-Racist Science: The 1950 UNESCO
Statement on Race and Economic Development in the Global South." *British
Journal for the History of Science* 51 (2): 281–303.

2023. *The Remnants of Race Science: UNESCO and Economic Development in the Global South*. New York: Columbia University Press.

Gill, Howard B. 1952. *Preliminary Report on the State of the Prisons of Puerto Rico*. San Juan: Department of Justice.

Gillingham, Paul. 2011. *Cuauhtémoc's Bones: Forging National Identity in Modern Mexico*. Albuquerque: University of New Mexico Press.

Ginsburg, Faye D., and Rayna Rapp. 1995. *Conceiving the New World Order: The Global Politics of Reproduction*. University of California Press.

Godreau, Isar P. 2015. *Scripts of Blackness: Race, Cultural Nationalism, and U.S. Colonialism in Puerto Rico*. Champaign: University of Illinois Press.

Goldstein, Jan E. 2015. "Toward an Empirical History of Moral Thinking: The Case of Racial Theory in Mid-Nineteenth-Century France." *The American Historical Review* 120 (1): 1–27.

González, Elisa M. 2016. "Food for Every Mouth: Nutrition, Agriculture, and Public Health in Puerto Rico, 1920s–1960s." PhD dissertation, Columbia University.

González, José Luis. 1993. *Puerto Rico: The Four-Storeyed Country and Other Essays*. Princeton: Markus Wiener.

Goodenough, Florence. 1926. *The Measurement of Intelligence by Drawings*. New York: World Book Company.

Gordillo, Gastón. 2016. "The Savage outside of White Argentina." In *Rethinking Race in Modern Argentina*, edited by Paulina Alberto and Eduardo Elena, 241–267. New York: Cambridge University Press.

Graham, Richard, ed. 1990. *The Idea of Race in Latin America, 1870–1940*. Austin: University of Texas Press.

Grandin, Greg. 2004. "Can the Subaltern Be Seen? Photography and the Affects of Nationalism." *Hispanic American Historical Review* 84 (1): 83–111.

Green, Bonnie A., and Harold Kiess. 2017. *Measuring Humans: Fundamentals of Psychometrics in Selecting and Interpreting Tests*. San Diego: Cognella.

Greenblatt, Stephen. 1991. *Marvelous Possessions: The Wonder of the New World*. Chicago: University of Chicago Press.

Greene, Jeremy, Victor Braitberg, and Gabriella Maya Bernadett. 2020. "Innovation on the Reservation: Information Technology and Health Systems Research among the Papago Tribe of Arizona, 1965–1980." *Isis* 111 (3): 443–470.

Griffith, Patricia. 1963. "Conscientious Non-Objectors to Important Medical Research." *Washington Post*, February 3.

Guerrero. 1950. "El hallazgo de Ichcateopan: Dictamen que rinde la comisión designada por acuerdo del c. Secretario de Educación Pública, en relación con las investigaciones y exploración realizadas en Ichcateopan." *Revista Mexicana de Estudios Antropológicos* 11: 197–295.

Gundersen, Edvard. 1927. "Is Diabetes of Infectious Origin?" *The Journal of Infectious Diseases* 41 (3): 197–202.

Gurvitz, Milton S. 1947. "Psychometric Procedure in Penal and Correctional Institutions." In *Handbook of Correctional Psychology*, edited by Robert M. Lindner and Robert V. Seliger, 58–71. New York: Philosophical Library.

Guzmán, Tracy Devine. 2013. *Native and National in Brazil: Indigeneity after Independence*. Chapel Hill: University of North Carolina Press.

Hacking, Ian. 2001. *The Social Construction of What?* Cambridge, MA: Harvard University Press.

Hale, Piers. 2014. *Political Descent: Malthus, Mutualism, and the Politics of Evolution in Victorian England*. Chicago: University of Chicago Press.

Hämäläinen, Pekka. 2008. *The Comanche Empire, The Lamar Series in Western History*. New Haven, CT: Yale University Press.

Haraway, Donna J. 1997. *Modest_Witness@Second_Millennium. "FemaleMan_ Meets_OncoMouse": Feminism and Technoscience.* New York: Routledge.

2016. *Staying with the Trouble: Making Kin in the Chthulucene.* Durham, NC: Duke University Press Books.

Harkness, Jon M. 1996. "Research behind Bars: A History of Nontherapeutic Research on American Prisoners." PhD dissertation, University of Wisconsin – Madison.

Harmon, Alexandra. 2010. *Rich Indians: Native People and the Problem of Wealth in American History Press*. Chapel Hill: University of North Carolina Press.

Harrison, Patti L., and Alan S. Kaufman. 2007. "History of Intelligence Testing." In *Encyclopedia of Special Education*, edited by Cecil R. Reynolds and Elaine Fletcher-Janzen, 1127–1130. New York: John Wiley & Sons.

Harrison, Simon. 2008. "Skulls and Scientific Collecting in the Victorian Military: Keeping the Enemy Dead in British Frontier Warfare." *Comparative Studies of Society and History* 50 (1): 285–303.

Hartman, Saidiya. 2008. "Venus in Two Acts." *Small Axe: A Journal of Criticism* 26: 1–14.

Hartmann, Heinrich, and Corinna Unger. 2014. *A World of Populations: Transnational Perspectives on Demography in the Twentieth Century*. New York: Berghann Books.

Head, Randolph C. 2019. *Making Archives in Early Modern Europe: Proof, Information, and Political Record-Keeping, 1400–1700*. Cambridge: Cambridge University Press.

Heaney, Christopher. 2011. *Cradle of Gold*. New York: St. Martin's Griffin.

2016. "A Peru of Their Own: English Grave Opening and Indian Sovereignty in Early America." *William and Mary Quarterly* 73 (4): 609–649.

2018. "How to Make an Incan Mummy: Andean Embalming, Peruvian Science, and the Collection of Empire." *Isis* 109 (1): 1–27.

Hecht, Gabrielle. 2012. *Being Nuclear: Africans and the Global Uranium Trade*. Cambridge, MA: MIT Press.

Henderson, Algo D. 1946. *Antioch College: Its Design for Liberal Education*. New York: Harper & Brothers.

Henley, Tracy B., and Michael B. Thorne. 2008. *Connections in the History and Systems of Psychology*, 3rd ed. Boston: Houghton Mifflin.

Herman, Ellen. 1995. *The Romance of American Psychology: Political Culture in the Age of Experts*. Berkeley: University of California Press.

Hesse, Frank G. 1959. "A Dietary Study of the Pima Indian." *The American Journal of Clinical Nutrition* 7 (5): 532–537.

Hines, Thomas S. 1973. "American Modernism in the Philippines: The Forgotten Architecture of William E. Parsons." *Journal of the Society of Architectural Historians* 32 (4): 316–326.

Hoofnagle, Jay H. 1984. "Viral Hepatitis." In *National Institute of Health: An Account of Research in Its Laboratories and Clinics*, edited by Dewitt Stetten, 141–153. Orlando: Academic Press.

Hornblum, Allen M. 1999. *Acres of Skin: Human Experiments at Holmesburg Prison*. New York: Routledge.

Horst, René Harder. 2007. *The Stroessner Regime and Indigenous Resistance in Paraguay*. Gainesville: University Press of Florida.

Horton, Benjamin J. 1934. *Report of the Attorney General of Puerto Rico for the Fiscal Year Ending June 30, 1934*, vol. 24. San Juan: Negociado de Materiales, Imprenta y Transporte.

1935. *Report of the Attorney General of Puerto Rico for the Fiscal Year Ending June 30, 1935*, vol. 26. San Juan: Bureau of Supplies, Printing, and Transportation.

Imada, Adria L. 2018. "Lonely Together: Subaltern Family Albums and Kinship during Medical Incarceration." *Photography and Culture* 11 (3): 297–321. https://doi.org/10.1080/17514517.2018.1465651.

2022. *An Archive of Skin, an Archive of Kin: Disability and Life-Making during Medical Incarceration*. Berkeley: University of California Press.

Immerwahr, Daniel. 2019. *How to Hide an Empire: A History of the Greater United States*. New York: Farrar, Straus, and Giroux.

Indigenous Peoples Council on Biocolonialism. 2017. "Human Genetics Issues." January 15. www.ipcb.org/issues/human_genetics/index.html.

Inglis, Kerri A. 2013. *Maʻi Lepera: Disease and Displacement in Nineteenth-Century Hawaiʻi*. Honolulu: University of Hawaiʻi Press.

Irvine-Rivera, Edith M. 1927. "Brief News Notes." *Porto Rico Health Review* 2 (12): 41.

Jackson, Zakiyyah Iman. 2020. *Becoming Human: Matter and Meaning in an Anti-Black World*. New York: New York University Press.

Jacobs, Margaret D. 2009. *White Mother to a Dark Race: Settler Colonialism, Maternalism, and the Removal of Indigenous Children in the American West and Australia, 1880–1940*. Lincoln: University of Nebraska Press.

Jacó-Vilela, Ana María. 2014. "Psychological Measurement in Brazil in the 1920s and 1930s." *History of Psychology* 17 (3): 237–248.

Jacyna, Stephen. 2006. "Medicine in Transformation, 1800–1849." In *The Western Medical Tradition, 1800–2000*, edited by W. F. Bynum, Anne Hardy,

Stephen Jacyna, Christopher Lawrence, and E. M. Tansey, 92–94. New York: Cambridge University Press.

2006. "Medicine in Transformation, 1800–50." In *The Western Medical Tradition, 1800–2000*, edited by W. F. Bynum, Anne Hardy, Stephen Jacyna, Christopher Lawrence, and E. M. Tansey, 11–97. New York: Cambridge University Press.

Janssen, Volker. 2005. "Convict Labor, Civic Welfare: Rehabilitation in California's Prisons, 1941–1971." PhD dissertation, University of California-San Diego.

Jiménez Moreno, Wigbert. 1962. "Los hallazgos de Ichcateopan." *Historia Mexicana* 12 (2): 161–181.

Jobe, Margaret M. 2004. "Native Americans and the U.S. Census: A Brief Historical Survey." *Journal of Government Information* 30 (1): 66–80.

Johnson, Christopher. 2003. *Claude Lévi-Strauss: The Formative Years*. Cambridge: Cambridge University Press.

Johnson, Courtney. 2009. "Understanding the American Empire: Colonialism, Latin Americanism, and Professional Social Science, 1898–1920." In *Colonial Crucible: Empire in the Making of the Modern American State*, by Francisco A. Scarano and Alfred W. McCoy. Madison: University of Wisconsin Press.

Johnson, Lyman L. 2004. "Digging up Cuauhtémoc." In *Death, Dismemberment, and Memory: Body Politics in Latin America*, edited by Lyman L. Johnson, 207–244. Albuquerque: University of New Mexico Press.

Johnson, Miranda. 2020. "Toward a Genealogy of the Researcher as Subject in Post/Decolonial Pacific Histories." *History and Theory* 59 (3): 421–429.

Jones, David. 2002. "The Health Care Experiments at Many Farms: The Navajo, Tuberculosis, and the Limits of Modern Medicine, 1952–1962." *Bulletin of the History of Medicine* 76 (4): 749–790.

2003. "Virgin Soils Revisited." *William and Mary Quarterly* 60 (4): 703.

Jones, David, and Robert Martensen. 2003. "Human Radiation Experiments and the Formation of Medical Physics at the University of California, San Francisco and Berkeley, 1937–1962." In *Useful Bodies: Humans in the Service of Medical Science in the Twentieth Century*, edited by Jordon Goodman, Anthony McElligott, and Lara Marks, 81–108. Baltimore: Johns Hopkins University Press.

Jones, David Shumway. 2004. *Rationalizing Epidemics: Meanings and Uses of American Indian Mortality since 1600*. Cambridge, MA: Harvard University Press.

Joseph, Gilbert, Catherine LeGrand, and Ricardo Salvatore. 1998. *Close Encounters of Empire: Writing the Cultural History of U.S.–Latin American Relations*. Durham, NC: Duke University Press.

Kauanui, Kēhaulani. 2016. "'A Structure, Not an Event': Settler Colonialism and Enduring Indigeneity." *Lateral* 5 (1). https://csalateral.org/issue/5-1/forum-alt-humanities-settler-colonialism-enduring-indigeneity-kauanui/.

Keller, Richard C. 2015. *Fatal Isolation: The Devastating Paris Heat Wave of 2003*, 1st ed. Chicago: University of Chicago Press.

Keme, Emil. 2018. "For Abiayala to Live, the Americas Must Die: Toward a Transhemispheric Indigeneity." *Native American and Indigenous Studies* 5 (1): 42–65.

Kerr, Ashley. 2017. "From Savagery to Sovereignty: Identity, Politics, and International Expositions of Argentine Anthropology (1878–1892)." *Isis* 108 (1): 62–81.

2020. *Sex, Skulls, and Citizens: Gender and Racial Science in Argentina (1860–1910)*. Nashville, TN: Vanderbilt University Press.

Kim, Hill, and A. Magdelena Hurtado. 1996. *Aché Life History: The Ecology and Demography of a Foraging People*. New York: Aldine de Gruyter.

Kimmerer, Robin Wall. 2013. *Braiding Sweetgrass: Indigenous Wisdom, Scientific Knowledge, and the Teachings of Plants*. Minneapolis, MN: Milkweed.

Kinsbruner, Jay. 1996. *Not of Pure Blood: The Free People of Color and Racial Prejudice in Nineteenth-Century Puerto Rico*. Durham, NC: University Press.

Kleinman, Arthur. 1993. "What Is Specific to Western Medicine?" *Companion Encyclopedia of the History of Medicine*, vol. 1, edited by William F. Bynum and Roy Porter, 15–23. New York: Routledge.

Klinenberg, Eric. 2003. *Heat Wave: A Social Autopsy of Disaster in Chicago*, 1st ed. Chicago: University Of Chicago Press.

Knight, Franklin W. 1996. *Race, Ethnicity, and Class: Forging the Plural Society in Latin America and the Caribbean*. Waco: Baylor University Press.

Knorr-Cetina, Karin D. 1981. *The Manufacture of Knowledge: An Essay on the Constructivist and Contextual Nature of Science*. Oxford: Pergamon Press.

Knorr-Cetina, Karin D., and Michael Mulkay. 1983. *Science Observed: Perspectives on the Social Study of Science*. Beverly Hills, CA: Sage.

Koel-Abt, Katrin, and Andreas Winkelmann. 2013. "The Identification and Restitution of Human Remains from an Aché Girl Named 'Damiana': An Interdisciplinary Approach." *Annals of Anatomy – Anatomischer Anzeiger* 195 (5): 393–400.

Kovach, Margaret. 2009. *Indigenous Methodologies: Characteristics, Conversations and Contexts*. Toronto: University of Toronto Press.

Kowal, Emma. 2013. "Orphan DNA: Indigenous Samples, Ethical Biovalue and Postcolonial Science." *Social Studies of Science* 43 (4): 577–597.

Kowal, Emma, Jenny Reardon, and Joanna Radin. 2013. "Indigenous Body Parts, Mutating Temporalities, and the Half-Lives of Postcolonial Technoscience." *Social Studies of Science* 43 (4): 465–483.

Kramer, Paul A. 2006. *The Blood of Government: Race, Empire, the United States, and the Philippines*. Chapel Hill: University of North Carolina Press.

Krawec, Patty. 2022. *Becoming Kin: An Indigenous Call to Unforgetting the Past and Reimagining Our Future*. Pine Bush, NY: Broadleaf Books.

Krieken, Robert. 1999. "The Barbarism of Civilization: Cultural Genocide and the 'Stolen Generations'." *The British Journal of Sociology* 50 (2): 297–315.

Kristeva, Julia. 1991. *Strangers to Ourselves*. New York: Columbia University Press.

Krmpotich, Cara, Joost Fontein, and John Harries. 2010. "The Substance of Bones: The Emotive Materiality and Affective Presence of Human Remains." *Journal of Material Culture* 15: 371–384.

Kuenzli, Gabrielle E. 2013. *Acting Inca: National Belonging in the Early Twentieth-Century Bolivia*. Pittsburgh: University of Pittsburgh Press.

Kuklick, Henrika. 1996. *The Savage Within: The Social History of British Anthropology, 1885–1945*. New York: Cambridge University Press.

Ladd-Taylor, Molly. 2017. *Fixing the Poor: Eugenic Sterilization and Child Welfare in the Twentieth Century*. Baltimore: Johns Hopkins University Press.

Lafitte, Paul. 1957. *The Person in Psychology: Reality or Abstraction*. London: Routledge.

Lamiell, James T. 1987. *The Psychology of Personality: An Epistemological Inquiry*. New York: Columbia University Press.

Landy, David. 1959. *Tropical Childhood: Cultural Transmission and Learning in a Rural Puerto Rican Village*. Chapel Hill: University of North Carolina Press.

Lapp, Michael. 1995. "The Rise and Fall of Puerto Rico as a Social Laboratory, 1945–1965." *Social Science History* 19 (2): 169–199.

Laqueur, Thomas W. 2015. *The Work of the Dead: A Cultural History of Mortal Remains*. Princeton: Princeton University Press.

Larson, Brooke. 2004. *Trials of Nation Making: Liberalism, Race, and Ethnicity in the Andes, 1810–1910*. Cambridge: Cambridge University Press.

Larson, Carolyne R. 2015. *Our Indigenous Ancestors: A Cultural History of Museums, Science, and Identity in Argentina, 1877–1943*. University Park: Penn State University Press.

Lasierra, Fray Iñigo Abbad y. 1966 [1778]. *Historia geográfica, civil y natural de la isla de San Juan Bautista de Puerto Rico*. San Juan: Editorial Universitaria.

Laurière, Christine. 2008. *Paul Rivet, le savant & le politique*. Paris: Publications Scientifiques du Muséum national d'Histoire naturelle.

2009. "La Société des Américanistes de Paris: une société savante au service de l'américanisme." *Journal de la Société des Américanistes* 95 (2): 93–115.

2010. "Anthropology and Politics, the Beginnings: The Relations between Franz Boas and Paul Rivet (1919–42)." *Histories of Anthropology Annual* 6 (1): 225–252.

2019. "Un lieu de synthèse de la science anthropologique: histoire du musée de l'Homme." In *Bérose – Encyclopédie internationale des histoires de l'anthropologie*. Paris. www.berose.fr/article1680.html?lang=fr.

Law, John, and Peter Lodge. 1984. *Science for Social Scientists*. London: Macmillan.

Lawrence, J. S., and P. H. Bennett. 1960. "Benign Polyarthritis." *Annals of the Rheumatic Diseases* 19 (1): 20–30.

LeBrón, Marisol. 2019. *Policing Life and Death: Race, Violence, and Resistance in Puerto Rico*. Berkeley: University of California Press.

Ledru, André Pierre. 1957. *Viaje a la isla de Puerto Rico en el año 1797, ejecutado por una comisión de sabios franceses, de orden de su gobierno bajo la dirección del capitán Nicolás Baudín*. Translated by Julio L. de Vizcarrondo. Río Piedras: Instituto de Literatura Puertorriqueña, Universidad de Puerto Rico.

Lehmann-Nitsche, Robert. 1908. Relevamiento antropológico de una india Guajaquí. Vol. 15, in *Revista del Museo de La Plata*, Tomo XV 91–99. UNLP.

Leite, Marcelo. 2005. "Projeto Genográfico e 'Projeto Vampiro'." *Folha de São Paulo*, April 17. www1.folha.uol.com.br/fsp/ciencia/fe1704200503.htm.

Lethabo King, Tiffany. 2019. *The Black Shoals: Offshore Formations of Black and Native Studies*. Durham, NC: Duke University Press.

Levine, Lawrence W. 1992. "The Folklore of Industrial Society: Popular Culture and Its Audiences." *The American Historical Review* 97 (5): 1369–1399.

Levine, Philippa. 2017. *Eugenics: A Very Short Introduction*. New York: Oxford University Press.

Levison, Julie H. 2003. "Beyond Quarantine: A History of Leprosy in Puerto Rico, 1898–1930s." *História, Ciências, Saúde-Manguinhos* 10 (1): 225–245.

Lévi-Strauss, Claude. 1952. *Race and History*, vol. 6. Paris: UNESCO.

1961. *Tristes Tropiques*. Translated by John Russel. New York: Criterion Books.

Leys, Ruth. 2011. "The Turn to Affect: A Critique." *Critical Inquiry* 73 (3): 434–472.

Livingston, Julie. 2014. "Figuring the Tumor in Botswana." *Raritan* 34 (1): 10–24.

Livingstone, David N., and Charles W. J. Withers. 2011. *Geographies of Nineteenth-Century Science*. Chicago: University of Chicago Press.

Livingstone Smith, David. 2016. "Paradoxes of Dehumanization." *Social Theory and Practice* 42 (4): 417–418.

Lonetree, Amy. 2012. *Decolonizing Museums: Representing Native America in National and Tribal Museums*. Chapel Hill: University of North Carolina Press.

López Lenci, Yazmín. 2007. *El Cusco, paqarina moderna: Cartografía de una modernidad e identidad en los Andes peruanos (1900–1935)*. Lima: Instituto Nacional de Cultura, Dirección Regional de Cultural de Cusco.

López Rey y Arrojo, Manuel, Toro Calder Jamie, and Ceferina Cedeño Zavala. 1975. *Extensión, características y tendencias de la criminalidad en Puerto Rico, 1964–70*. Río Piedras: Editorial de la Universidad de Puerto Rico.

Lossio Chávez, Jorge. 2012. *El peruano y su entorno: Aclimatándose a las alturas andinas*. Lima: Instituto de Estudios Peruanos.

Loveland, Frank. 1950. *Classification in the Prison System*. Washington, DC: Bureau of Prisons.

Lovell, George D., Gloria Laurie, and Doris Marvin. 1947. "A Comparison of the Minnesota Personality Scale and the Bell Adjustment Inventory for Student Counseling." *Proceedings of the Iowa Academy of Science* 54: 247–251.

Loveman, Mara. 2014. *National Colors: Racial Classification and the State in Latin America*. New York: Oxford University Press.

Lucero, José Antonio. 2020. "'To Articulate Ourselves': Trans-Indigenous Reflections on Film and Politics in Amazonia." *Native American and Indigenous Studies* 7 (2): 1–28.

Lyons, Scott Richard. 2011. "Actually Existing Indian Nations: Modernity, Diversity, and the Future of Native American Studies." *The American Indian Quarterly* 35 (3): 294–312.

Mabuchi, Kiyohiko, and Arthur B. Schneider. 2014. "Do Nuclear Power Plants Increase the Risk of Thyroid Cancer?" *Nature Reviews Endocrinology* 10 (7): 385–387.

MacDonald, David B. 2015. "Canada's History Wars: Indigenous Genocide and Public Memory in the United States, Australia and Canada." *Journal of Genocide Research* 17 (4): 411–431. https://doi.org/10.1080/14623528.2015 .1096583.

Macquarrie, Thomas W. 1924. "A Measure of Mechanical Ability." PhD dissertation, Stanford University.

Malavet, Pedro A. 2004. *America's Colony: The Political and Cultural Conflict between the United States and Puerto Rico*. New York: New York University Press.

Mallon, Florencia E. 2002. "Editor's Introduction." In Rosa Isolde Reuque Paillalef, *When a Flower Is Reborn: The Life and Times of a Mapuche Feminist*, 1–34. Durham, NC: Duke University Press.

——— ed. 2012. *Decolonizing Native Histories: Collaboration, Knowledge, and Language in the Americas*. Durham, NC: Duke University Press.

Mandler, Peter. 2009. "One World, Many Cultures: Margaret Mead and the Limits to Cold War Anthropology." *History Workshop Journal* 68 (1): 149–172.

Mann, Charles. 2018. *The Wizard and the Prophet*. New York: Knopf.

Marín, Rosa Celeste, Awilda Paláu de López, and Gloria P Barbosa de Chardón. 1959. *La efectividad de la rehabilitación de los delincuentes en Puerto Rico*. San Juan: Universidad de Puerto Rico.

Martinez-Novo, Carmen. 2006. *Who Defines Indigenous? Identities, Development, Intellectuals, and the State in Northern Mexico*. New Brunswick, NJ: Rutgers University Press.

Mason, John Alden. 1950. *The Languages of South America. Vol. 6*. In *Handbook of South American Indians*, edited by Julian H. Steward, 157–317. Washington, DC: Government Printing Office.

Massin, Benoit. 1998. "From Virchow to Fischer: Physical Anthropology and 'Modern Race Theories' in Wilhelmine Germany." In *History of Anthropology: Volksgeist as Method and Ethic*, edited by George Stocking, 79–154. Madison: University of Wisconsin Press.

Mayer, Ruth. 2008. "The Things of Civilization, the Matters of Empire: Representing Jemmy Button." *New Literary History* 39 (2): 193–215.

Mayo Santana, Raúl, Annette B. Ramírez de Arellano, and José G. Rigau Pérez. 2008. *A Sojourn in Tropical Medicine: Francis W. O'Connor's Diary of a Porto Rican Trip, 1927*. San Juan: Editorial de la Universidad de Puerto Rico.

Maza, Sarah. 1996. "Stories in History: Cultural Narratives in Recent Works in European History." *The American Historical Review* 101 (5): 1493–1515.

Mbembe, Achille, Olivier Mongin, Nathalie Lempereur, Jean-Louis Schlegel, and John Fletcher. 2006. "What Is Postcolonial Thinking?" *Esprit* 12: 117–133.

McCook, Stuart George. 2002. *States of Nature: Science, Agriculture, and Environment in the Spanish Caribbean, 1760–1940*. Austin: University of Texas Press.

2013. "Focus: Global Currents in National Histories of Science." *Isis* 104 (4): 773–776.

McCoy, Alfred W., Francisco A. Scarano, and Courtney Johnson. 2009. "On the Tropic of Cancer: Transitions and Transformations in the U.S. Imperial State." In *Colonial Crucible: Empire in the Making of the Modern American State*, edited by Alfred W. McCoy and Francisco A. Scarano, 3–33. Madison: University of Wisconsin Press.

McGranahan, Carole, and John F. Collins, eds. 2018. *Ethnographies of U.S. Empire*. Durham, NC: Duke University Press.

McHugh, Gelolo. 1945. "Relationship between the Goodenough Drawing a Man Test and the 1937 Revision of the Stanford-Binet Test." *Journal of Educational Psychology* 36 (2): 119–124.

McMillen, Christian W. 2015. *Discovering Tuberculosis: A Global History, 1900 to the Present*. New Haven, CT: Yale University Press.

Menchú, Rigoberta, and Elisabeth Burgos-Debray. 1993. *Rigoberta Menchú: An Indian Woman in Guatemala*. London: Verso.

Méndez, Cecilia. 1996. "Incas Sí, Indios No: Notes on Peruvian Creole Nationalism and Its Contemporary Crisis." *Journal of Latin American Studies* 28 (1): 197–225.

Méndez, José Luis. 2005. *Las ciencias sociales y el proceso político puertorriqueño*. San Juan: Ediciones Puerto.

Mendoza, Zoila S. 2008. *Creating Our Own: Folkore, Performance, and Identity in Cuzco, Peru*. Durham, NC: Duke University Press.

Mendoza Lizasuain, Gil G. 2017. "Desarrollo del sistema de salud pública de Puerto Rico desde el 1900 al 1957: 'una visión salubrista hacia las comunidades aisladas'." PhD dissertation, Universidad Interamericana de Puerto Rico-Recinto Metropolitano.

Métraux, Alfred. 1950. "An Indian Girl with a Lesson for Humanity." *UNESCO Courier* 3 (8): 8.

Métraux, Alfred, and Herbert Baldus. 1946. *The Guayaki*, vol. 1. In *Handbook of South American Indians*, by Julian H. Steward. Washington, DC: United States Government Printing Office.

Meyerowitz, Joanne. 2010. "'How Common Culture Shapes the Separate Lives': Sexuality, Race, and Mid-Twentieth-Century Social Constructionist Thought." *The Journal of American History* 96 (4): 1057–1084.

Mignolo, Walter D., and Catherine E. Walsh. 2018. *Decoloniality: Concepts, Analytics, Praxis.* Durham, NC: Duke University Press.

Milam, Erika Lorraine. 2019. *Creatures of Cain.* Princeton: Princeton University Press.

Miles, Tyla. 2019. "Beyond a Boundary: Black Lives and the Settler–Native Divide." *William and Mary Quarterly* 76 (3): 417–426.

Miller, Elizabeth. 1938. "Follow-up Study of Fifty Former Waialee Training School Boys." MA thesis, University of Hawai'i.

Million, Dian. 2009. "Felt Theory: An Indigenous Feminist Approach to Affect and History." *Wicazo Sa Review* 24 (2): 53–76.

Mintz, Sidney. 1960. *Worker in the Cane: A Puerto Rican Life History.* New Haven, CT: Yale University Press.

Moeller, H. G., ed. 1981. *The Selected Papers of Frank Loveland,* vol. ix. College Park: American Correctional Association.

Moreno, Francisco. 1874. "Cimetiéres et paraderos préhistoriques de Patagonie." *Revue d'anthropologie* 3 (1): 72–90.

1878. "Apuntes sobre las tierras patagónicas." *Anales de la Sociedad Científica Argentina* 5: 189–224.

1879. *Viaje a la Patagonia Austral.* Buenos Aires: La Nacion.

2002. *Perito Moreno's Travel Journal: A Personal Reminiscence.* Edited by Eduardo V. Moreno. Translated by Victoria Barcelona. Buenos Aires: Elephante Blanco.

Moreno Fraginals, Manuel. 1999. *La historia como arma y otros estudios sobre esclavos, ingenios, y plantaciones.* Barcelona: Crítica.

Moreno Toscano, Alejandra. 1980. *Los hallazgos de Ichcateopan 1949–1951.* Mexico City: UNAM.

Moreton-Robinson, Aileen. 2015. *The White Possessive: Property, Power, and Indigenous Sovereignty.* Minneapolis: University of Minnesota Press.

Morgan, Lewis Henry. 2020. "Ancient Society Or, Researches in the Lines of Human Progress from Savagery, through Barbarism to Civilization." Project Gutenberg, May 20. www.gutenberg.org/ebooks/45950.

Morrow, Daniel, and Barbara Brookes. 2013. "The Politics of Knowledge: Anthropology and Maori Modernity in Mid-Twentieth-Century New Zealand." *History and Anthropology* 24 (4): 453–471.

Moure-Eraso, Rafael. 1999. "Observational Studies as Human Experimentation: The Uranium Mining Experience in the Navajo Nation (1947–66)." *New Solutions: A Journal of Environmental and Occupational Health Policy* 9 (2): 163–178.

Moyn, Samuel. 2004. "Of Savagery and Civil Society: Pierre Clastres and the Transformation of French Political Thought." *Modern Intellectual History* 1 (1): 55–80.

Mülberger, Annette. 2014. "The Need for Contextual Approaches to the History of Mental Testing." *History of Psychology* 17 (3): 177–186.

Müller-Wille, Staffan. 2010. "Claude Lévi-Strauss on Race, History and Genetics." *BioSocieties* 5 (3): 330–347.

Muñoz Marín, Luis. 2003. *Historia del Partido Popular Democrático*. San Juan: Fundación Muñoz Marín.

Murphy, Michelle. 2017. *The Economization of Life*. Durham, NC: Duke University Press.

——— 2017. "What Cannot a Body Do?" *Catalyst: Feminism, Theory, Technoscience* 3 (1): 1–15.

Murphy, Robert F. 1991. "Memoir: Anthropology at Columbia: A Reminiscence." *Dialectical Anthropology* 16 (1): 65–81.

Nadesan, Majia Holmer. 2008. *Governmentality, Biopower, and Everyday Life*. New York: Routledge.

Navarro, José-Manuel. 2010. *Creating Tropical Yankees: Social Science Textbooks and U.S. Ideological Control in Puerto Rico, 1898–1908*. New York: Routledge.

Neale, Timothy, and Emma Kowal. 2020. "'Related' Histories: On Epistemic and Reparative Decolonization." *History and Theory* 593: 403–412.

Negrón Fernández, Luis. 1950. *Report of the Attorney General to the Governor of Puerto Rico for the Fiscal Year Ended June 30, 1947*. San Juan: Real Hermanos.

Nelson, Cary. 2007. "Antioch: An Education in the Real World." *The Chronicle of Higher Education* 53 (43): B5.

Nemser, Daniel. 2017. *Infrastructures of Race: Concentration and Biopolitics in Colonial Mexico*. Austin: University of Texas Press.

Noll, Steven. 1994. *Feeble-Minded in Our Midst: Institutions for the Mentally Retarded in the South, 1900–1940*. Chapel Hill: University of North Carolina Press.

Norton, Marcy. 2017. "Subaltern Technologies and Early Modernity in the Atlantic World." *Colonial Latin American Review* 26 (1): 18–38.

Novoa, Adriana, and Alex Levine. 2010. *From Man to Ape: Darwinism in Argentina*. Chicago: University of Chicago Press.

O'Brien, Anne. 2015. *Philanthropy and Settler Colonialism*. Basingstoke: Palgrave Macmillan.

Ocasio, Rafael. 2020. *Race and Nation in Puerto Rican Folklore: Franz Boas and John Alden Mason in Porto Rico*. New Brunswick, NJ: Rutgers University Press.

Ocasio Meléndez, Marcial E. 1985. *Río Piedras: ciudad universitaria, notas para su historia*. San Juan: Comité Historia de los Pueblos.

Offner, Amy C. 2019. *Sorting Out the Mixed Economy: The Rise and Fall of Welfare and Developmental States in the Americas*. Princeton: Princeton University Press.

Oreskes, Naomi. 2013. "Why I Am a Presentist." *Science in Context* 26 (4): 595–609.

Ortega, Julio. 2008. *Adiós Ayacucho*. Lima: Grupo Cultural Yuyachkani and Fondo Editorial de la UNMSM.

Ortiz Díaz, Alberto. 2017. "Redeeming Bodies and Souls: Penitentiary Science and Spirituality in Twentieth-Century Puerto Rico and the Dominican Republic." PhD dissertation, University of Wisconsin-Madison.

———. 2020. "Pathologizing the Jíbaro: Mental and Social Health in Puerto Rico's Oso Blanco (1930s to 1950s)." *The Americas* 77 (3): 409–441.

———. 2023. *Raising the Living Dead: Rehabilitative Corrections in Puerto Rico and the Caribbean*. Chicago: University of Chicago Press.

Osborn, Fairfield. 1948. *Our Plundered Planet*. London: Faber and Faber.

Otis, Arthur S. 1937. *Otis Quick-Scoring Mental Ability Tests*. New York: World Book Company.

Outes, Feliz, and Carlos Bruch. 1910. *Los aborígenes de la República Argentina*. Buenos Aires: Estrada Y Cia.

Padilla, Tanalís. 2016. "Memories of Justice: Rural Normales and the Cardenista Legacy." *Mexican Studies/Estudios Mexicanos* 32 (1): 111–143. https://doi .org/10.1525/mex.2016.32.1.111.

Paisley, Fiona. 2007. "Childhood and Race: Growing Up in the Empire." In *Gender and Empire*, edited by Philippa Levine, 240–259. Oxford: Oxford University Press.

Pantojas García, Emilio. 1989. "Puerto Rican Populism Revisited: The PPD during the 1940s." *Journal of Latin American Studies* 21 (3): 521–557.

Parahym, Orlando. 1940. *O Problema alimentar no sertao*. Recife: Imprensa Industrial.

Parezo, Nancy, ed. 1993. *Hidden Scholars: Women Anthropologists and the Native American Southwest*. Albuquerque: University of New Mexico Press.

Parker, James D. A. 1986. "From the Intellectual to the Non-Intellectual Traits: A Historical Framework for the Development of American Personality Research." MA thesis, York University, Toronto.

Paul, Herman. 2011. "Distance and Self-Distanciation: Intellectual Virtue and Historical Method Around 1900." *History and Theory* 50 (4): 104–116.

Pavlich, George. 2009. "The Subjects of Criminal Identification." *Punishment & Society* 11 (2): 171–190.

Pearl, Josh. 2016. *Native Americans and the Census*. January 25. https://journeys .dartmouth.edu/censushistory/2016/01/25/native-americans-and-the- census/.

Peláez, Severo Martínez. 2009 [1970]. *La patria del criollo: An Interpretation of Colonial Guatemala*. Durham, NC: Duke University Press.

Peñaloza, Fernanda. 2010. "On Skulls, Orgies, Virgins and the Making of Patagonia as a National Territory: Francisco Pascasio Moreno's Representation of Indigenous Tribes." *The Bulletin of Hispanic Studies* 87 (4): 455–472.

Penny, H. Glenn. 2003. "The Politics of Anthropology in the Age of Empire: German Anthropologists, Brazilian Indians, and the Case of Alberto Vojtech Frič." *Comparative Studies in Society and History* 45 (2): 249–280.

———. 2013. *Kindred by Choice: Germans and American Indians Since 1800.* Chapel Hill: University of North Carolina Press.

———. 2017. "Material Connections: German Schools, Things, and Soft Power in Argentina and Chile from the 1880s through the Interwar Period." *Comparative Studies in Society and History* 59 (3): 519–549.

Pérez, José G. 1976. *Puerto Rico: U.S. Colony in the Caribbean.* New York: Pathfinder Press.

Pérez Jr., Louis A. 1998. *The War of 1898: The United States and Cuba in History and Historiography.* Chapel Hill: University of North Carolina Press.

Pérez, Pilar. 2016. *Archivos del silencio: Estado, indígenas y violencia en Patagonia Central, 1878–1941.* Buenos Aires: Prometeo Libros.

Pérez Martínez, Héctor. 1951. *La supervivencia de Cuauhtémoc: Hallazgo De los Restos Del Heroe.* Mexico City: Ediciones Criminalia.

Pero, Alejandra. 2003. "The Tehuelche of Patagonia as Chronicled by Travelers and Explorers in the Nineteenth Century." In *Anthropological Perspectives on the Native Peoples of Pampa, Patagonia, and Tierra del Fuego to the Nineteenth Century,* edited by Claudia Briones and José Luis Lanata, 103–120. Westport, CT: Bergin and Garvey.

Pfefferbaum, Betty, Rennard J. Strickland, Everett R. Rhoades, and Rose L. Pfefferbaum. 1995. "Learning How to Heal: An Analysis of the History, Policy, and Framework of Indian Health Care." *American Indian Law Review* 20: 365.

Picart, Juan B. 1960. "Validation of the WBIS for a Disabled Puerto Rican Veteran College Population." *Vocational Rehabilitation and Education Quarterly Information Bulletin* 1: 14–15.

Picó, Fernando. 1994. *El día menos pensado: historia de los presidiarios en Puerto Rico, 1793–1993.* Río Piedras: Ediciones Huracán.

———. 2006. *History of Puerto Rico: A Panorama of its People.* Princeton: Markus Wiener.

Pietsch, Tamson. 2013. *Empire of Scholars: Universities, Networks, and the British Academic World, 1850–1939.* Manchester: Manchester University Press.

Piñero, Jesús T. 1948. *Forty-Seventh Annual Report of the Governor of Puerto Rico for the Fiscal Year 1946–1947.* San Juan: Service of the Government of Puerto Rico Printing Division.

Pintner, Rudolph. 1924. "Results Obtained with the Non-Language Group Test." *Journal of Educational Psychology* 15 (8): 473–483.

———. 1949. *Intelligence Testing: Methods and Results.* New York: Henry Holt.

Podgorny, Irina. 2005. "Bones and Devices in the Constitution of Paleontology in Argentina at the End of the Nineteenth Century." *Science in Context* 18 (2): 249–283.

2008. "De ángeles, gigantes y megaterios. El intercambio de fósiles de las provincias del Plata en la primera mitad del siglo XIX." In *Los lugares del saber*, edited by Ricardo Salvatore, 125–157. Argentina: Beatriz Viterbo Editora

2016. "Human Origins in the New World? Florentino Ameghino and the Emergence of Prehistoric Archaeology in the Americas, 1875–1912." *Paleoamerica* 1 (1): 68–80.

Podgorny, Irina, and María Margaret Lopes. 2008. *El desierto en una vitrina: museos e historia natural en la Argentina*. Mexico City: UNAM.

Poole, Deborah. 1997. *Vision, Race, and Modernity*. Princeton: Princeton University Press.

2004. "An Image of 'Our Indian': Type Photographs and Racial Sentiments in Oaxaca, 1920–1940." *Hispanic American Historical Review* 84 (1): 37–82.

Porteus, Elizabeth Dole. 1991. *Let Us Go Exploring: The Life of Stanley D. Porteus, Hawaii's Pioneer Psychologist*. Honolulu: Ku Pa'a.

Porteus, Stanley. 1948. *The Institutions of the Territory of Hawaii and Their Policies, Plans and Needs For Sound Institutional Practices*. Honolulu: University of Hawai'i.

Potchen, E. James. 2000. "Reflections on the Early Years of Nuclear Medicine." *Radiology* 214 (3): 623–629.

Potchen, E. James, and Alexander Gottschalk. 1976. *Diagnostic Nuclear Medicine, Golden's Diagnostic Radiology Series; Section 20*. Baltimore: Williams & Wilkins.

Povinelli, Elizabeth A. 2002. *The Cunning of Recognition: Indigenous Alterities and the Making of Australian Multiculturalism*. Durham, NC: Duke University Press.

2016. *Geontologies: A Requiem to Late Liberalism*. Durham, NC: Duke University Press.

Presidential Commission for the Study of Bioethical Issues. 2011. *"Ethically Impossible": STD Research in Guatemala from 1946 to 1948*. Washington, DC: Presidential Commission for the Study of Bioethical Issues.

Prewitt, Kenneth. 2013. *What Is Your Race?: The Census and Our Flawed Efforts to Classify Americans*. Princeton: Princeton University Press.

Pribilsky, Jason. 2015. "Developing Selves: Photography, Cold War Science and 'Backwards' People in the Peruvian Andes, 1951–1966." *Visual Studies* 30 (2): 131–150.

Priegue, Celia N. 2002. "Mortuary Rituals among the Southern Tehuelche." In *Contemporary Perspectives on the Native Peoples of Pampa, Patagonia, and Tierra del Fuego*, edited by Claudia Briones and José Luis Lanata, 47–56. Westport, CT: Bergin and Garvey.

Puerto Rico Anemia Commission. 1916. *Uncinariasis en Puerto Rico: un problema médico y económico*. Edited by Bailey Ashford and Pedro

Gutiérrez Igaravídez. San Juan: Bureau of Supplies, Printing, and Transportation.

Quevedo Báez, Manuel. 1946–1949. *Historia de la medicina y cirugía en Puerto Rico.* 2 vols. San Juan: Asociación Médica de Puerto Rico.

Quijada, Mónica. 1998. "Ancestros, ciudadanos, piezas de museo. Francisco P. Moreno y la articulación del indígena en la construcción nacional argentina." *Estudios Interdisciplinarios de América Latina* 9 (2): 21–46.

Quijano, Aníbal. 2000. "Coloniality of Power and Eurocentrism in Latin America." *International Sociology* 15 (2): 215–232.

———. 2000. "Coloniality of Power, Eurocentrism, and Latin America." *Nepantla: Views from South* 1 (3): 533–580.

———. 2007. "Questioning 'Race'." *Socialism and Democracy* 21 (1): 45–53.

Quinn, Sandra Crouse. 2006. "Hurricane Katrina: A Social and Public Health Disaster." *American Journal of Public Health* 96 (2): 204.

Raby, Megan. 2017. *American Tropics: The Caribbean Roots of Biodiversity Science.* Chapel Hill: University of North Carolina Press.

Radin, Joanna. 2017. *Life on Ice: A History of New Uses for Cold Blood.* Chicago: University of Chicago Press.

Ramírez de Arellano, Annette B., and Conrad Seipp. 2011. *Colonialism, Catholicism, and Contraception: A History of Birth Control in Puerto Rico.* Chapel Hill: University of North Carolina Press.

Ramos, Ana, and Claudia Briones, eds. 2016. *Parentesco y política: Topologías indígenas en Patagonia.* Viedma, Argentina: Editorial Universidad Nacional de Río Negro.

Ramos, Rodríguez. 2010. *Rethinking Puerto Rican Precolonial History.* Tuscaloosa: University of Alabama Press.

Razack, Sherene, ed. 2002. *Race, Space, and the Law: Unmapping a White Settler Society.* Toronto: Between the Lines.

Reardon, Jenny. 2004. *Race to the Finish.* Princeton: Princeton University Press.

Reardon, Jenny, and Kim TallBear. 2012. "'Your DNA Is Our History': Genomics, Anthropology, and the Construction of Whiteness as Property." *Current Anthropology* 54 (5): 233–245.

Reardon, Jenny, Jacob Metcalf, Martha Kenney, and Karen Barad. 2015. "Science & Justice: The Trouble and the Promise." *Catalyst: Feminism, Theory, Technoscience* 1 (1): 1–49.

Redman, Samuel J. 2010. *Bone Rooms: From Scientific Racism to Human Prehistory in Museums.* Chicago: University of Chicago Press.

Reed, Richard, and John Renshaw. 2012. "The Aché and Guaraní: Thirty Years after Maybury-Lewis and Howe's Report on Genocide in Paraguay." *Tipití: Journal of the Society for the Anthropology of Lowland South America* 10 (1): 1–18.

Reid, Joshua L. 2019. "Introduction: Indigenous Agency and Colonial Law." *American Historical Review* 124 (1): 20–27.

Rembis, Michael A. 2011. *Defining Deviance: Sex, Science, and Delinquent Girls, 1890–1960.* Urbana: University of Illinois Press.

Rénique, José Luis. 1991. *Los sueños de la sierra: Cusco en el siglo XX.* Lima: CEPES.

Reverby, S. M. 1999. "Rethinking the Tuskegee Syphilis Study. Nurse Rivers, Silence and the Meaning of Treatment." *Nursing History Review: Official Journal of the American Association for the History of Nursing* 7 (1): 3–28.

Rice, Mark. 2018. *Making Machu Picchu.* Chapel Hill: University of North Carolina Press.

Rich, Elizabeth. 2004. "'Remember Wounded Knee': AIM's Use of Metonymy in 21st Century Protest." *College Literature* 31 (3): 70–91.

Richardson, Jamie. 1940. "Keiki o ka Aina." Salvation Army Girls' Home Agency Report, Community Chests and Councils, Honolulu, Hawai'i.

    2005. "Keiki o Ka 'Aina: Institutional Care for Hawaii's Dependent Children." PhD dissertation, University of Hawai'i.

Rieber, Robert W. 2012. *Encyclopedia of Psychological Theories,* 1st ed. New York: Springer.

Rifkin, Mark. 2014. *Settler Common Sense.* Minnesota: University of Minnesota Press.

Rigau Pérez, José G. 1980. "El Dr. Francisco Oller y el inicio de la salud pública moderna en Puerto Rico, 1790–1831." *XXVII Congreso Internacional de Historia de la Medicina* 199–202.

    1989. "The Introduction of Smallpox Vaccine in 1803 and the Adoption of Immunization as a Government Function in Puerto Rico." *Hispanic American Historical Review* 69 (3): 393–423.

    2000. "Historia de la medicina: la salud en Puerto Rico en el siglo XX." *Puerto Rico Health Sciences Journal* 19 (4): 357–368.

Rivera Cusicanqui, Silvia. 2012. "Ch'ixinakax utxiwa: A Reflection on the Practices and Discourses of Decolonization." *South Atlantic Quarterly* 111 (1): 95–109.

    2020. *Ch'ixinakax utxiwa: On Practices and Discourses of Decolonization.* Boston: Polity Press.

Robertson, Thomas. 2012. *Malthusian Moment: Global Population Growth & the Birth of American Environmentalism.* New Brunswick, NJ: Rutgers University Press.

    2014. "Revisiting the Early 1970s Commoner-Ehrlich Debate about Population and Environment: Dueling Critiques of Production and Consumption in a Global Age." In *A World of Populations: Transnational Perspectives on Demography in the Twentieth Century,* edited by Heinrich Hartmann and Corinna Unger, 108–126. New York: Berghann Books.

Rodriguez, Julia. 2006. *Civilizing Argentina: Science, Medicine, and the Modern State.* Chapel Hill: University of North Carolina Press.

Rodríguez, Mariela Eva. 2016. "'Invisible Indians,' 'Degenerate Descendants': Idiosyncrasies of Mestizaje in Southern Patagonia." In *Rethinking Race in Modern Argentina*, edited by Paula L. Alberto and Eduardo Elena, 126–154. New York: Cambridge University Press.

Rodríguez Ramos, Reniel, Jaime Pagán Jiménez, Jorge Santiago Blay, Joseph B. Lambert, and Patrick R. Craig. 2013. "Some Indigenous Uses of Plants in Pre-Columbian Puerto Rico." *Life: The Excitement of Biology* 1 (1): 83–90.

Rodríguez-Silva, Ileana. 2012. *Silencing Race: Disentangling Blackness, Colonialism, and National Identities in Puerto Rico*. Basingstoke: Palgrave Macmillan.

Roesch, Jennifer, Walker MacKenzie, Kyra Millard, and Parker Parker. n.d. "Howard B. Gill, Architect of the Fallen Community Prison Model." *Architecture and Planning: States of Incarceration*. https://statesofincarceration.org/story/howard-b-gill-architect-fallen-community-prison-model.

Rorschach, Hermann. 1942. *Psychodiagnostics: A Diagnostic Test Based on Perception*. Translated by Paul Lemkau et al. Berne: Huber.

Rosemblatt, Karin Alejandra. 2018. *The Science and Politics of Race in Mexico and the United States 1910–1950*. Chapel Hill: University of North Carolina Press.

Ross, Dorothy, ed. 1994. *Modernist Impulses in the Human Sciences, 1870–1930*. Baltimore: John Hopkins University Press.

Rotker, Susana. 2002. *Captive Women: Oblivion and Memory in Argentina*. Minneapolis: University of Minnesota Press.

Rubin, Sol. 1951. *Planning for Correctional Progress in Puerto Rico*. Washington, DC: National Probation and Parole Association.

Rueda, Salvador. 1997–1998. "De conspiradores y mitógrafos: Entre el mito, la historia y el hecho estético." *Historias* 24: 17–26.

Ruiz-Serna, Daniel. 2023. *When Forests Run Amok: War and Its Afterlives in Indigenous and Afro-Colombian Territories*. Durham, NC: Duke University Press.

Sabol, Steven. 2017. *"The Touch of Civilization": Comparing American and Russian Internal Colonization*. Boulder: University Press of Colorado.

Saldaña Portillo, María Josefina. 2016. *Indian Given: Racial Geographies across Mexico and the United States*. Durham, NC: Duke University Press.

Salesa, Damon. 2013. "The Pacific in Indigenous Time." In *Pacific Histories: Ocean, Land, People*, edited by David Armitage and Alison Bashford, 31–52. New York: Bloomsbury.

Salomon Tarquini, Claudia. 2019. "Academic Knowledge about Indigenous Peoples in the Americas: A Comparative Approach about the Conditions of Its International Circulation." *Tapuya: Latin American Science, Technology, and Society* 2 (1): 269–294.

Salvatore, Ricardo D. 2016. *Disciplinary Conquest: U.S. Scholars in South America, 1900–1945*. Durham, NC: Duke University Press.

2020. "Live Indians in the Museum: Connecting Evolutionary Anthropology with the Conquest of the Desert." In *The Conquest of the Desert: Argentina's Indigenous Peoples and the Battle for History*, edited by Carolyne Larson, 97–121. Albuquerque: University of New Mexico Press.

Salzano, Francisco M. 2015. "Bioethics, Population Studies, and Geneticophobia." *Journal of Community Genetics* 6 (3): 197–200.

Salzano, Francisco M., and Maria C. Bortolini. 2005. *The Evolution and Genetics of Latin American Populations*. Cambridge: Cambridge University Press.

Salzano, Francisco M., and Sidia M. Callegari-Jacques. 1988. *South American Indians: A Case Study in Evolution*. Oxford: Oxford University Press.

Samson, Alice V. M., Lucy J. Wrapson, Caroline R. Cartwright, Diana Sahy, Rebecca J. Stacey, and Jago Coopere. 2017. "Artists before Columbus: A Multi-Method Characterization of the Materials and Practices of Caribbean Cave Art." *Journal of Archaeological Science* 88: 24–36.

Sandburg, Carl. 2002. *The Family of Man*. Reissue. Edited by Edward Steichen. New York: The Museum of Modern Art, New York.

Sanders, Karin. 2009. *Bodies in the Bog and the Archaeological Imagination*. Chicago: University of Chicago Press.

Santos, Ricardo Ventura. 2006. "Indigenous Peoples, Bioanthropological Research, and Ethics in Brazil: Issues in Participation and Consent." In *The Nature of Difference: Science, Society and Human Biology*, edited by George Ellison and Alan H. Goodman, 181–202. London: Taylor & Francis Books.

Saranillio, Dean Itsuji. 2018. *Unsustainable Empire: Alternative Histories of Hawai'i Statehood*. Durham, NC: Duke University Press.

Saunt, Claudio. 2020. *Unworthy Republic: The Dispossession of Native Americans and the Road to Indian Territory*, 1st ed. New York: W. W. Norton.

Sauvy, Alfred. 1952. "Trois mondes, une planète." *L'Observateur* (118): 14.

Sayre, Nathan F. 2008. "The Genesis, History, and Limits of Carrying Capacity." *Annals of the Association of American Geographers* 98 (1): 120–134.

Schell, Patience. 2013. *The Sociable Sciences: Darwin and His Contemporaries in Chile*. New York: Palgrave Macmillan.

Schilling, Rebecca, and Stephen T. Casper. 2015. "Of Psychometric Means: Starke R. Hathaway and the Popularization of the Minnesota Multiphasic Personality Inventory." *Science in Context* 28 (1): 77–98.

Schilling Dowd, Patrick. 1999. "Rudolf Virchow and the Science of Humanity." PhD thesis, University of Pittsburgh.

Schneider, Peter B. 1964. "Falsely Abnormal Responses of Normal Subjects to the Metopirone (Su-4885) Test for Pituitary ACTH Reserve." *The Journal of Clinical Endocrinology & Metabolism* 24 (2): 218–221.

Schneider, William H. 1986. "Puericulture, and the Style of French Eugenics." *History and Philosophy of the Life Sciences* 8 (2): 265–277.

Scott, Linda H. 1981. "Measuring Intelligence with the Goodenough-Harris Drawing Test." *Psychological Bulletin* 89 (3): 483–505.

Searis, Damion. 2017. *The Inkblots: Hermann Rorschach, His Iconic Test, and the Power of Seeing*. New York: Crown.

Selcer, Perrin. 2009. "The View from Everywhere: Disciplining Diversity in Post–World War II International Social Science." *Journal of the History of the Behavioral Sciences* 45 (4): 309–329.

——— 2012. "Beyond the Cephalic Index: Negotiating Politics to Produce UNESCO's Scientific Statements on Race." *Current Anthropology* 53 (S5): S173–S184.

Sen, Amartya. 1981. *Poverty and Famines: An Essay on Entitlement and Deprivation*. Oxford: Clarendon Press.

Seth, Suman. 2009. "Putting Knowledge in Its Place: Science, Colonialism, and the Postcolonial." *Postcolonial Studies* 12 (4): 373–388.

——— 2014. "Focus: Relocating Race." *Isis* 105 (4): 759–763.

——— 2016. "Darwin and the Ethnologists: Liberal Racialism and the Geological Analogy." *Historical Studies in the Natural Sciences* 46 (4): 490–527.

——— 2018. *Difference and Disease: Medicine, Race, and the Eighteenth-Century British Empire*, 1st ed. Cambridge: Cambridge University Press.

Shapin, Steven. 1992. "Discipline and Bounding: The History and Sociology of Science as Seen through the Externalism-Internalism Debate." *History of Science* 30 (4): 333–369.

Shoemaker, Nancy. 2004. *A Strange Likeness: Becoming Red and White in Eighteenth-Century North America*. Oxford: Oxford University Press.

Shohat, Ella, and Robert Stam. 2012. *Race in Translation: Culture Wars around the Postcolonial Atlantic*. New York: New York University Press.

Shumway, Nicolás. 1991. *The Invention of Argentina*. Berkeley: University of California Press.

Siegel, Eleanor. 1992. "Arthur Sinton Otis and the American Mental Testing Movement." PhD dissertation, University of Miami.

Sievers, Maurice L., and James R. Marquis. 1962. "The Southwestern American Indian's Burden: Biliary Disease." *JAMA* 182 (5): 570–572.

Silman, Alan. 1996. "OBITUARY: Dr John Lawrence." *The Independent*, May 14. www.independent.co.uk/incoming/obituary-dr-john-lawrence-5616071.html.

Silva, Guilherme C. 2022. "The South as a Laboratory (Again)? Dealing with Calls for 'Alternatives' in the North." *4S Reflections*.

Silva, Noenoe K. 2004. *Aloha Betrayed: Native Hawaiian Resistance to American Colonialism*. Durham, NC: Duke University Press.

Silverblatt, Irene Marsha. 2004. *Modern Inquisitions: Peru and the Colonial Origins of the Civilized World*. Durham, NC: Duke University Press.

Simpson, Audra. 2007. "On Ethnographic Refusal: Indigeneity, 'Voice' and Colonial Citizenship." *Junctures: The Journal for Thematic Dialogue* 9: 67–80.

——— 2014. *Mohawk Interruptus: Political Life across the Borders of Settler States*. Durham, NC: Duke University Press.

Sintenis, Paul Ernst Emil, G. Bresadola, P. Hennings, and P. Magnus. 1893. *Die Von Herrn P. Sintenis auf der Insel Portorico, 1884–1887, gesammelten Pilze*. Leipzig: W. Engelmann.

Sivasundaram, Sujit. 2010. "Focus: Global Histories of Science." *Isis* 101 (1): 95–97.

Skurski, Julie, and Fernando Coronil. 2006. "Introduction: States of Violence and the Violence of States." In *States of Violence*, edited by Julie Skurski and Fernando Coronil, 1–31. Ann Arbor: University of Michigan Press.

Sleeper-Smith, Susan et al., eds. 2019. *Why You Cannot Teach United States History without American Indians*. Chapel Hill: University of North Carolina.

Smallwood, Stephanie. 2019. "Reflections on Settler Colonialism, the Hemispheric Americas, and Chattel Slavery." *William and Mary Quarterly* 76 (3): 407–416.

Smith, Andrea. 2014. "Native Studies on the Horizon of Death: Theorizing Ethnographic Entrapment and Settler Self-Reflexivity." In *Theorizing Native Studies*, edited by Andrea Smith and Audra Simpson, 207–234. Durham, NC: Duke University Press.

Smith, Carl. 2006. *The Plan of Chicago: Daniel Burnham and the Remaking of the American City*. Chicago: University of Chicago Press.

Smith, Linda Tuhiwai. 2012. *Decolonizing Methodologies: Research and Indigenous Peoples*. New York: Zed Books.

Smith, Roger. 1997. *The Fontana History of the Human Sciences*. London: Fontana Press.

Sokal, Michael M., ed. 1987. *Psychological Testing and American Society, 1880–1930*. New Brunswick, NJ: Rutgers University Press.

Soto Laveaga, Gabriela. 2009. *Jungle Laboratories: Mexican Peasants, National Projects, and the Making of the Pill*. Durham, NC: Duke University Press.

2018. "Largo Dislocare: Connecting Microhistories to Remap and Recenter Histories of Science." *History and Technology* 34 (1): 21–30.

Speed, Shannon. 2017. "Structures of Settler Capitalism in Abya Yala." *American Quarterly* 69 (4): 783–790.

Spiegel, Gabrielle. 2014. "The Future of the Past: History, Memory, and the Ethical Imperatives of Writing History." *Journal of the Philosophy of History* 8: 149–179.

Stahl, Agustín. 1883–1888. *Estudios sobre la flora de Puerto Rico*. San Juan: Tip. "El Asimilista."

Stark, Laura. 2007. "Victims in Our Own Minds? IRBs in Myth and Practice." *Law & Society Review* 41 (4): 777–786.

2012. *Behind Closed Doors: IRBs and the Making of Ethical Research*. Chicago: University of Chicago Press.

2016. *The Normals: A People's History*. Chicago: University of Chicago Press.

2018. "Contracting Health: Procurement Contracts, Total Institutions, and Problem of Virtuous Suffering in Post-War Human Experiment." *Social History of Medicine* 31 (4): 818–846.

2019. "Emergence." *Isis* 110 (2): 332–336.

Stark, Laura, and Nancy D. Campbell. 2014. "Stowaways in the History of Science: The Case of Simian Virus 40 and Clinical Research on Federal Prisoners at the US National Institutes of Health, 1960." *Studies in History and Philosophy of Biological and Biomedical Sciences* 48 Pt B: 218–230.

Stegman, Stephanie. 2010. "Taking Control: Fifty Years of Diabetes in the American Southwest 1940–1990." PhD Dissertation, Arizona State University. *ProQuest Dissertations Publishing.* http://search.proquest.com/docview/504799320/?pq-origsite=primo.

Stepan, Nancy Leys. 1991. *"The Hour of Eugenics": Race, Gender, and Nation in Latin America.* Ithaca: Cornell University Press.

Stern, Alexandra M. 2005. *Eugenic Nation: Faults and Frontiers of Better Breeding in Modern America.* Berkeley: University of California Press.

2006. "An Empire of Tests: Psychometrics and the Paradoxes of Nationalism in the Americas." In *Haunted by Empire: Geographies of Intimacy in North American History,* edited by Ann L. Stoler, 328. Durham, NC: Duke University Press.

2007. "Yellow Fever Crusade: U.S. Colonialism, Tropical Medicine, and the International Politics of Mosquito Control, 1900–1920." In *Medicine at the Border: Disease, Globalization and Security, 1850 to the Present,* edited by Alison Bashford, 41–59. Basingstoke: Palgrave Macmillan.

Stevenson, Lisa. 2020. "Looking Away." *Cultural Anthropology* 35 (1): 6–13.

Steward, Julian H. 1956. *The People of Puerto Rico: A Study in Social Anthropology.* Urbana: University of Illinois Press.

Stocking, George. 1987. *Victorian Anthropology.* New York: New York Free Press.

Stocking Jr., George W. 1989. *A Franz Boas Reader: The Shaping of American Anthropology, 1883–1911.* Chicago: University of Chicago Press.

1995. *After Tylor: British Social Anthropology, 1888–1951.* Madison: University of Wisconsin Press.

Stoczkowski, Wiktor. 2008. "Claude Lévi-Strauss: The View from Afar – UNESCO Digital Library." *The UNESCO Courier.* https://unesdoc.unesco.org/ark:/48223/pf0000162711.

Stoler, Ann Laura. 2009. *Along the Archival Grain: Epistemic Anxieties and Colonial Common Sense.* Princeton: Princeton University Press.

2010. "Intimidations of Empire: Predicaments of the Tactile and Unseen." In *Haunted by Empire: Geographies of Intimacy in North American History,* edited by Ann Laura Stoler, 1–22. Durham, NC: Duke University Press.

Strang, Cameron. 2014. "Violence, Ethnicity, and Human Remains during the Second Seminole War." *Journal of American History* 100 (4): 973–994.

Swanson, Drew. 2019. "In Living Color: Early 'Impressions' of Slavery and the Limits of Living History." *The American Historical Review* 124 (5): 1732–1748.

Swartz, Rebecca. 2019. *Education and Empire: Children, Race and Humanitarianism in the British Settler Colonies, 1833–1880.* Basingstoke: Palgrave Macmillan.

Sweet, James. 2022. "Is History History? Identity Politics and Teleologies of the Present." www.historians.org/publications-and-directories/perspectives-on-history/september-2022/is-history-history-identity-politics-and-teleologies-of-the-present.

Szreter, Simon. 1993. "The Idea of Demographic Transition and the Study of Fertility Change: A Critical Intellectual History." *Population and Development Review* 19 (4): 659–701.

Taiz, Lillian. 2001. *Hallelujah Lads & Lasses: Remaking the Salvation Army in America, 1880–1930.* Chapel Hill: University of North Carolina Press.

Takaki, Ronald T. 1983. *Pau Hana: Plantation Life and Labor in Hawaii, 1835–1920.* Honolulu: University of Hawai'i Press.

TallBear, Kim. 2011. "Why Interspecies Thinking Needs Indigenous Standpoints." *Cultural Anthropology Online.* https://culanth.org/fieldsights/why-interspecies-thinking-needs-indigenous-standpoints.

    2013. "Genomic Articulations of Indigeneity." *Social Studies of Science* 43 (4): 509–533.

    2013. *Native American DNA: Tribal Belonging and the False Promise of Genetic Science.* Minneapolis: University of Minnesota Press.

    2014. "Standing with and Speaking as Faith: A Feminist-Indigenous Approach to Inquiry." *Journal of Research Practice* 10 (2): 1–7.

    2015. "An Indigenous Reflection on Working beyond the Human/Not Human." In "Dossier: Theorizing Queer Inhumanisms." *GLQ: A Journal of Lesbian and Gay Studies* 21 (2–3): 230–235.

    2019. "Caretaking Relations, Not American Dreaming." *Kalfou* 6 (1): 24–41.

Telles, Edward E. 2014. *Pigmentocracies: Ethnicity, Race, and Color in Latin America.* Chapel Hill: University of North Carolina Press.

Terman, Lewis M. 1916. *The Measurement of Intelligence.* Boston: Houghton Mifflin.

Thurner, Mark. 1997. *From Two Republics to One Divided: Contradictions of Postcolonial Nationmaking in Andean Peru.* Durham, NC: Duke University Press.

Tobbell, Dominique A. 2011. *Pills, Power, and Policy: The Struggle for Drug Reform in Cold War America and Its Consequences,* 1st ed. Berkeley: University of California Press.

Todd, Zoe. 2014. "Fish Pluralities: Human–Animal Relations and Sites of Engagement in Paulatuuq, Arctic Canada." *Études/Inuit/Studies* 38 (1&2): 217–238.

    2017. "Fish, Kin, and Hope: Tending to Water Violations in Amiskwaciwaskahikan and Treaty Six Territory." *Afterall* 43 (1): 102–107.

Tortorici, Zeb. 2018. *Sins against Nature: Sex and Archives in Colonial New Spain.* Durham, NC: Duke University Press.

Trent, James W. 1995. *Inventing the Feeble Mind: A History of Mental Retardation in the United States.* Berkeley: University of California Press.

Treuer, David. 2019. *The Heartbeat of Wounded Knee: Native America from 1890 to the Present*. New York: Riverhead.

Trouillot, Michel-Rolph. 1995. *Silencing the Past: Power and the Production of History*. Boston: Beacon Press.

2004. *Global Transformations: Anthropology and the Modern World*, 1st ed. New York: Palgrave Macmillan.

Trujillo-Pagán, Nicole E. 2014. *Modern Colonization by Medical Intervention: U.S. Medicine in Puerto Rico*. Chicago: Haymarket.

Tsing, Anna. 2005. *Friction: An Ethnography of Global Connection*. Princeton: Princeton University Press.

Tuck, Eve. 2009. "Suspending Damage: A Letter to Communities." *Harvard Educational Review* 79 (3): 409–427.

Tuck, Eve, and C. Ree. 2013. "A Glossary of Haunting." In *Handbook of Autoethnography*, by Stacey Holman Jones, Tony E. Adams, and Carolyn Ellis, 639–658. Oakland, CA: Left Coast Press.

Tuck, Eve, and K. Wayne Yang. 2012. "Decolonization Is Not a Metaphor." *Decolonization: Indigeneity, Education & Society* 1 (1): 1–40.

2014. "R-Words: Refusing Research." In *Humanizing Research: Decolonizing Qualitative Inquiry with Youth and Communities*, edited by Django Paris and Maisha T. Winn, 223–248. Thousand Oaks, CA: Sage.

Turda, Marius, and Aaron Gillette. 2014. *Latin Eugenics in Comparative Perspective*. London: Bloomsbury.

United States Atomic Energy Commission. 1964–1965. *Isotopes and Radiation Technology* 2 (2): 192–193.

United States Environmental Protection Agency. 2017. *Case Summary: $600 Million Settlement to Clean up 94 Abandoned Uranium Mines on the Navajo Nation*. www.epa.gov/enforcement/case-summary-600-million-settlement-clean-94-abandoned-uranium-mines-navajo-nation.

University of Puerto Rico-Río Piedras Agricultural Experiment Station. 1952. *The Story of the Agricultural Experiment Station of the University of Puerto Rico*. Río Piedras: University of Puerto Rico.

Unknown. 1949. "Tierra de todas las zonas indígenas en el monumento a Cuauhtémoc." *Excélsior*, October 15.

1963. "Dr. MacQuarrie's Funeral Today." *Spartan Daily*, November 18: 1. https://scholarworks.sjsu.edu/cgi/viewcontent.cgi?article=4476&context=spartandaily

2014. "O'Brien, William." *The Daily Progress Newspaper*, March 11. www.dailyprogress.com/obituaries/o-brienwilliam/article_73a7cd0c-3a1a-53a2-8dab-edeb93a25c04.html.

U.S. Department of Health, Education and Welfare. 1959. *Bulletin No. 11: Research Relating to Children August 1959–January 1960*. Bulletin, U.S. Department of Health, Education, and Welfare. Washington, DC: Government Printing Office.

U. V. Administration. 1948. "A Study of the Bell Adjustment Inventory and Deviant Scores on the MMPI." *Minnesota Counselor* 3: 1–7.

Vaughan, Alden T. 1982. "From White Man to Redskin: Changing Anglo-American Perceptions of the American Indian." *The American Historical Review* 87 (4): 917–953.

Vega Ramos, Luis. 2006. *Pan, tierra, y libertad: historia y filosofía del Partido Popular Democrático.* San Juan: EMS Editores.

Vellard, Jehan Albert. 1932. "Exploration du Dr Vellard au Paraguay." *Journal de la Société des Américanistes* 24 (1): 215–218.

1933. "Une mission scientifique au Paraguay (15 juillet 1931–16 janvier 1933)." *Journal de la Société des Américanistes* 25 (2): 293–334.

1954. *Une Civilisation Du Miel: Les Indiens Guayakis Du Paraguay.* Paris: Libr. Gallimard.

Venator-Santiago, Charles R. 2015. *Puerto Rico and the Origins of U.S. Global Empire: The Disembodied Shade.* New York: Routledge.

Verran, Helen. 2002. "A Postcolonial Moment in Science Studies: Alternative Firing Regimes of Environmental Scientists and Aboriginal Landowners." *Social Studies of Science* 32 (5–6): 729–762.

Vetö, Silvana. 2019. "Child Delinquency and Intelligence Testing at Santiago's Juvenile Court, Chile, 1929–1942." *History of Psychology* 22 (3): 244–265.

Villar, Diego. 2017. "Les Expéditions du Doctor Vellard." In *Les Années folles de l'ethnographie: Trocadéro 28–37,* edited by André Delpuech, Christine Laurière, and Carine Peltier-Caroff, 536–579. París: Muséum national d'histoire naturelle.

Virchow, Rudolf. 1879. "Drei Patagonier." *Zeitschrift für Ethnologie* XI: 198–204.

1892. *Crania ethnica americana. Sammlung Auserlesener amerikanischer Schädeltypen herausgegeben von Rudolf Virchow.* Berlin: A. Asher & Co.

1992. *Sämtliche Werke in 71 Bänden.* Edited by Christian Andree. Berlin: Blackwell Wissenschafts-Verlag.

Vizenor, Gerald, ed. 2008. *Survivance: Narratives of Native Presence.* Lincoln: University of Nebraska Press.

Vogt, William. 1948. *The Road to Survival.* New York: William Sloane Association.

Wade, Peter. 1997. *Race and Ethnicity in Latin America.* Sterling, VA: Pluto Press.

2017. *Degrees of Mixture, Degrees of Freedom: Genomics, Multiculturalism, and Race in Latin America.* Durham, NC: Duke University Press.

Wagner, Kim. 2010. "Confessions of a Skull: Phrenology and Colonial Knowledge in Early Nineteenth-Century India." *History Workshop Journal* 69 (1): 27–51.

Wald, Priscilla. 1995. *Constituting Americans: Cultural Anxiety and Narrative Form.* Durham, NC: Duke University Press.

Walker, Alexis K., and Elizabeth L. Fox. 2018. "Bioethics, 'Vulnerability' and Marginalization." *AMA Journal of Ethics* 20 (10): E941–E947.

Walsh, Sarah. 2021. *The Religion of Life: Eugenics, Race, and Catholicism in Chile.* Pittsburgh: University of Pittsburgh Press.

Warren, Adam. 2018. "Collaboration and Discord in International Debates about Coca Chewing, 1949–1950." *Medicine Anthropology Theory* 5 (2): 35–51.

Washington, Harriet A. 2006. *Medical Apartheid: The Dark History of Medical Experimentation on Black Americans from Colonial Times to the Present 244–270.* New York: Harlem Moon.

Weheliye, Alexander. 2014. *Habeas Viscus: Racializing Assemblages, Biopolitics, and Black Feminist Theories of the Human.* Durham, NC: Duke University Press.

Welch, Teresa J. C. 1972. *Fundamentals of the Tracer Method.* Philadelphia: Saunders.

Whitley, Richard. 1984. *The Intellectual and Social Organization of the Sciences.* Oxford: Clarendon Press.

Williams, Ronald Jr. 2015. "Race, Power, and the Dilemma of Democracy: Hawai'i's First Territorial Legislature, 1901." *Hawaiian Journal of History* 49 (1): 1–45. https://doi.org/10.1353/hjh.2015.0017.

Wilson, John P. 2014. *Peoples of the Middle Gila: A Documentary History of the Pimas and Maricopas 1500s–1945,* 1st ed. Sacaton, AZ: Gila River Indian Community.

Wilson, Shawn. 2008. *Research Is Ceremony: Indigenous Research Methods,* 1st ed. Black Point, Nova Scotia: Fernwood Publishing.

2009. *Research Is Ceremony: Indigenous Research Methods.* Winnipeg: Fernwood Publishing.

Wilson, William H. 1989. *The City Beautiful Movement.* Baltimore: John Hopkins University Press.

Winter, Charles E. 1934. *Report of the Attorney General of Puerto Rico for the Fiscal Year Ending June 30, 1933,* vol. 18. San Juan: Negociado de Materiales, Imprenta y Transporte.

Wolfe, Patrick. 1999. *Settler Colonialism and the Transformation of Anthropology: The Politics and Poetics of an Ethnographic Event.* New York: Cassell.

2006. "Settler Colonialism and the Elimination of the Native." *Journal of Genocide Research* 8 (4): 387–409.

Woodson, M. Kyle. 2016. *The Social Organization of Hohokam Irrigation in the Middle Gila River Valley, Arizona,* 1st ed. Sacaton, AZ: Gila River Indian Community.

Wynter, Sylvia. 1994. "No Humans Involved: An Open Letter to My Colleagues." *Forum NHI: Knowledge for the 21st Century* I (1): 42–71.

2003. "Unsettling the Coloniality of Being/Power/Truth/Freedom: Towards the Human, after Man, Its Overrepresentation – An Argument." *CR: The New Centennial Review* 3 (3): 257–337.

Zamorano, Gabriela. 2011. "Traitorous Physiognomy: Photography and the Racialization of Bolivian Indians by the Créqui-Montfort Expedition

(1903)." *Journal of Latin American and Caribbean Anthropology* 16 (2): 425–455.

Zenderland, Leila. 1987. "The Debate over Diagnosis: Henry Herbert Goddard and the Medical Acceptance of Intelligence Testing." In *Psychological Testing and American Society, 1880–1930*, edited by Michael M. Sokal, 46–74. New Brunswick, NJ: Rutgers University Press.

———. 1998. *Measuring Minds: Henry Goddard and the Intelligence Testing Movement.* New York: Cambridge University Press.

Ziman, John M. 1968. *Public Knowledge: An Essay concerning the Social Dimension of Science.* Cambridge: Cambridge University Press.

Zimmerman, Andrew. 2001. *Anthropology and Antihumanism in Imperial Germany.* Chicago: University of Chicago Press.

Zulawski, Ann. 2007. *Unequal Cures: Public Health and Political Change in Bolivia, 1900–1950.* Durham, NC: Duke University Press.

INDEX

349

Printed in the United States
by Baker & Taylor Publisher Services